FIRE SERVICE GROUND LADDER PRACTICES

EIGHTH EDITION

VALIDATED BY
THE INTERNATIONAL FIRE SERVICE TRAINING ASSOCIATION

PUBLISHED BY
FIRE PROTECTION PUBLICATIONS
OKLAHOMA STATE UNIVERSITY

Dedication

This manual is dedicated to the members of that unselfish organization of men and women who hold devotion to duty above personal risk, who count sincerity of service above personal comfort and convenience, who strive unceasingly to find better ways of protecting the lives, homes and property of their fellow citizens from the ravages of fire and other disasters . . . **The Firefighters of All Nations.**

Dear Firefighter:

The International Fire Service Training Association (IFSTA) is a nonprofit organization that exists for the sole purpose of serving firefighters' training needs. IFSTA is a member of the Joint Council of National Fire Organizations. Fire Protection Publications is the publisher of IFSTA materials. Fire Protection Publications' staff members participate in the National Fire Protection Association and the International Society of Fire Service Instructors.

If you need help in locating additional information concerning these organizations or assistance with manual orders, contact:

Customer Services
Fire Protection Publications
Oklahoma State University
Stillwater, Oklahoma 74078
(800)654-4055 in Continental United States

For assistance with training materials, recommended materials for inclusion in a manual, or questions on manual content, contact:

Technical Services
Fire Protection Publications
Oklahoma State University
Stillwater, Oklahoma 74078
(405)624-5723

First Printing - October 1984

ISBN 0-87939-055-7
Library of Congress 84-61707
Eighth Edition
Printed in the United States of America

Table of Contents

GLOSSARY **11**

INTRODUCTION **1**

LADDER TYPES AND LADDER TERMS **9**
Ladder Types 9
Single Ladders 9
Roof Ladders 10
Folding Ladders 11
Extension Ladders 12
Pole Ladders 12
Combination Ladders 12
Pompier Ladders 14
Ladder Terms 14
Angle of Inclination 14
Beam 14
Bed Section 15
Bedded Position 15
Butt 15
Designated Length 16
Dogs 16
Fly Section 16
Halyard 16
Heel 16
Identification Number 17
Inside Width 18
Maximum Extended Length 18
Nesting 18
Outside Width 18
Pawls 19
Rail 19
Retracted 19
Rungs 20
Side Rail 20
Stripping Ladder 20
Tip 21

DESIGN AND CONSTRUCTION, MAINTENANCE, SERVICE TESTING **27**
Design and Construction 28
Workmanship 28
Materials Used 31

Features of Construction . . . 38
Design Verification Testing . . . 74
Inspection and Maintenance . . . 77
Inspection . . . 78
Maintenance . . . 79
Service Testing Ground Ladders . . . 84
When Should Ladders be Service Tested? . . . 85
What Constitutes Failure? . . . 85
How to Service Test Ground Ladders . . . 85
Records . . . 93

HANDLING LADDERS . . . 99
Location and Methods of Mounting on Apparatus . . . 99
Ladders Carried on Pumpers . . . 100
Ladders Carried on Aerial Apparatus . . . 102
Ladders Carried on Other Apparatus . . . 105
Selecting the Correct Ladder for the Job . . . 107
Removing Ladders from Apparatus . . . 110
Proper Lifting and Lowering Methods . . . 113
Carrying Ladders . . . 114
One-Firefighter Carries . . . 114
Two-Firefighter Carries . . . 130
Three-Firefighter Carries . . . 145
Four-Firefighter Carries . . . 157
Five-Firefighter Carries . . . 164
Six-Firefighter Carries . . . 167
Carrying Other Ladders . . . 169
Special Carry for Narrow Passageways . . . 171
Positioning (Placement) . . . 172
Responsibility for Positioning . . . 172
Objectives . . . 172

RAISING AND CLIMBING/LADDER SAFETY . . . 183
Raising . . . 183
Miscellaneous Procedures . . . 183
One-Firefighter Raises . . . 194
Two-Firefighter Raises . . . 201
Three-Firefighter Raises . . . 204
Four-Firefighter Raises . . . 207
Pole Ladder Raises . . . 209
Combination Ladder Raises . . . 231
Extension/A-Frame Combination Ladder . . . 231
Single/A-Frame Combination Ladder . . . 233
Telescoping Beam: Single/A-Frame Combination Ladder . . . 233
Extending A-Frame Combination Ladder . . . 236
Pompier Ladder Raise . . . 237
Special Raises . . . 239
Dome (Auditorium) Raise . . . 239
Hotel or Factory Raise . . . 242

Raising Ladders Under Obstructions 245
Two-Firefighter Single or Roof Ladder Obstructed Raise 245
Three-Firefighter Single or Roof Ladder Obstructed Raise 248
Alternate Three-Firefighter Single or Roof Ladder Obstructed Raise 249
Four-Firefighter Obstructed Raise for an Extension Ladder . . 252
Securing the Ladder 252
Heeling 252
Tying the Ladder In 254
Tying the Halyard 255
Climbing 259
Ladder Climbing Skills 259
Climbing to Place a Roof Ladder 270
Climbing a Pompier Ladder 280
Ladder Safety 283

SPECIAL USES 291
Ladder Rescue Operations 292
Assisting Conscious and Physically Able Persons Down Ground Ladders 292
Bringing Unconscious Persons Down Ground Ladders 293
Removing an Unconscious Firefighter Leg Locked on a Ground Ladder 299
Lowering Extension Ladders to Below-Grade Locations 301
Bridging with Ladders for Rescue 305
Using a Ladder as a Fulcrum for Lowering Injured Persons 315
Using a Ladder as a Slide for Removing Injured Persons 319
Using a Ladder Leaning Against a Building to Support Lowering of an Injured Person (Ladder Sling) 321
Using a Ladder as a Stretcher 323
Using a Ladder for Ice Rescue 326
Using a Ladder Float Drag for Ice and Water Rescue 328
Prying with a Ground Ladder 328
Using Ground Ladders for Shoring and Other Uses During Trench Rescue 328
Making Devices for Hoisting 333
Extending Reach from Aerial Platforms 340
Using a Roof Ladder as a Pompier Ladder for Rescue 343
Using a Ladder for Emergency Ventilation 345
Fire Fighting Operations and Uses 346
Directing Fire Streams from Ground Ladders 346

Hoisting Ladders 349
Bridging for Fire Fighting 353
Improvising a Long Roof Ladder 359
Using Ladders to Support Smoke Ejectors 361
Keeping a Suction Strainer off the Bottom by Using a Ladder 363
Making a Water Chute with a Tarp and a Ladder 364
Using a Ladder as a Battering Ram 366
Using a Roof Ladder and Rope for Remote Control and Advancement of a Nozzle (Also Called an Improvised Cellar Pipe) 367
Making a Ladder Pipe Applicator 370
Making an Under-a-Pier Applicator 373
Using Ladders to Construct a Catch Basin 374
Making a Dam Across a Stream Using a Ladder and a Salvage Cover 375

APPENDIX A
Ground Ladder Testing and Repair Form 380

INDEX 381

REVIEW ANSWERS 386

List of Tables

2.1 Weight Comparison of Most Common Types of Ladders . . . 32
2.2 Minimum Inside Clear Width 39
2.3 Maximum Ladder Loading 39
2.4 Summary of Design Verification Testing 77
2.5 Hardness Test Minimum Readings 91
2.6 Additional Hardness Test Readings 91
3.1 Maximum Working Heights for Ladders set at Proper Climbing Angle 109

THE INTERNATIONAL FIRE SERVICE TRAINING ASSOCIATION

The International Fire Service Training Association is an educational alliance organized to develop training material for the fire service. The annual meeting of its membership consists of a workshop conference which has several objectives —

... to develop training material for publication
... to validate training material for publication
... to check proposed rough drafts for errors
... to add new techniques and developments
... to delete obsolete and outmoded methods
... to upgrade the fire service through training

This training association was formed in November 1934, when the Western Actuarial Bureau sponsored a conference in Kansas City, Missouri, to determine how all agencies that were interested in publishing fire service training material could coordinate their efforts. Four states were represented at this conference and it was decided that, since the representatives from Oklahoma had done some pioneering in fire training manual development, other interested states should join forces with them. This merger made it possible to develop nationally recognized training material which was broader in scope than material published by an individual state agency. This merger further made possible a reduction in publication costs, since it enabled each state to benefit from the economy of relatively large printing orders. These savings would not be possible if each individual state developed and published its own training material.

From the original four states, the adoption list has grown to forty-four American States; six Canadian Provinces; the British Territory of Bermuda; the Australian State of Queensland; the International Civil Aviation Organization Training Centre in Beirut, Lebanon; the Department of National Defence of Canada; the Department of the Army of the United States; the Department of the Navy of the United States; the United States Air Force; the United States Bureau of Indian Affairs; The United States General Services Administration; and the National Aeronautics and Space Administration (NASA). Representatives from the various adopting agencies serve as a voluntary group of individuals who govern policies, recommend procedures, and validate material before it is published. Most of the representatives are members of other international fire protection organizations and this meeting brings together individuals from several related and allied fields, such as:

... key fire department executives and drillmasters,
... educators from colleges and universities,
... representatives from governmental agencies,
... delegates of firefighter associations and organizations, and
... engineers from the fire insurance industry.

This unique feature provides a close relationship between the International Fire Service Training Association and other fire protection agencies, which helps to correlate the efforts of all concerned.

The publications of the International Fire Service Training Association are compatible with the National Fire Protection Association's Standard 1001, "Fire Fighter Professional Qualifications (1981)," and the International Association of Fire Fighters/International Association of Fire Chiefs "National Apprenticeship and Training Standards for the Fire Fighter." The standards are an effort to attain professional status through progressive training. The NFPA and IAFF/IAFC Standards were prepared in cooperation with the Joint Council of National Fire Service Organizations of which IFSTA is a member.

The International Fire Service Training Association meets each July at Oklahoma State University, Stillwater, Oklahoma. Fire Protection Publications at Oklahoma State University publishes all IFSTA training manuals and texts. This department is responsible to the executive board of the association. While most of the IFSTA training manuals can be used for self-instruction, they are best suited to group work under a qualified instructor.

Preface

This is the eighth edition of **Fire Service Ground Ladder Practices.** The greatly expanded text and illustrations detailing all phases of ladder terminology, construction, testing, carries, raises, climbing, and special uses make it a virtual pictorial encyclopedia of ground ladders.

Acknowledgment and grateful thanks are extended to the members of the validating committee, who contributed their time, wisdom, and knowledge to this manual in the memory of Earl Hood the former committee chairman.

Chairman
Wayne Nelson
New Mexico State Fire Marshal's Office
Santa Fe, New Mexico

1981 Secretary
Ron Graw, Captain
Rockford Fire Department
Rockford, Illinois

1982 Secretary
Phillip Schwab, President
Duo-Safety Ladder Co.
Oshkosh, Wisconsin

Other persons serving on the committee during its tenure were:

Richard Calhoun
Louisiana State Fire Marshal's Office

Joe Gage
Midwest City, (OK) Fire Department

Sam Cramer
Aluminum Ladder Co.

Robert Hasbrook
Neosho, (MO) Fire Department

Larry Northcutt
Norman, (OK) Fire Department

Special acknowledgement and thanks are extended to Joel Woods, Regional Coordinator, Upper Eastern Shore Office, Maryland Fire and Rescue Institute and Don Davis, Coordinator, Publications Production, Fire Protection Publications for taking most of the hundreds of photographs appearing in this manual.

A book of this scope would be impossible to publish were it not for the assistance of many persons and organizations. To the following and others as noted in the captions we owe a great debt as they gave freely of time, advice, equipment, and/or personnel.

ALACO Ladder Company
Aluminum Ladder Company
Community Volunteer Fire Company, Inc. of Rising Sun, Maryland
Duo-Safety Ladder Company
Howard County Fire Department
    Columbia Station
Maryland Fire and Rescue Institute
Montgomery County, Maryland Fire and Rescue Training Academy
    Captain Jerry Eiler
    Training Academy Staff
Montgomery County, Maryland Fire and Rescue Service
    Gaithersburg-Washington Grove Volunteer Fire Department, Inc.
    Hillandale Volunteer Fire Department, Inc.
    Rockville Volunteer Fire Department, Inc.
    Silver Spring Volunteer Fire Department, Inc.
National Bureau of Standards Fire Department
Oklahoma City, Oklahoma Fire Department
    Deputy Chief Jerry Smith, Chief Training Officer
    Major John Paraich, Assistant Training Officer
    Truck Co. 1
Prince George County, Maryland Fire Department
    Chief James Estepp
    Station 12-College Park
    Station 20-Marlboro
    Station 31-Beltsville
    Station 45-Marlboro
Pierre Thibault Truck, Inc.
Singerly Fire Company, Elkton, Maryland
Stillwater, Oklahoma Fire Department
    Jim Smith, Fire Chief
    Headquarters Station Personnel
    Station 3 Personnel
Tahlequah, Oklahoma Fire Department
    Sam Pinson, Fire Chief
    Steve Smith, Assistant Fire Chief
Tulsa, Oklahoma Fire Department
    Stanley Hawkins, Fire Chief
    Michael Conley, Deputy Drill Master
    Training Academy Staff

Cover Photo By: William F. Noonan, N. Weymouth, Massachusetts.

We also express our thanks to the following Oklahoma Fire Service Training staff members for their assistance in demonstrating certain evolutions described in this manual.

David Fischer
Eric Haussermann
Fred Myers

Gratitude is also extended to the following members of the Fire Protection Publications staff whose contributions made the final publication of the manual possible.

Charles Donaldson	Associate Editor
Lynne Murnane	Associate Editor
Teresa Tackett	Marketing Associate
Gary Courtney	Research Technician
Scott Kerwood	Research Technician
Kevin Roche	Research Technician
Scott Tyler	Research Technician
Karen Murphy	Phototypesetter Operator II
Desa Porter	Phototypesetter Operator II
Ann Moffat	Graphic Designer
Lynda Halley	Graphic Artist
Carol Smith	Publications Validation Assistant
Cindy Brakhage	Unit Assistant

Gene P. Carlson
Editor

Glossary

A

Alloy — A substance composed of two or more metals fused together and dissolved in each other when molten.

C

Certification — Refers to the manufacturer's certification that the ladder has been constructed to meet requirements of NFPA 1931.

Checks — Cracks or breaks in wood.

Conductor — A substance that transmits electrical or thermal energy.

Corrugated — Formed into ridges or grooves.

D

Deformation — An alteration of form or shape.

Dimpled — Depressed or indented, as on a metal surface to aid in gripping.

Dry Hoseline — A hoseline without water in it; an uncharged hoseline.

E

Eave — The lower border of a roof that projects beyond the face of a building.

Electrical Service — The conductor and equipment for delivering energy from the electrical supply system to the wiring system of the premises.

Engulf — To flow over and enclose. In this text it refers to being enclosed in flames.

F

Fascia — The broad flat surface over a storefront or below a cornice.

Flat Roof — A roof that has a pitch not exceeding 20 degrees.

G

Ground Ladders — Ladders specifically designed for fire service use that are not mechanically or physically attached permanently to fire apparatus, and not requiring mechanical power from the apparatus for ladder use and operation.

Growth Ring — A layer of wood (as an annual ring) produced during a single period of growth.

H

Heat Treatment — A controlled cooling or quenching of heated metals, usually by immersion in a liquid quenching medium; its purpose is to harden the metal.

Heel — The act of preventing the butt end of a ladder from slipping while in use.

Heelman — The firefighter who carries the butt end of the ladder and/or who subsequently heels or secures it from slipping during operations.

K

Knurled — Having a series of small ridges or beads as on a metal surface to aid in gripping.

L

Leg Lock — A method of entwining a leg around a ladder rung to free the climber's hands for working while insuring that the individual cannot fall from the ladder.

Life Belt — A wide adjustable belt with a snap hook that can be fastened to the rungs of a ladder leaving the hands free for working.

M

May — A term used in NFPA standards that denotes voluntary or optional compliance.

Mortise — A hole, groove, or slot cut into a wood ladder beam to receive a rung tenon.

N

NFPA — An abbreviation for National Fire Protection Association, an international nonprofit technical organization which develops fire safety standards by a system of more than 150 committees which make proposals to the association for adoption as standards.

NFPA 1901 — Refers to National Fire Protection Association: Standard for Automotive Fire Apparatus, 1979 Edition.

NFPA 1931 — Refers to National Fire Protection Association: Standard on Design, and Design Verification Tests for Fire Department Ground Ladders, 1984 Edition.

NFPA 1932 — Refers to National Fire Protection Association: Standard on Use, Maintenance, and Service Testing of Fire Department Ground Ladders, 1984 Edition.

Nondestructive Test — A method of testing that does not damage the ladder structure or its components.

O

OSHA — An abbreviation for Occupational Safety and Health Administration.

P

Permanent Deformation — That deformation remaining in any part of a ladder or its components after all test loads have been removed.

Protective Clothing — As used in this text means turnout coat, bunker pants and boots, or three-quarter length boots, helmet, and gloves.

Pitch — There are two meanings as used in this text:

1. The slope of a roof.
2. Resin present in certain woods.

R

Rack — 1. A framework used to support ladders while being carried on fire apparatus.
2. The act of placing a ladder on apparatus.

Rhythm — Handling and climbing ladders with smooth motion.

Ridge — The peak or sharp edge along the very top of the roof of a building.

Roof — The outside top covering of a building.

S

Scuttle — An opening in the roof or ceiling providing access to the roof or attic.

Serrated — Notched or toothed edge.

Set — See deformation.

Shall — A term used in NFPA standards denoting compulsory compliance.

Spanish Windlass — An apparatus such as a stick or dowel used for post-tensioning lines by twisting.

Spotting — Positioning a ladder to reach an object or person.

Splice — To join two ropes or cables by weaving the strands together.

Standard — A document containing requirements and specifications.

Story — That portion of a building between the upper surface of a floor and the floor or roof that is next above it.

T

Throw A Ladder — Raise a ladder quickly.

Tongue — The rib on one edge of a ladder beam that fits into a corresponding groove or channel attached to the edge of another ladder beam; its purpose is to hold the two sections together while allowing the sections to move up and down.

Torque — Something that produces or tends to produce rotation; a turning or twisting force.

Tying In — Securing a ladder to a building or object.

V

Ventilation — Controlled removal of smoke and other gases and replacement with fresh air.

Visible — That which is clearly evident by visual inspection without recourse to optical measuring devices.

W

Wind Shakes — Damage done to timber by repeated flexing in the wind.

Wood Grain — Stratification of wood fibers in a piece of wood.

Introduction

The major objectives of fireground operations are rescue, confinement, and extinguishment. In order to meet these objectives firefighters are required to perform many tasks using a variety of tools and appliances, primary among these being ground ladders.

Many fire service ground ladders resemble commercial ladders in shape and appearance but the potential uses, frequently under very adverse conditions, demand that a margin of safety not usually expected of commercial ladders be provided. For this reason, fire department ground ladders are specially constructed.

As with most fire department tools and appliances, frequent concentrated training is required to develop the individual skills and teamwork necessary for efficient use of ground ladders. This is especially true in the case of ground ladders because the ground ladder is an item of equipment upon which the life of both the firefighter and the public may depend. Even when aerial apparatus is part of the fire fighting attack team ground ladders will still be needed to gain access to locations inaccessible to apparatus.

The uniqueness of ground ladders among the collection of items carried on fire apparatus (there is nothing else carried that is similar) and their immediate general availability make them subject to special and exotic uses. Therefore this edition has been considerably expanded to present evolutions detailing these uses.

The NFPA recently updated the standards for fire department ground ladders, making two standards where there was previously one. This edition presents the highlights of these new standards.

Other editions contained only descriptions of evolutions which began with ladders being raised up from a position flat on the ground. Further, the evolutions were based on the premise that a ladder would always be placed on the ground prior to raising. This edition recognizes that it is often quicker and more logical to begin a ladder carry directly from the apparatus and to proceed directly from the carry to the raise without first placing the ladder flat on the ground. Evolutions are now detailed with this in mind.

The style of this text has been enhanced by the use of color. The number of photographs and drawings has been increased from less than 100 to nearly 750; all are up-to-date illustrations. Finally, study and review questions have been provided to emphasize and assist in understanding important points.

HISTORY AND DEVELOPMENT

The use of ladders goes back to early civilizations. The earliest evidence of their use in North America is not certain, but records indicate that crudely built ladders were being used as early as 1200 AD by cliff dwellers in what is now Arizona, Colorado, New Mexico, and Utah. These primitive people built cavelike dwellings that were often three or four stories high. At each level the rooms were set back into the cliff a few feet from the rooms beneath, thus forming a ledge at each level.

Ladders, rather than steps, were used to reach each level, probably because they could be drawn up when necessary to keep out intruders. These ladders consisted of two poles to which rungs were bound with cords or strips of rawhide. A single beam ladder was also used, which consisted of a pole with notches cut into it to serve as steps. Some rope ladders were also used. They were hand-woven vegetable fibers, usually yucca and milkweed.

Early ladders were single section ladders. Available records do not identify when or where the extension ladder originated. Records do show that in the late 1700s it was the custom to have buckets and a ladder hung at a convenient location in a village for use by whoever could be mustered when a fire occurred.

Hand drawn hook and ladder trucks such as that pictured below were evident by the mid 1800s. They consisted of rackings of single and extension ladders on a specially constructed wagon. Thus, it appears that fire service ladders as we know them today began to emerge at this time.

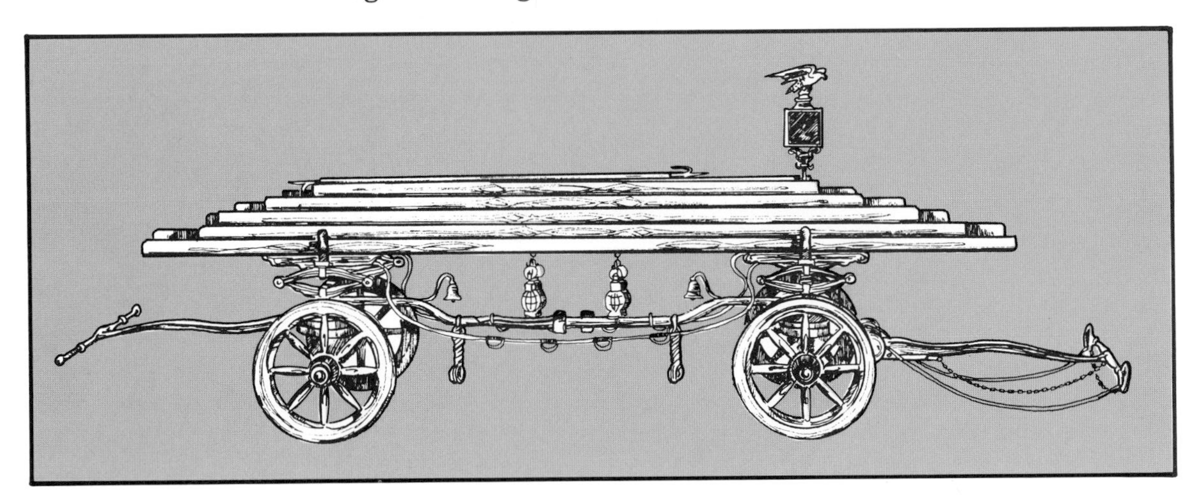

PURPOSE

A training manual is more useful when it presents a variety of methods of accomplishing a task, allowing the user to select the one best suited adaptation to a particular locality. To achieve this purpose, a selection of methods and techniques for handling, raising, and climbing ground ladders have been recommended by a committee of the International Fire Service Training Association.

Other IFSTA training manuals present only basic information on ground ladders because they are broader in scope and so have limited space. The purpose of this manual is to complement and expand this information by providing more detailed in-depth material.

The National Fire Protection Association has adopted and published a set of guidelines of proficiency for a firefighter: NFPA 1001 Fire Fighter Professional Qualifications. Sections 3-9, 4-9, and 5-5 of this standard are requirements for working with ground ladders. This manual presents all the necessary information for the firefighter to achieve NFPA Fire Fighter III level requirements for working with ground ladders.

SCOPE

The scope of this manual is the presentation of both the *need to know* and the *nice to know* information on ground ladders. The information provided ranges from technical to practical and covers everything from design and construction through carries, raises, and climbing to special uses.

FIRE SERVICE LADDER REQUIREMENTS

The fire service needs ladders

- To gain access to upper stories of buildings for rescue and for fire fighting operations
- To provide a vantage point from which a fire stream may be directed onto a fire
- For a variety of special uses

The general requirements for ladders are similar in most communities. However, for a number of reasons, the number, design, and types of ladders and the apparatus carrying them vary considerably. Hydraulically-operated ladders are necessary where there are taller structures but access to many remaining structures is quicker, easier, and more practical using ground ladders.

The minimum lengths, types, and numbers of ground ladders to be carried on apparatus are specified by NFPA Standard 1931: *Standard on Design and Design Verification Tests for Fire Department Ground Ladders*. The use, care, and service testing of ground ladders are specified by NFPA Standard 1932: *Standard on the Use, Maintenance, and Service Testing of Fire Department Ground Ladders*.

LADDERS

Chapter 1

Ladder Types and Ladder Terms

NFPA STANDARD 1001
Fire Fighter I

3-9.1 The fire fighter shall identify each type of ladder and define its use.

Chapter 1

Ladder Types and Ladder Terms

No single type of ground ladder meets all fire service needs. The number of types varies from five to seven depending on how one classifies variations of two types. This manual discusses ground ladders in considerable detail, so it is more desirable to define seven types: single, roof, folding, extension, pole, combination, and pompier. It is important that the firefighter be familiar with the various types because an understanding of configuration in relation to designation is the first step toward understanding ladder uses.

Specialized equipment usually has a language of its own that is used in its day-to-day operation. Ladders are no exception. The section on ladder terms details words and phrases that will be helpful to the firefighter. Both the designations of types and terms may vary with local use. Where possible these variations are identified.

LADDER TYPES

Single Ladders (Wall Ladders)

Single ladders have only one section and are of a fixed length (Figure 1.1). Lengths vary from six feet (2 m) to 32 feet (10 m) with

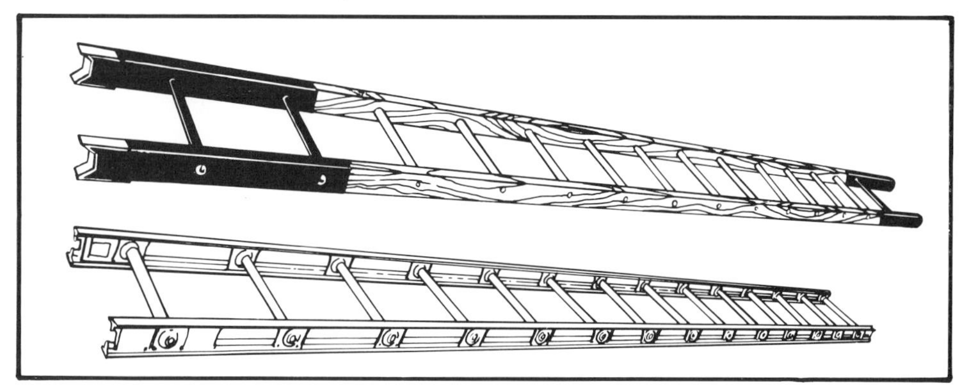

Figure 1.1 Single ladder.

the more common lengths ranging from 12 feet (4 m) to 24 feet (7 m). They are mainly used for operations involving one-and two-story buildings. The primary advantage of this type ladder is that it can be quickly placed by a single firefighter.

Roof Ladders

Roof ladders are single ladders that have hooks attached to the tip end. The hooks are nested between the beams when the ladder is used as a single ladder (Figure 1.2). They are swiveled out (Figure 1.3) only when needed to provide a means of anchoring the ladder when it is used on a sloped roof (Figure 1.4). Lengths vary from 12 feet (4 m) to 24 feet (7 m).

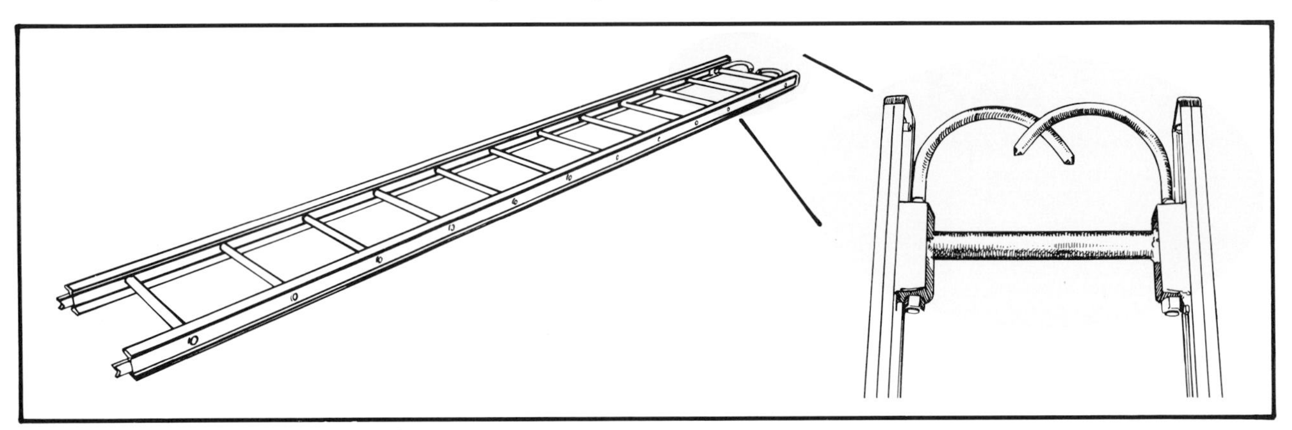

Figure 1.2 Roof ladder with hooks nested.

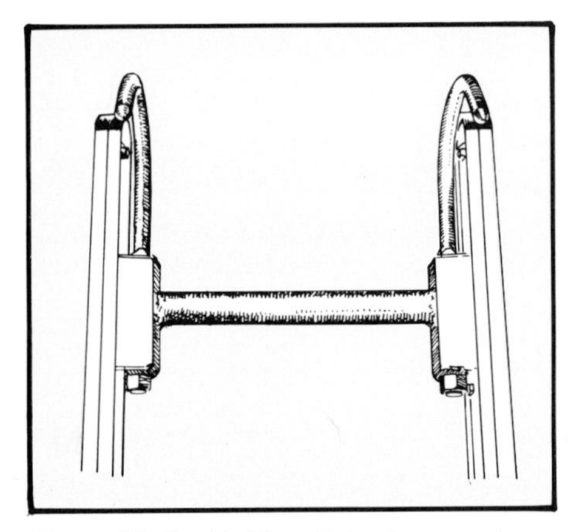

Figure 1.3 Roof ladder with hooks opened.

Figure 1.4 Roof ladder in place on a sloped roof.

Folding Ladders

Folding ladders are a special type of single ladder that have hinged rungs so that they can be folded into a compact assembly with one beam resting against the other (Figure 1.5). This feature allows the ladder to be carried in narrow hallways and aisles and to be taken around corners not possible with regular single ladders. When there are low ceilings, as in residential structures, the ladder's compactness makes it much easier to get into attic scuttle openings because it can be inserted while still folded and then opened in place.

Folding ladders are narrower when open than regular single ladders (Figure 1.6). This feature allows them to be used in narrow scuttle openings common to residential structures. Lengths range from eight feet (2 m) to 16 feet (5 m) with the most common being 10 feet (3 m). Most have foot pads attached to the butt to prevent slipping on floor surfaces, and NFPA 1931 now requires foot pads for folding ladders. The disadvantages of this type ladder are limited weight loading (less than half a regular single ladder) and its narrowness, which makes climbing awkward and leg locking impractical.

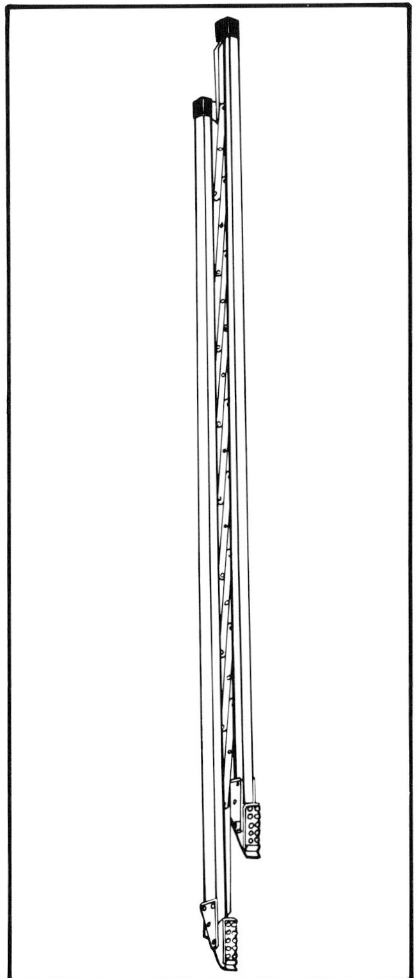
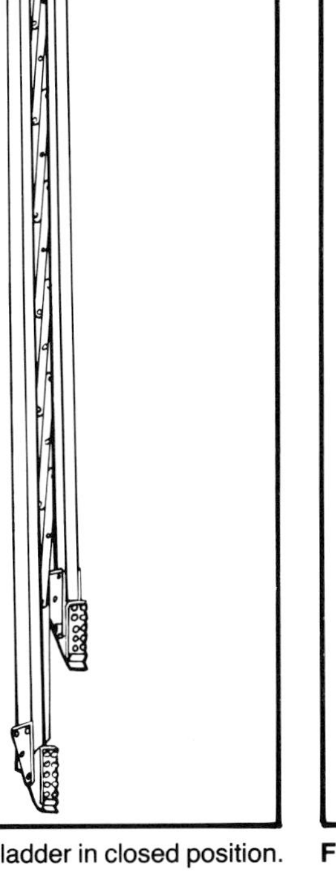

Figure 1.5 Folding ladder in closed position.

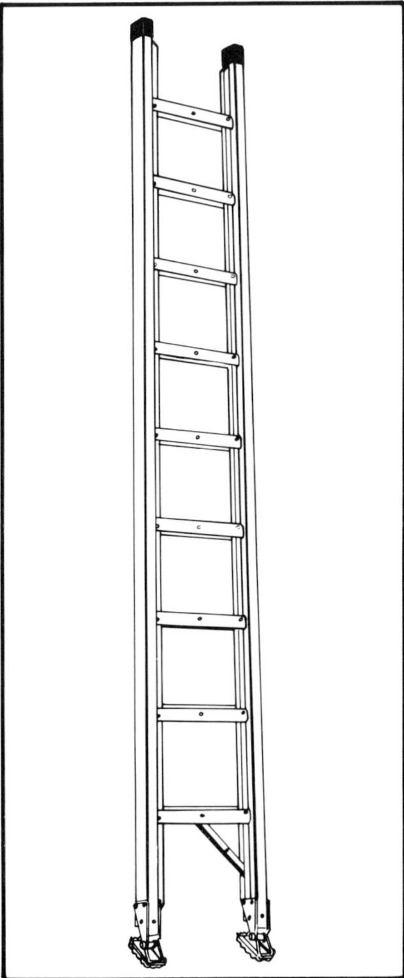

Figure 1.6 Folding ladder in open position.

Extension Ladders

Extension ladders have two or three sections. Upper sections are manually raised and lowered by rope or rope and cable to permit length adjustment. They are referred to by the fully extended length. Lengths normally range from 12 feet (4 m) to 35 feet (11 m) (Figure 1.7).

This design makes it possible to carry longer ladders. For example, it would not be practical to carry a 20 foot (6 m) single ladder on the side of a pumper but a 35 foot (11 m) three section extension ladder can easily be accommodated.

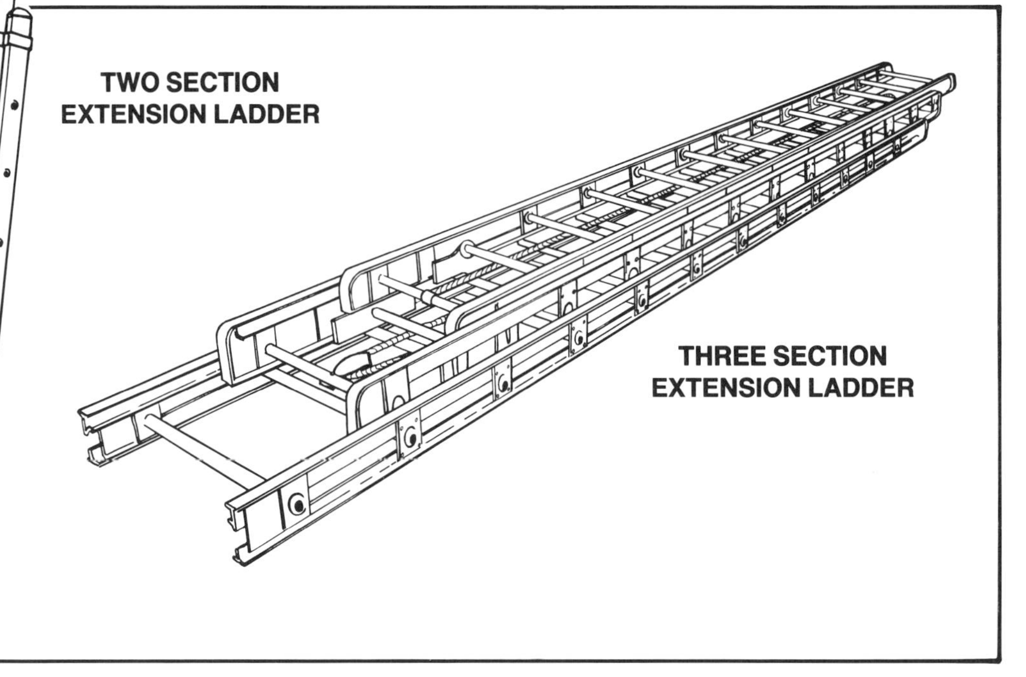

Figure 1.7 Two-and three-section extension ladders.

Pole Ladders (Bangor Ladders)

Pole ladders are extension ladders that have staypoles added for stability (Figure 1.8). They are manufactured with two to four sections. Lengths vary from 35 feet (11 m) to 65 feet (20 m); however, most do not exceed 50 feet (15 m). Use of pole ladders is limited due to manpower shortages in many fire departments. However, they do allow ladder operations on buildings of up to five stories when aerial apparatus is not available or cannot gain access.

Combination Ladders (A-Frame Ladders)

Combination ladders are ladders which can be used both as a step ladder (A-Frame) and either a single or an extension ladder (Figures 1.9 and 1.10). Lengths range from eight feet (2 m) to 14 feet (4 m) with the most popular being the 10 foot (3 m) model. They are usually employed for inside operations and are very useful in reaching ceiling-mounted light fixtures.

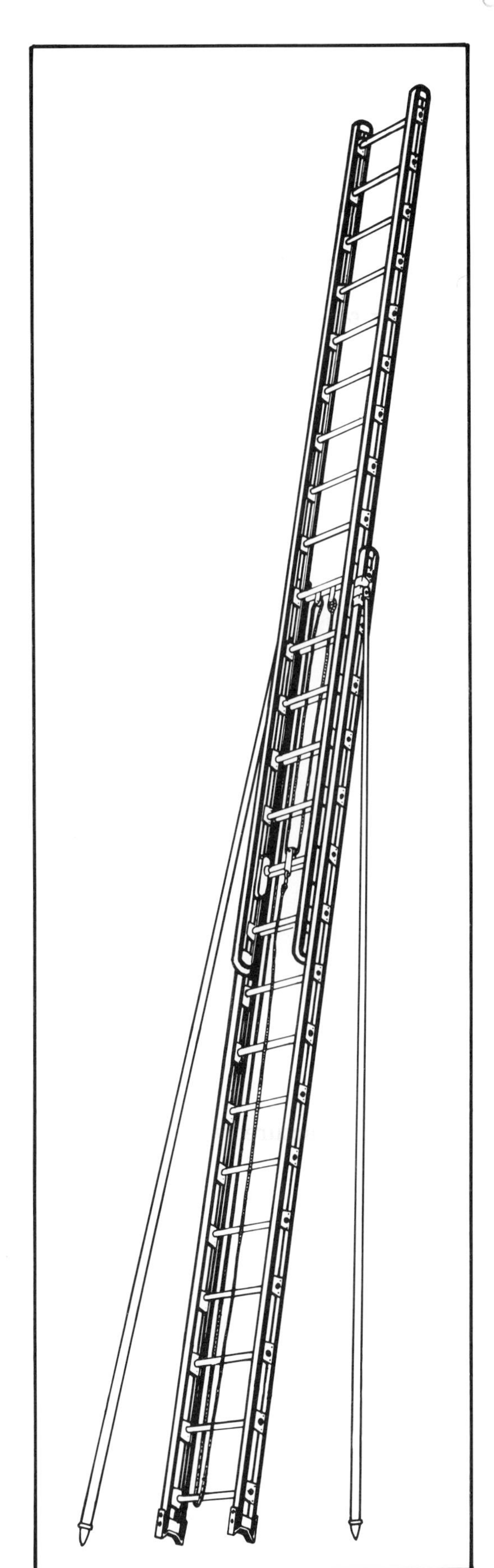

Figure 1.8 Pole ladder.

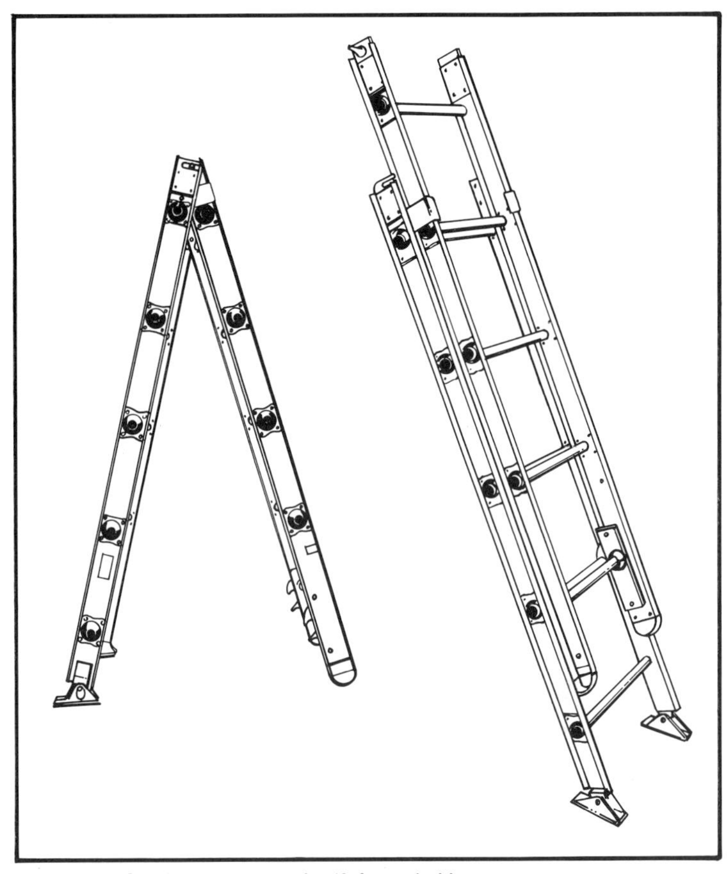

Figure 1.9 Combination extension/A-frame ladder.

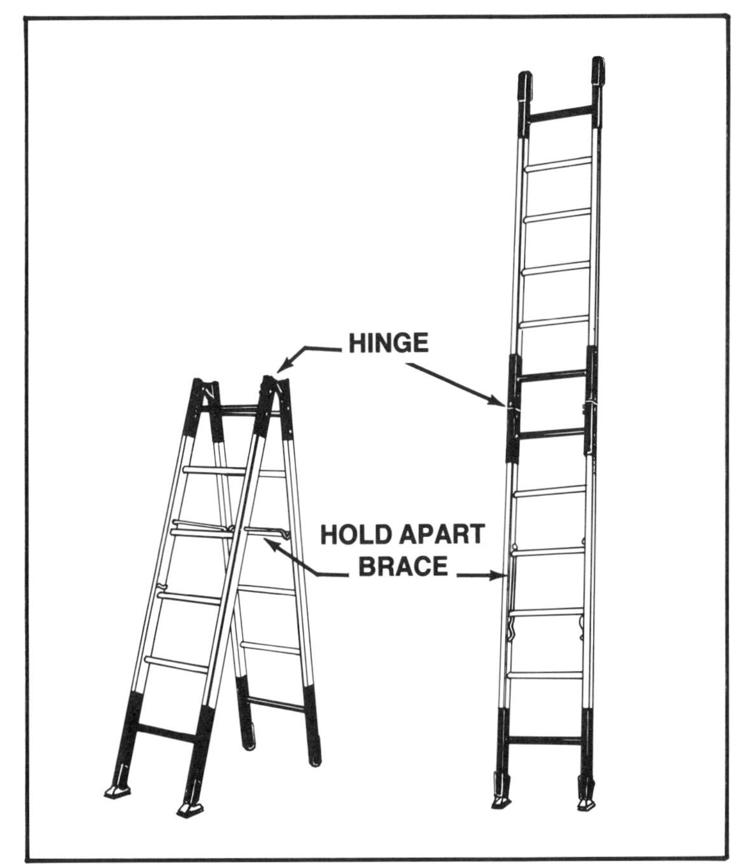

Figure 1.10 Combination single/A-frame ladder.

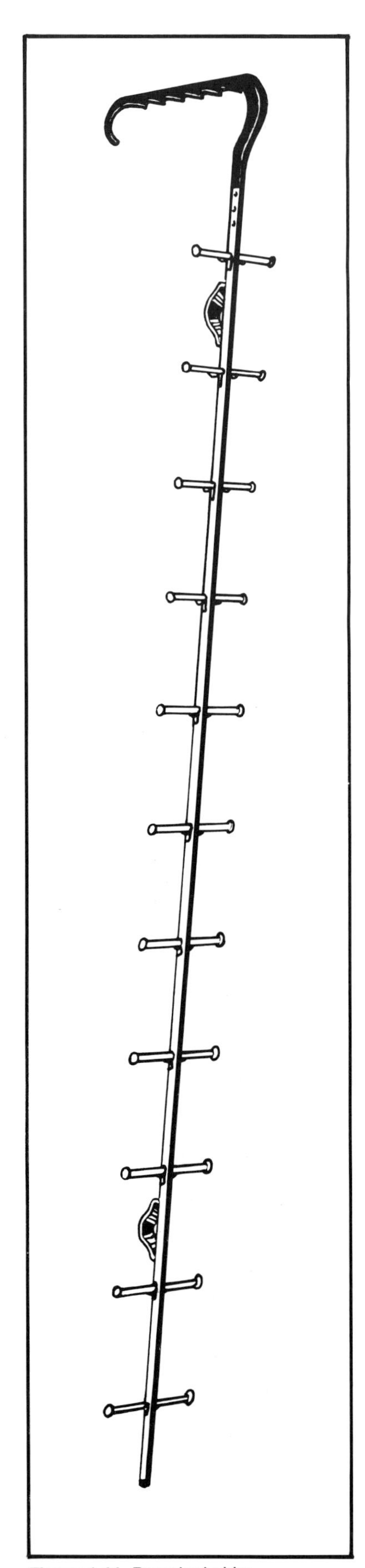

Figure 1.11 Pompier ladder.

Pompier Ladders

Pompier ladders are single beam ladders with rungs projecting from both sides. They have a large metal "gooseneck" or hook projecting at the top for insertion into windows or other openings (Figure 1.11). Lengths vary from 10 feet (3 m) to 20 feet (6 m). They are mainly used for training as a ladder climbing confidence builder, but some fire departments still use pompier ladders in scaling operations to reach points beyond the reach of other ground ladders and aerial apparatus.

LADDER TERMS

Angle of Inclination

Refers to the angle of an in-place ground ladder in relation to horizontal (Figure 1.12).

Beam

The side rail of a ladder (Figure 1.13).

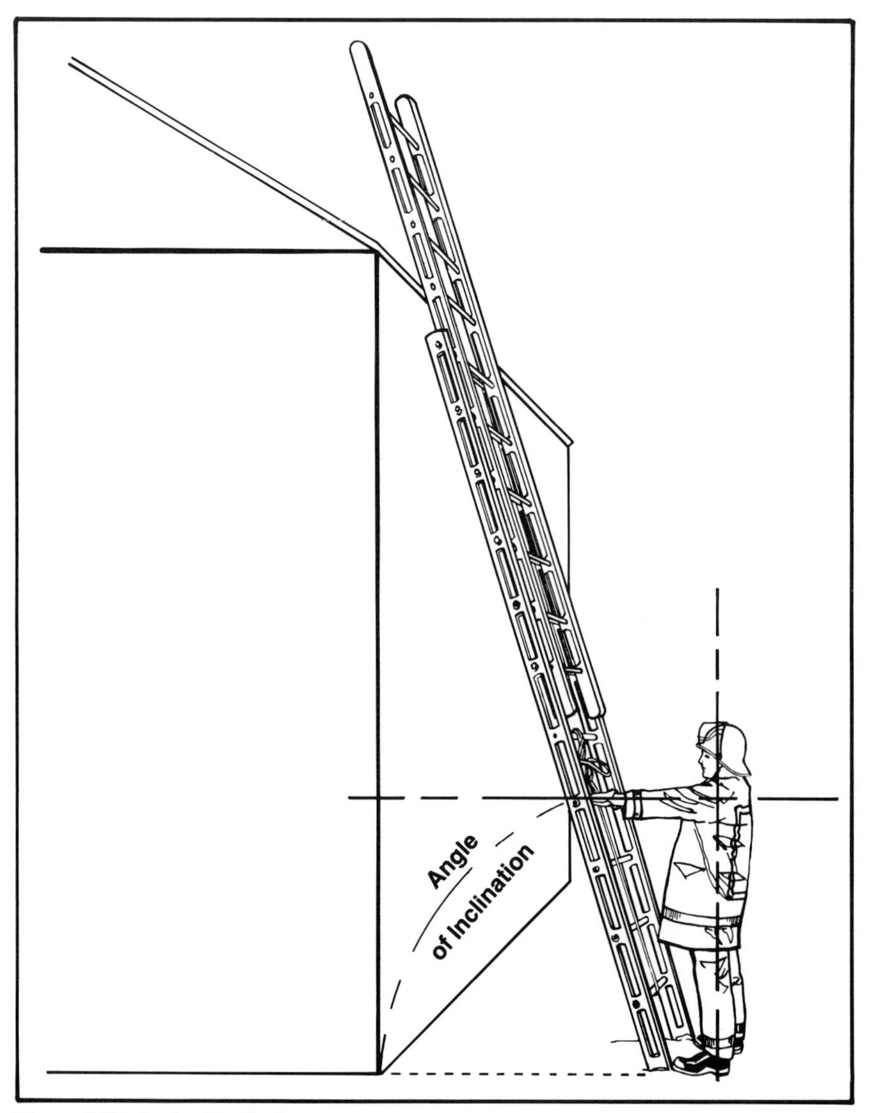

Figure 1.12 Angle of inclination.

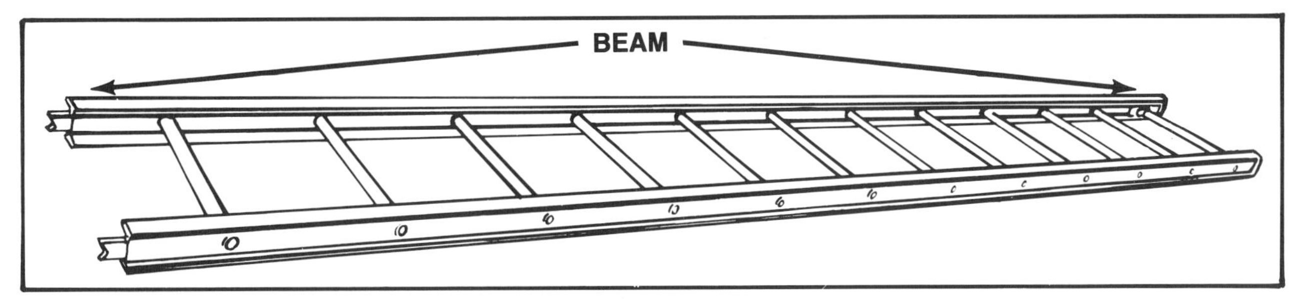

Figure 1.13 A ladder beam or side rail.

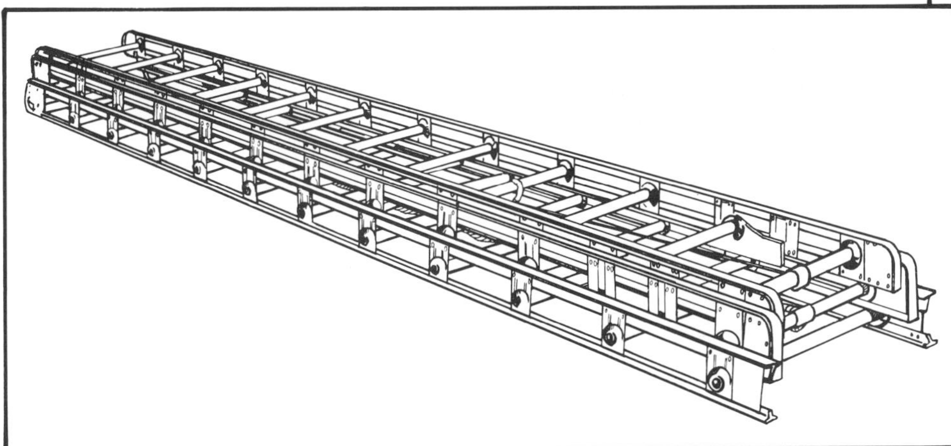

Figure 1.14 Ladder in a bedded position.

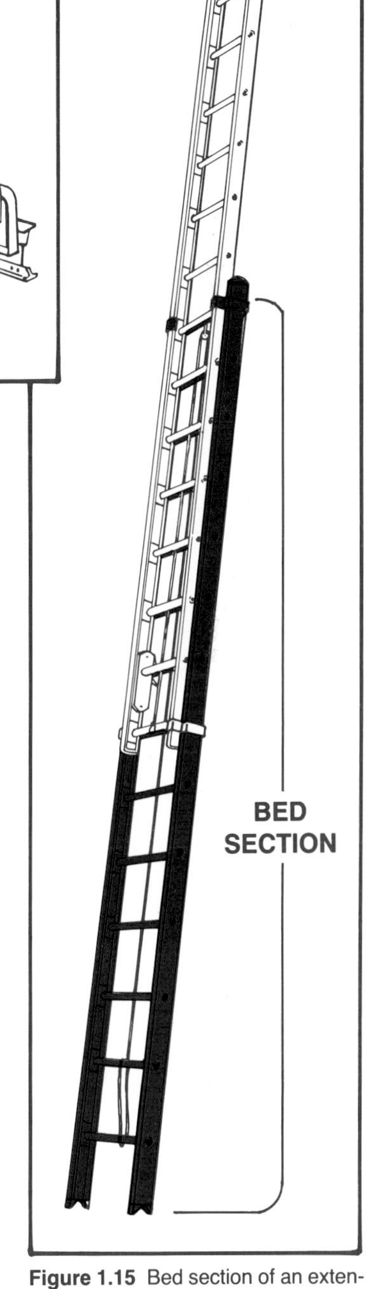

Figure 1.15 Bed section of an extension ladder.

Bedded Position

When the fly(s) of an extension ladder is (are) fully retracted it is said to be in the bedded position; the position in which the ladder is carried on the apparatus (Figure 1.14).

Bed Section (Base Section)

The lowest or widest section of a ground ladder (Figure 1.15).

Butt (Heel)

The bottom end of the ladder; the end which will be placed on the ground or other supporting surface when the ladder is raised (Figure 1.16).

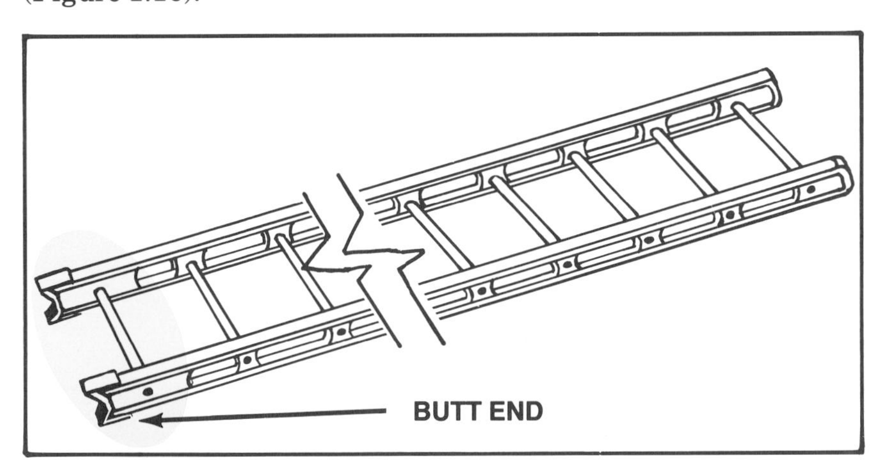

Figure 1.16 The bottom of the ladder is known as the butt.

Designated Length

The length marked on the ladder. NFPA 1931 requires this marking; it specifies that it be within 12 inches (300 mm) of the butt of each side rail of a single ladder and the base section of extension and pole ladders. These markings are also required to be visible when the ladder is in the bedded position and stored on the apparatus (Figure 1.17).

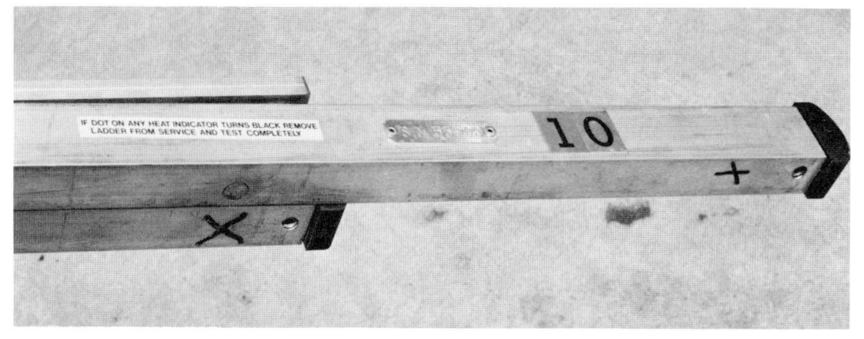

Figure 1.17 Ladders marked to show their designated length.

Dogs

See Pawls

Fly Section

Upper section(s) of extension, pole, or some combination ladders (Figure 1.18).

Halyard

A rope or cable used for hoisting and lowering extension and pole ladder fly sections (Figure 1.19).

Heel

See Butt

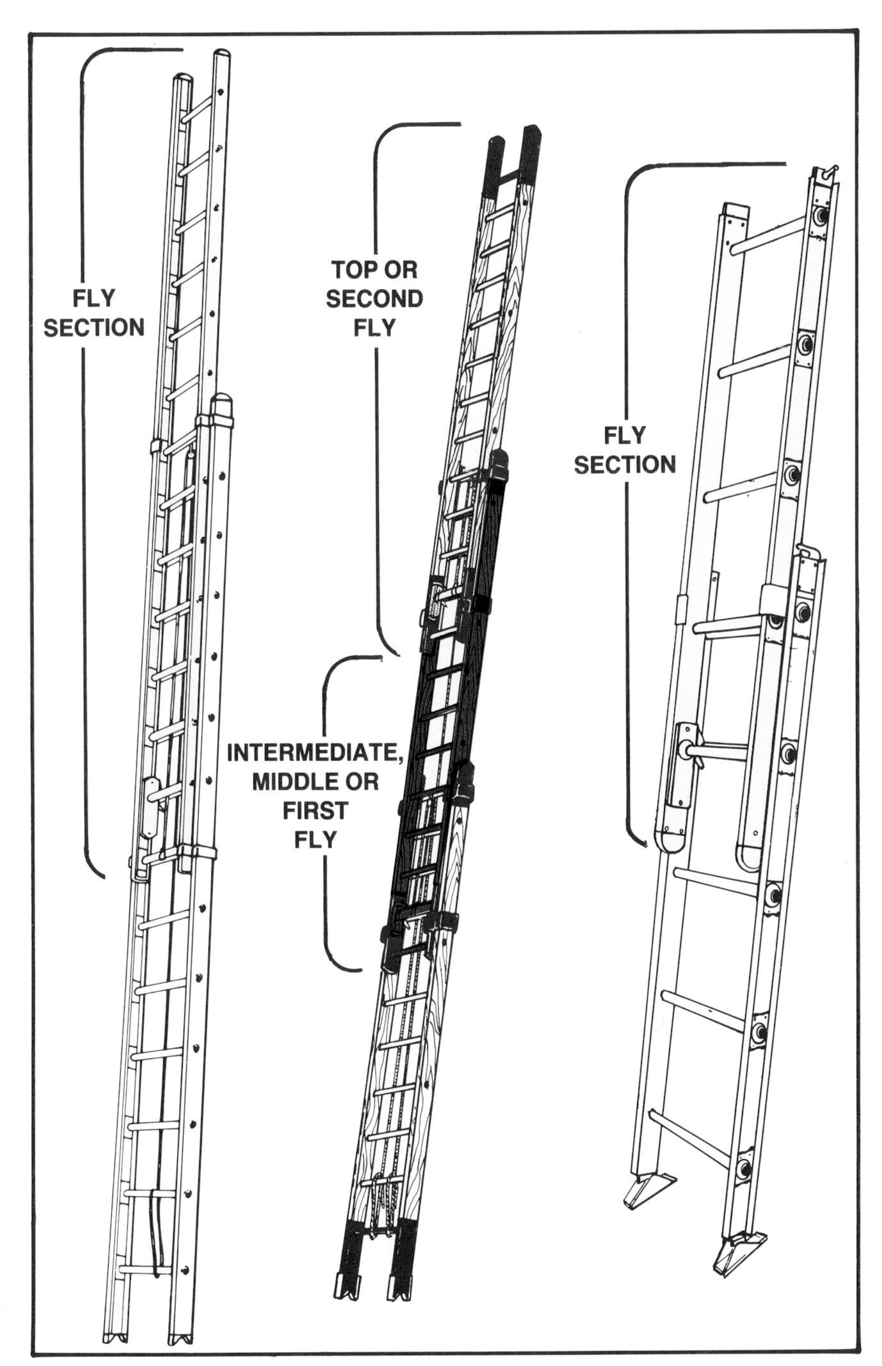

Figure 1.18 Fly sections of extension and combination ladders.

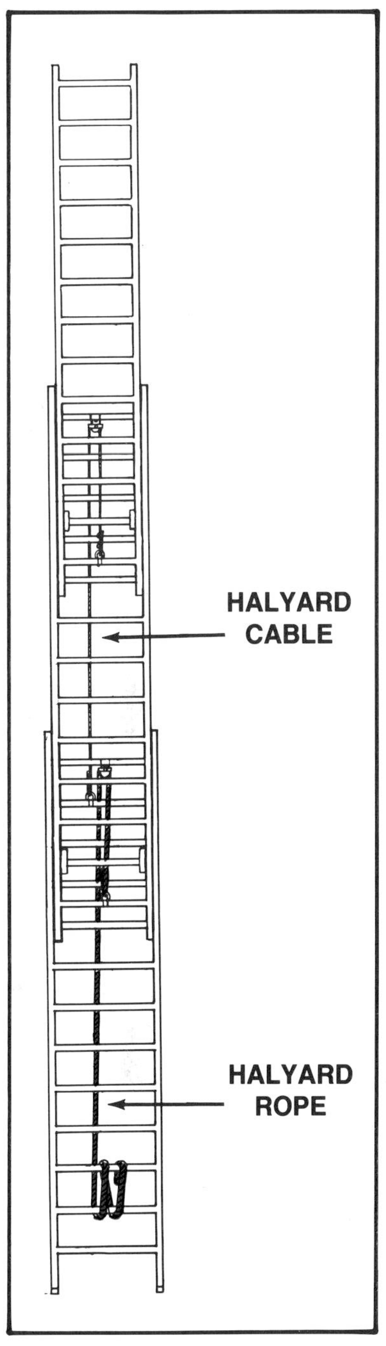

Figure 1.19 Halyard rope and cable on a three-section extension ladder.

Identification Number

NFPA 1931 now requires that ground ladders bear a unique individual identification number (Figure 1.20).

Figure 1.20 Identification or serial number.

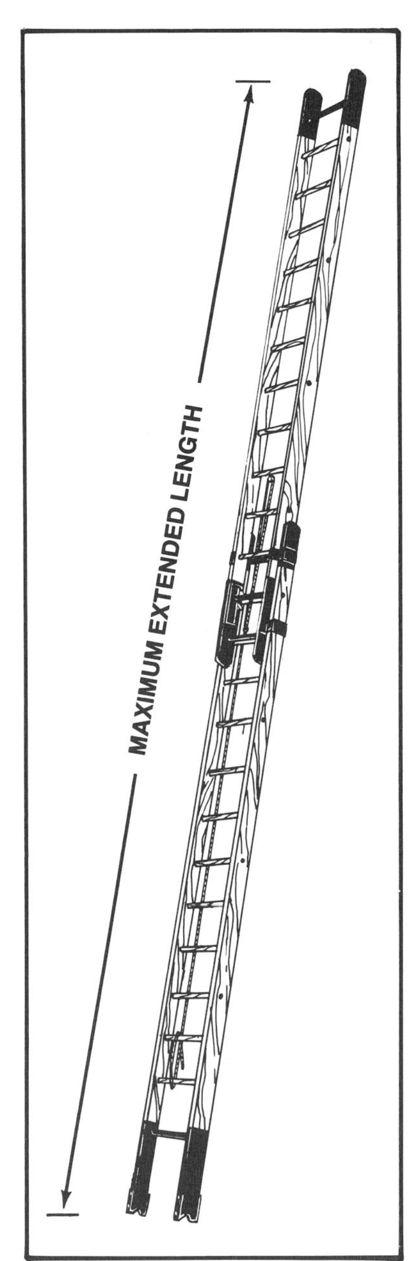

Figure 1.22 Maximum extended length.

Inside Width

The distance measured from the inside of one beam to the inside of the opposite beam (Figure 1.21).

Maximum Extended Length

The total length of the extension, pole, or some combination ladders when all fly sections are fully extended and the pawls are engaged (Figure 1.22).

Nesting

The procedures whereby ladders of different sizes and/or types are racked partially within one another to reduce the space required for storage on the apparatus. The most common arrangement is to nest the roof ladder with an extension ladder on the side of a pumper (Figure 1.23).

Outside Width

The distance measured from the outside of one ladder beam to the outside of the opposite ladder beam (Figure 1.24).

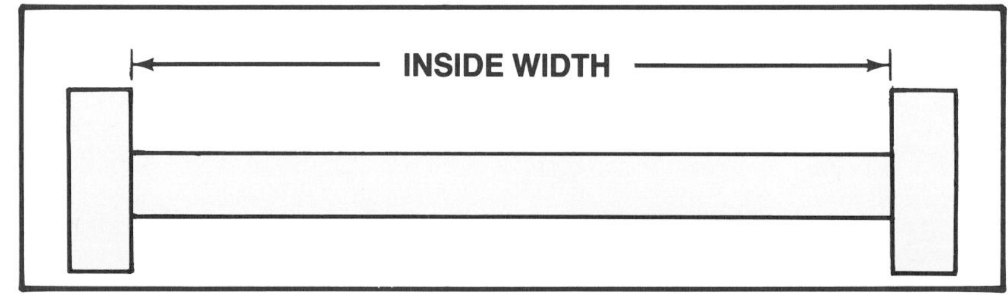

Figure 1.21 Inside width.

Figure 1.23 Roof ladder nested with an extension ladder on the side of a pumper.

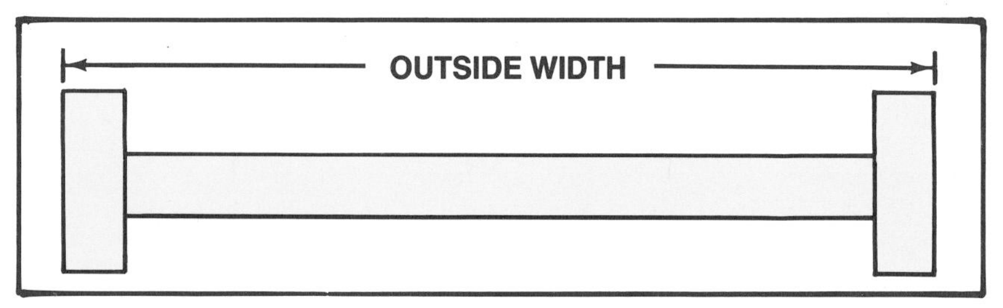

Figure 1.24 Outside width.

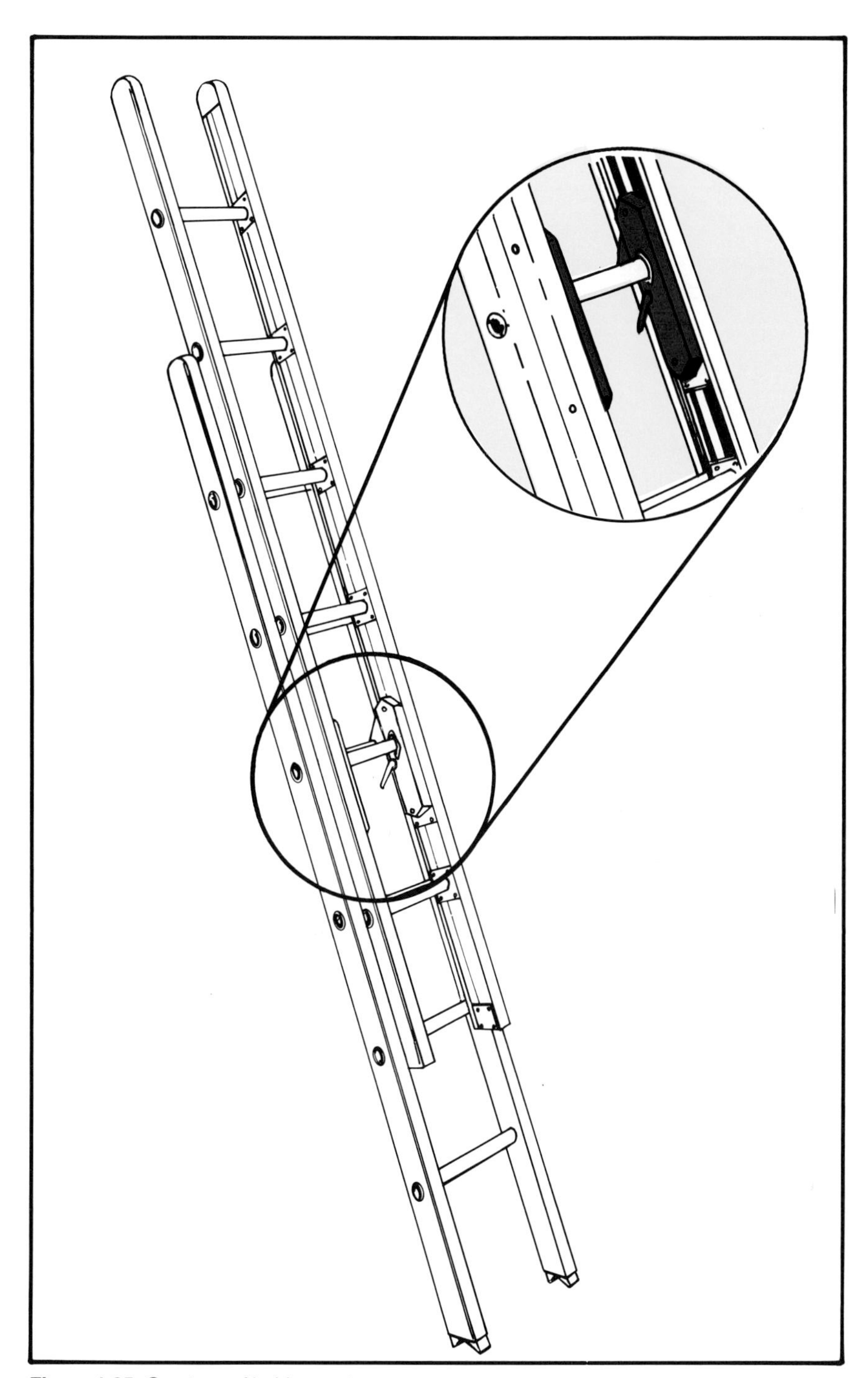

Figure 1.25 One type of ladder pawls.

Pawls

Devices used to hold fly sections at the desired height during use (Figure 1.25).

Rail

See Beams

Retracted

See Bedded Position

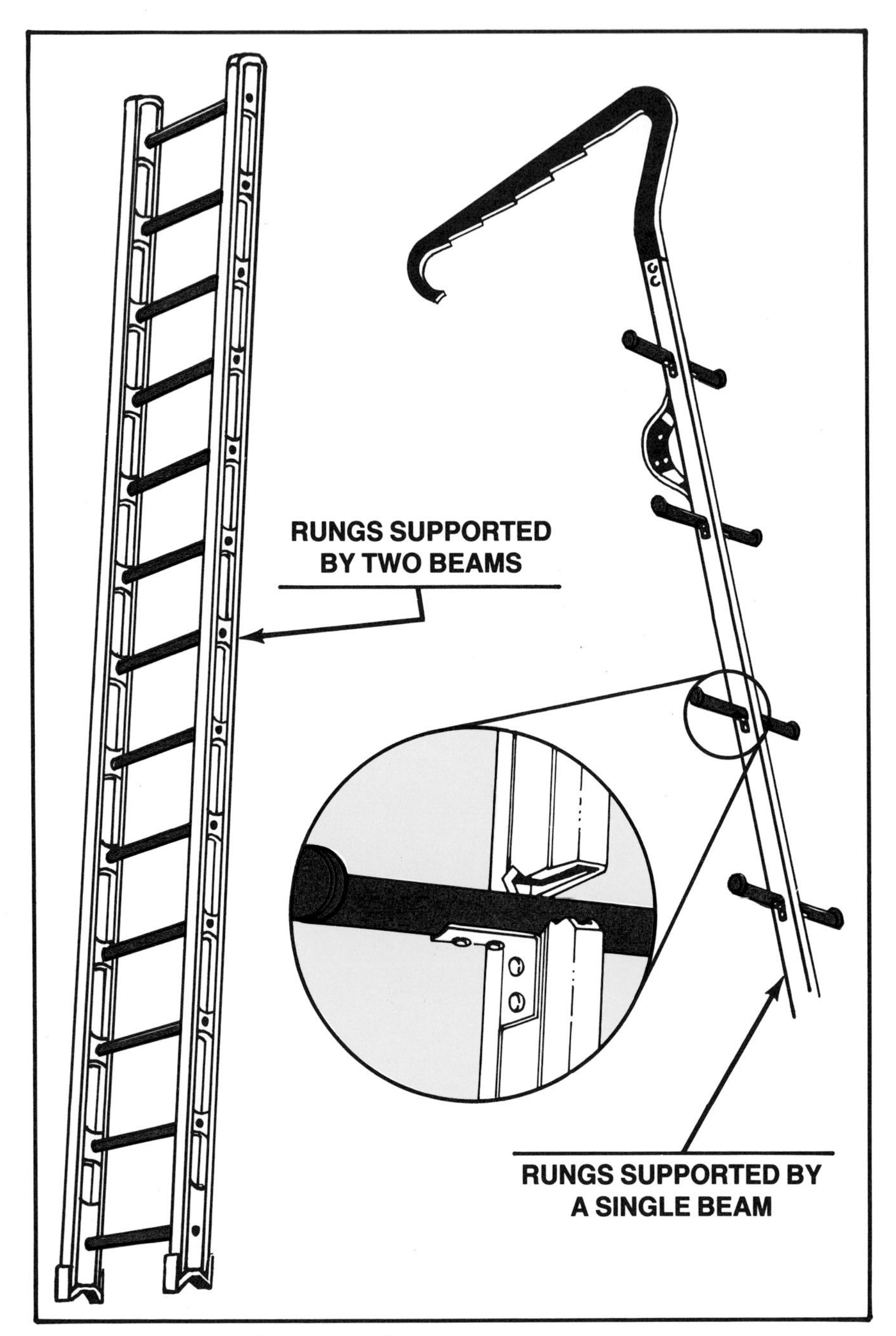

Figure 1.26 Regular and pompier ladder rungs.

Rungs

Cross members which provide the foothold for climbing. In all except pompier ladders, the rungs extend from one beam to the other; on a pompier ladder the rungs pierce the single beam (Figure 1.26).

Side Rail

See Beam

Stripping Ladder

A device which resembles a short, beefed up roof ladder but which actually is a tool to strip roofs of shingles, paper, and roof boards; it is NOT used for climbing. Stripping ladders are usually shorter and narrower than roof ladders and have fixed open hooks of sturdier construction. The butt is reinforced so that it may be used for removing ceilings from above after clearing the roof opening (Figure 1.27).

Tip (Top)

The extreme top of the ladder (Figure 1.28).

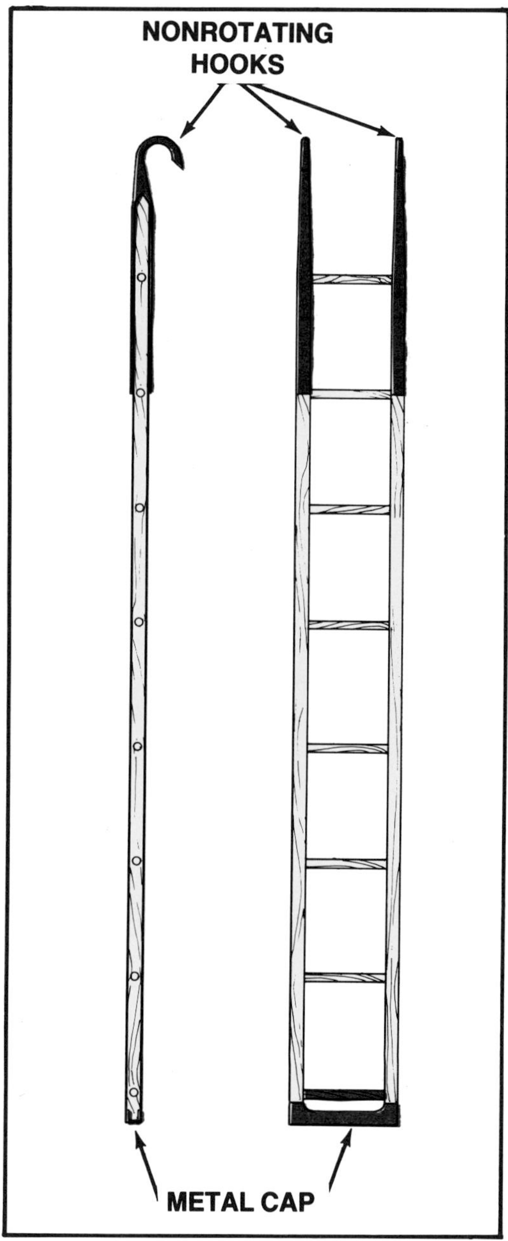

Figure 1.27 Stripping ladder.

Figure 1.28 The top of the ladder is known as the tip.

Review

Answers on page 386

Check the correct response.

1. A ladder that has one section and is of fixed length is a ________.
- ☐ A. Pole ladder.
- ☐ B. Combination ladder.
- ☐ C. Single ladder.

2. A ladder designed for use in narrow hallways or aisles is a _____.
- ☐ A. Folding ladder.
- ☐ B. Pompier ladder.
- ☐ C. Roof ladder.

3. A ladder that can be used as either a single ladder or an A-frame ladder is one type of __________.
- ☐ A. Jacobs ladder.
- ☐ B. Combination ladder.
- ☐ C. Pompier ladder.

4. A ladder having a single beam with rungs protruding from both sides of it is a ______________.
- ☐ A. Folding ladder.
- ☐ B. Jacobs ladder.
- ☐ C. Pompier ladder.

5. In the sketch below, line A shows ____________.
- ☐ A. Rung diameter.
- ☐ B. Inside width of a ladder.
- ☐ C. Outside width of a ladder.

A

True or False.

	True	False
6. Pole ladders are long extension ladders that have staypoles attached.	☐	☐
7. Roof ladders are single ladders with a fixed gooseneck shaped hook attached to the tip.	☐	☐
8. The term *LADDER NESTING* refers to the construction feature of extension ladders where the fly section fits between the beams of the bed section.	☐	☐

	True	False
9. When referring to the position of fly sections of extension ladders, *RETRACTED* is the opposite of extended.	☐	☐
10. A *STRIPPING LADDER* is not really a ladder. It is a ventilation tool.	☐	☐

Fill in the blank.

11. ______________________________ is the term that refers to the way a ladder is placed in relation to a horizontal plane.

True or False.

	True	False
12. The bed section of a ladder is the middle section of a three-section extension ladder.	☐	☐
13. The bedded position of an extension ladder is when the flys are fully retracted.	☐	☐

Chapter 2

Design and Construction, Maintenance, Service Testing

NFPA STANDARD 1001
Fire Fighter I

3-9.8 The fire fighter shall demonstrate the techniques of cleaning ladders.

4-9.1 The fire fighter shall identify the materials used in ladder construction.

4-9.2 The fire fighter shall identify the load safety features of all ground and aerial ladders.

4-9.3 The fire fighter shall demonstrate inspection and maintenance techniques for different types of ground and aerial ladders.

Fire Fighter III

5-5.1 The fire fighter shall conduct an annual service test for ground ladders.

Chapter 2

Design and Construction, Maintenance, Service Testing

Considering that there are seven types of ground ladders and that there are a number of different manufacturers, it is not surprising to find differences in design, construction, and materials used for construction. Choosing between solid and truss design or between wood, metal, or fiber glass is frequently a matter of personal preference because all are constructed to meet the same standards. These are identified and discussed in the Design and Construction section of this chapter.

Fire department ground ladders, the same as any tool or appliance, require unique maintenance procedures. These will vary according to the types of construction and materials used. The importance of maintenance is critical in the case of ground ladders because of their direct involvement in lifesaving situations. Most of this maintenance is performed by the firefighter who is, after all, one of the main beneficiaries of the effort involved. The maintenance section of this chapter deals with this aspect.

Ground ladders, like apparatus pumps, aerial apparatus, fire hose, rope, breathing apparatus, portable extinguishers, etc., are required to be service tested at specified intervals. The Service Testing section enumerates the procedures, most of which the firefighter will be involved in at least to the extent that the individual will be assisting with testing.

PROPER MAINTENANCE OF LADDERS IS CRITICAL DUE TO THEIR DIRECT RELATIONSHIP TO LIFE SAFETY.

DESIGN AND CONSTRUCTION

Workmanship

Fire service ground ladders have to be able to withstand considerable abuse: they may be overloaded (Figure 2.1), they may be subjected to temperature extremes (Figure 2.2), they must withstand flames suddenly rolling out to engulf them (Figure 2.3), and parts of collapsing structures may fall on them (Figure 2.4). Since they are a basic piece of equipment that is widely available at the fire scene, ladders are sometimes utilized for tasks other than what they were designed for (Figure 2.5).

Figure 2.1 Example of an overloaded ladder.

Figure 2.2 Ice encrusted ground ladders. *Courtesy of Bill Noonan, N. Weymouth, Mass.*

Figure 2.3 Flame impinging on a ladder. *Courtesy of Bob Norman Elkton, Md.*

Figure 2.4 Debris falling on a ladder.

Figure 2.5 Using a ladder for a purpose other than what it was designed for.

2.1 True or False.

	True	False
1. Due to the nature of fire service operations, fire service ground ladders have to be able to withstand considerable abuse. Because of this they must be stronger and of a higher quality workmanship than ladders built for public or commercial use.	☐	☐

Nearly every use of a fire service ladder has lives depending on it, most frequently those of the firefighters. Because of this it is very important that the ladder have no structural defects or design weaknesses. Workmanship should also be such that other defects that may cut or tear clothing or skin do not exist.

It was for these reasons that NFPA 1931 *Standard on Design and Design Verification Tests For Fire Department Ground Ladders,* 1984 Edition, was developed and adopted. Fire service ground ladders should be purchased from specifications which adopt NFPA 1931 by reference.

How does the firefighter or fire official know that a particular ladder has been constructed in accordance with the standard? NFPA 1931 answers this by requiring that the manufacturer provide certification attesting that a particular ladder was manufactured in accordance with the requirements of the standard. This certification is in the form of a label affixed to the ladder (Figure 2.6).

Figure 2.6 The label on the ladder in this photograph is an example of the way a ladder manufacturer provides certification that the ladder was constructed in accordance with NFPA Standard 1931.

2.2 Fill in the blank.

1. NFPA Standard ______ prescribes performance specifications for construction of fire service ground ladders.

2.1 True

Materials Used

Obviously the materials used to build fire service ladders have to be of first quality. This requirement is one of the factors that led to major changes in the materials used for construction of most fire service ladders.

Wood for the construction of fire service ladders must be carefully selected before the manufacturing process begins. Only a few types of lumber meet the requirements of weight per cubic inch, bending strength, stiffness, hardness, and resistance to shock. The most desirable of these are clear straight-grained Douglas fir for beam construction and hickory for rungs.

Up to the time of World War II (1941-1945) nearly all fire service ladders were constructed of wood.A problem developed when high-grade Douglas fir became hard to find and ladder manufacturers were forced to look for new materials. Although it was not liked at first, the metal ladder is now the most widely used ladder in the fire service. However, wood ladders are still being manufactured and many more are still in use.

> **2.3** Fill in the blanks.
> **1.** Prior to 1945 nearly all fire service ground ladders were constructed of ______. This is no longer true. ______ is now the most commonly used material.

Fiber glass is just entering the picture as a material used for construction of fire service ground ladders. Its acceptance has been slow and it may never be really popular because it is heavier than corresponding metal and wood ladders.

Ladders constructed of a combination of wood beams and metal rungs have also been marketed, primarily to meet the objections that some have to the electrical conductivity hazard of metal ladders.

> **2.4** Fill in the blank.
> **1.** ______ is the newest material used for construction of fire department ground ladders.

METAL

Most fire service metal ground ladders are built of heat-treated aluminum alloy. Other than the fact that it is more available than wood, the metal ladder has been found to have some definite advantages, as detailed below.

Advantages

- In most lengths and models metal ladders are lighter in weight than their wood and fiber glass counterparts. But there are a few exceptions. Table 2.1 provides a comparison of the weights of the more popular lengths and types carried on fire department pumpers; the information comes from major manufacturer's catalogs. These catalogs provide complete weight information on each ladder made by the particular manufacturer.

TABLE 2.1
Weight Comparison of Most Common Types of Ladders

U.S.

		METAL									WOOD			FIBER GLASS	COMPOSITION
		SOLID BEAM						TRUSS BEAM			SOLID BEAM	TRUSS BEAM		SOLID BEAM	TRUSS BEAM
		MFG 1		MFG 2		MFG 3		MFG 1	MFG 2		MFG 1	MFG 1	MFG 4	MFG 1	MFG 1
LENGTH IN FEET	TYPE	Regular	X-Duty	Regular	X-Duty	Regular	X-Duty	Regular	Regular	X-Duty	Regular	Regular	Regular	Regular	Regular
14'	Roof	31	40	30	41	33	40	49	30	43	37	45	42	38	40
24'	Two Section Extension		74	57	80	79	100	97	92	102	75	95	110	85	90
35'	Three Section Extension		130		116	132	159	163	121	138	185	149			145

WEIGHT IN POUNDS

METRIC

		METAL									WOOD			FIBER GLASS	COMPOSITION
		SOLID BEAM						TRUSS BEAM			SOLID BEAM	TRUSS BEAM		SOLID BEAM	TRUSS BEAM
		MFG 1		MFG 2		MFG 3		MFG 1	MFG 2		MFG 1	MFG 1	MFG 4	MFG 1	MFG 1
LENGTH IN METERS	TYPE	Regular	X-Duty	Regular	X-Duty	Regular	X-Duty	Regular	Regular	X-Duty	Regular	Regular	Regular	Regular	Regular
4 m	Roof	14	18	14	19	15	18	22	14	20	17	20	19	17	18
7 m	Two Section Extension		34	26	36	36	45	44	42	46	34	43	50	39	41
11 m	Three Section Extension		59		53	60	72	74	55	66	84	68			66

WEIGHT IN KILOGRAMS

- Metal ladders are tougher than either wood or fiber glass, so they show less wear and tear from everyday use.
- They will not chip or crack when subject to impact as can wood and fiber glass ladders.
- Users are not subject to injury from splinters, as is sometimes the case with wood ladders.
- Metal ladders are easier to care for than wood ladders because they do not need to be sanded down and refinished periodically. Since this is a time-consuming process, metal ladders have considerably less downtime.
- Finally, metal ladders do not deteriorate with age to the extent that wood ladders do. (Wood ladders are subject to water absorption and dry rotting.) So in the long run, metal ladders can be a better investment.

2.2 1931

2.3 Wood, Metal

2.4 Fiber Glass

2.5 True or False.

	True	False
1. In most models a metal ladder is heavier than a corresponding wood ladder.	☐	☐
2. Metal ladders are tougher than wood ladders.	☐	☐
3. Metal ladders do not crack or chip easily as do wood and fiber glass ladders.	☐	☐
4. Metal ladders require less maintenance than wood ladders.	☐	☐

Disadvantages

There are also some distinct disadvantages to using metal ladders.

- First and foremost - they are good conductors of electricity. EXTREME CAUTION IS NECESSARY WHENEVER METAL LADDERS ARE USED NEAR ELECTRICAL POWER SOURCES.
- Second and only slightly less important, metal ladders are subject to sudden failure when exposed to heat or flame temperatures of 600°F (316°C) or more, even for short periods. Even though they may not fail immediately, the metal loses its strength when so heated and does not regain it when it cools. These kinds of temperatures are routinely encountered at fires and so a second safety precaution is necessary:
  ANY METAL LADDER SUBJECT TO DIRECT FLAME CONTACT OR HEAT HIGH ENOUGH TO CAUSE WATER CONTACTING IT TO SIZZLE OR TURN TO

STEAM, OR WHOSE HEAT SENSOR LABEL HAS CHANGED COLOR, SHOULD BE REMOVED FROM SERVICE AND SUBJECTED TO A HARDNESS TEST.*

- Metal ladders may become very cold in winter or hot in summer because of the good conductivity of the aluminum alloy.

2.6 Check the correct responses.

From the list below select those statements which are the DISADVANTAGES of metal ladders.

- ☐ 1. They expose the user to the possibility of injury from splinters.
- ☐ 2. They deteriorate with age more than other types of ladders.
- ☐ 3. They are good conductors of electricity even when they are dry.
- ☐ 4. They are subject to sudden failure when exposed to temperatures of 600°F (316°C) or more even for short periods.
- ☐ 5. They take on the characteristics of the surrounding environment: hot when exposed to heat, cold when exposed to winter temperatures.

WOOD

As pointed out earlier, a select grade of clear straight-grained Douglas fir is used for construction of ladder beams. Truss blocks are also usually Douglas fir although some white spruce was used. Douglas fir is used because it is relatively free from knots, checks (cracks), pitch pockets, and wind shakes (damage due to wind flexing the growing tree), and otherwise has the strength and flexibility needed.

Douglas fir is not used when it is freshly cut because it is first necessary to reduce the moisture content of the wood. Otherwise, excessive shrinkage would occur after the ladder was constructed, with resulting warping and looseness. The drying period is usually two years.

Not just any Douglas fir can be used; the lumber selected has to have annual growth rings relatively close together. Figure 2.7 shows characteristic growth rings sawed so that maximum strength is provided. Figure 2.8 shows a section of ladder stock side rail with a flat grain which makes it unsuitable because it does not wear well. Such stock may separate at the grain when

---

*A detailed description of the Hardness Test and of the Hardness Tester may be found in the testing section of this chapter.

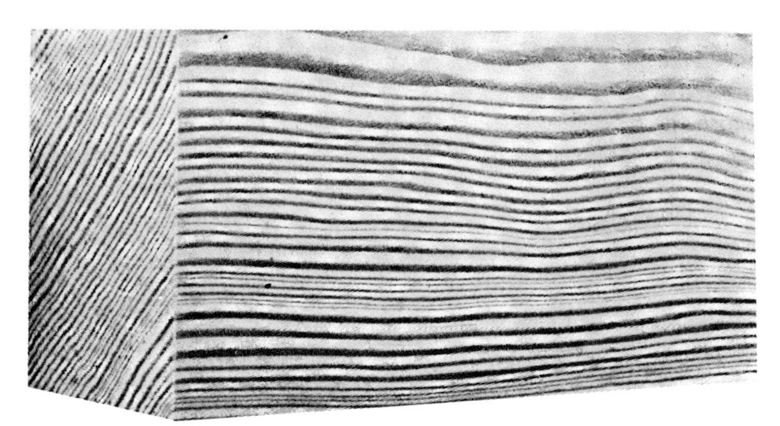

Figure 2.7 Ladder side rail stock sawed so that the grain is angular. Sawing in this manner provides maximum strength.

2.5 1. False
2. True
3. True
4. True

Figure 2.8 Ladder side rail stock sawed with the grain flat. The wood is not as strong when sawed this way and it does not wear as well.

under stress. A sectional view of a trussed beam for a wood ladder shown in Figure 2.9 indicates the proper grain angle for hard edge bearing.

An enlargement of a piece of Douglas fir is shown in Figure 2.10. Note the cellular structure of the wood. Although all wood is

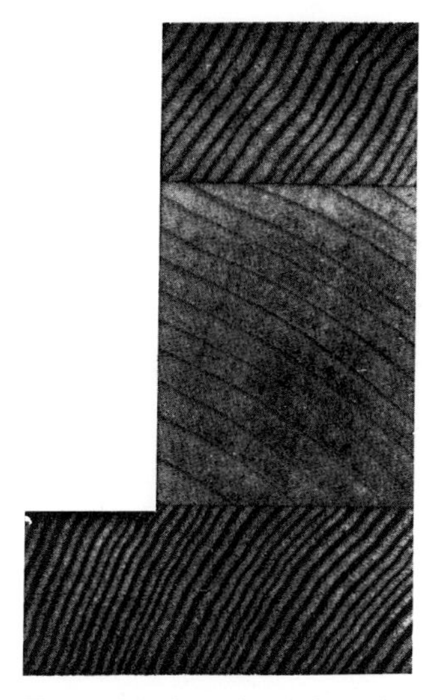

Figure 2.9 A sectional view of a wood truss beam.

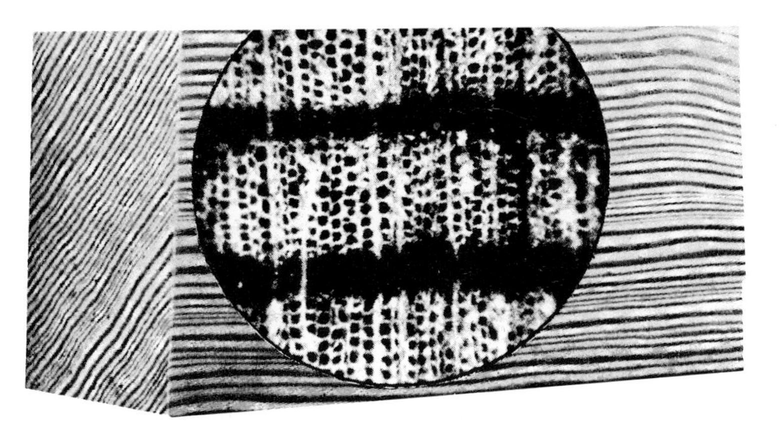

Figure 2.10 Enlargement of a piece of Douglas fir. Note the cellular structure and the porous honeycomb-like composition.

not the same, the enlargement shows that wood is porous like a honeycomb. The dark diagonal streaks, which are the dense summer wood, are generally termed the growth rings. It is not difficult to understand how excessive loss of moisture, resins, and oils can leave empty cells which result in shrinkage and loss of strength.

Wood shrinkage is particularly noticeable after a wood ladder has been subjected to low humidity and artificial heat, conditions that are common in fire stations. This slow drying process is a primary cause of loose rungs and cracked rails in wood ladders.

Rungs for wood ladders are usually made from second-growth hickory. Second-growth wood is derived from a stump of a tree that has been felled previously and which has grown back. The wood is split before turning so that the grain will run the full length of the rung.

2.7 True or False.

	True	False
1. The commonly used materials for wood ladder construction are Douglas fir for beams and second growth hickory for rungs.	☐	☐
2. Douglas fir with wide annual growth rings should be used for wood ladder beam construction.	☐	☐
3. Low humidity and artificial heat are common causes of wood ladder shrinkage.	☐	☐
4. Second growth wood is obtained from trees that grow from the stump of a tree that has been previously felled.	☐	☐

Advantages

- When dry they are nonconductors of electricity, which makes them much safer to use around live electrical power sources.
- Wood ladders are not subject to sudden failure when subjected to heat or flames. When flames engulf a wood ladder it may ignite, but only the surface will burn; the center remains solid and strong.
- Wood is a good insulator; it does not transmit temperature extremes like metal ladders do.

Disadvantages

The big drawbacks to the user are

- A finish, such as a clear varnish, is required to prevent

aging. This must be kept intact, which often requires considerable time and labor.

- A wood ladder will deteriorate with age.

2.6 3, 4, 5

2.8 Fill in the blank.

1. When dry, wood ladders are _______ of electricity.
2. _______ ladders are not subject to sudden failure when exposed to heat and flame.
3. A coating of _______ is required to prevent aging of wood ladders.

FIBER GLASS

Ladders made of fiber glass are relatively new to the fire service. These ladders are not actually all fiber glass; the beams are fiber glass and the remaining parts are metal.

Advantages

- Fiber glass is a nonconductor of electricity when dry, making it comparable to wood in this aspect.
- Fiber glass can be exposed to heat and flame and will regain its original load-carrying capacity when it cools.
- It is a tough material which will take considerable abuse and requires little maintenance.
- It does not require any type of preservative finish like wood ladders do.
- Areas exposed to flame will discolor, providing a positive indication of this occurrence.
- Fiber glass is a poor conductor of heat and is much like wood in this respect.

Disadvantages

The disadvantages of fiber glass construction are

- It is a dense man-made material and so is relatively heavy.
- It tends to chip and crack with severe impact.
- When overloaded it may suddenly crack and fail.

FIBER GLASS LADDERS COMBINE THE LOW MAINTENANCE ADVANTAGES OF METAL LADDERS WITH THE NONCONDUCTIVITY OF WOODEN LADDERS.

2.9 True or False.

	True	False
1. Fiber glass ladder beams exposed to high temperatures will regain their load-carrying capacity when the beams cool to normal temperatures.	☐	☐
2. Fiber glass ladders require more maintenance than wood ladders.	☐	☐
3. Fiber glass ladders are lighter in weight than comparable wood or metal ladders.	☐	☐
4. Fiber glass ladders are designed not to fail or crack when overloaded.	☐	☐

COMPOSITE WOOD AND METAL

This new ladder attempts to combine the best features of wood and metal. It uses wood truss beams with metal rungs.

Advantages

- It is a nonconductor of electricity.
- It is lighter than comparable all-wood ladders.

Disadvantages

- The beams have all the wear and maintenance disadvantages of a wood ladder.

2.10 True or False.

	True	False
1. A composite ladder is heavier than comparable wood ladders.	☐	☐
2. The primary advantages of the composite ladder are that when dry it is a nonconductor of electricity and the use of metal rungs reduces the maintenance necessary for a comparable wood ladder.	☐	☐

Features of Construction

NFPA 1931 has a major influence in this area because the standard prescribes widths and loading limits that have to be met.

WIDTH

The specified widths are based on the type of ladder except in one case where it is based on the length of the ladder. Table 2.2 shows the minimum inside clear widths required.

TABLE 2.2
Minimum Inside Clear Width

Type of Ladder	*Inside Width Between Side Rails*
Folding ladders in open position	7½ inches (190 mm)
Combination ladders	7½ inches (190 mm)
Single and Roof ladders	16 inches (410 mm)
Extension ladders 16 feet (5 m) and under	7½ inches (190 mm)
Extension ladders over 16 feet (5 m) and Pole ladders	16 inches (410 mm)
	Width of Rungs
Pompier ladders	12 inches (300 mm)

LADDER LOADING

CAUTION: Fire service personnel should be alert to the fact that the maximum loading figures shown in Table 2.3 are for ladders that have been raised at the proper angle and placed with the top supported.

The load is the TOTAL WEIGHT ON THE LADDER including persons, their equipment, and any other weight such as charged hoselines.

TABLE 2.3
Maximum Ladder Loading

Type of Ladder	**Load** *US*	*Metric*
Folding	300 pounds	(136 kg)
Pompier	300 pounds	(136 kg)
Single and Roof	750 pounds	(340 kg)
Extension and Pole	750 pounds	(340 kg)
Combination	750 pounds	(340 kg)

NOTE: These figures are based on the 1984 edition of NFPA 1932. A SIGNIFICANT WEIGHT LOAD INCREASE is allowed by this new standard. For example, the weight load increase allowed for the two most commonly encountered ladders, the 14 foot (4 m) roof ladder and the 24 foot (7 m) extension ladder is 250 pounds (113 kg). (These ladders were formerly rated for a 500 pound [227 kg].)
Before using this standard for ladders manufactured prior to adoption of the new standard fire departments are cautioned to:

- Check with the manufacturers for their particular ladders and if approved by the manufacturer, service test the ground ladders to the new standard.

2.7
1. True
2. False
3. True
4. True

2.8
1. Nonconductors
2. Wood
3. Varnish

MAJOR COMPONENTS

The major components of ladders are, of course, the beams and the rungs.

Single, Roof, Extension, and Pole Ladders

There are two basic designs used: SOLID BEAM and TRUSS BEAM construction. Both types have to meet the same specifications. Then how does one decide which ladder to purchase? As mentioned earlier, individual preference is often the answer. However, weight is a legitimate criterion.

2.11 Fill in the blanks.

1. The major components of any ladder are the ______ and the ______.
2. The two basic designs of ladders are: ______ beam and ______ beam.

For lengths of 24 feet (7 m) or less, solid beam construction provides a lighter weight ladder in most cases (See Table 2.1 on page 32).

For intermediate lengths of 25 feet (8 m) to 35 feet (11 m), solid beam metal ladders will usually be lightest in weight. However, in wood construction the lightest weight ladder usually is obtained by using truss construction.

For larger ladders, over 35 feet (11 m), truss construction provides the lightest weight ladder.

2.12 Fill in the blank.

1. For lengths of 24 feet (7 m) or less ______ beam construction provides the lightest weight ladder.
2. For lengths of over 35 feet (11 m) ______ beam construction provides the lightest weight ladders.

METAL LADDERS

Metal ladders are available in both solid and truss beam construction. The term "solid beam" when used to refer to a metal ladder is actually a misnomer because metal "solid beams" are not solid in the sense that wood beams are. One design uses a C-channel. Rung plates are riveted across the open side of the "C" to provide support and to serve as a mounting for the rungs (Figure 2.11). Some extension ladders made with this type beam construction have modified rung plates for bed and intermediate fly sections (Figure 2.12). The beam rail of the fly section is modified by adding a "tongue" (Figure 2.13). The resulting assembly is called tongue and groove construction (Figure 2.14). Others use the

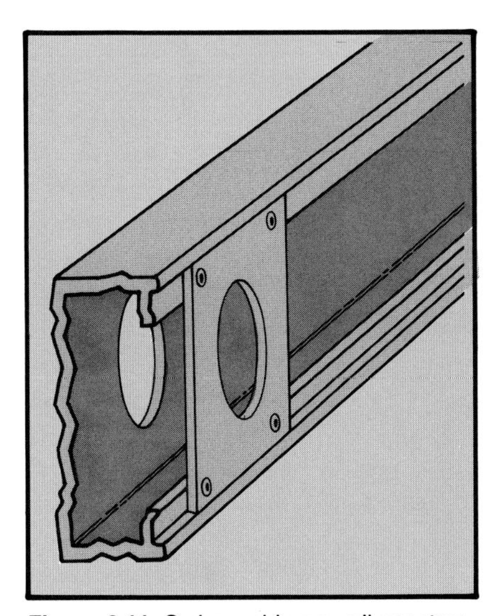

Figure 2.11 C-channel beam rail construction. Rung plates are riveted across the open side of the C.

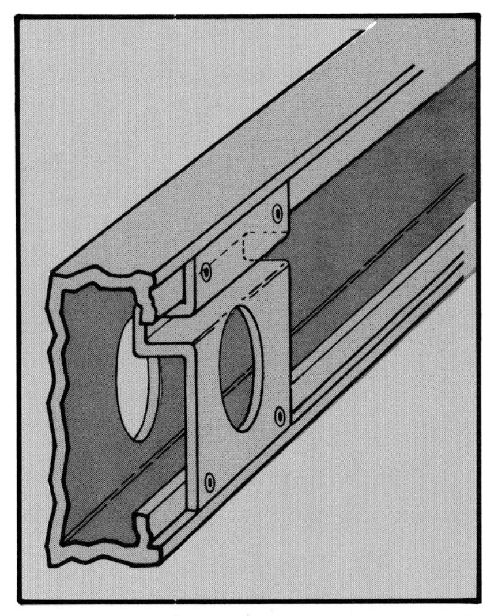

Figure 2.12 Rung plate modification for C-channel beam rail of the bed section of an extension ladder. The difference being that it is notched to receive a tongue located on the outside of the fly section beam rail.

2.9 1. True
2. False
3. False
4. False

2.10 1. False
2. True

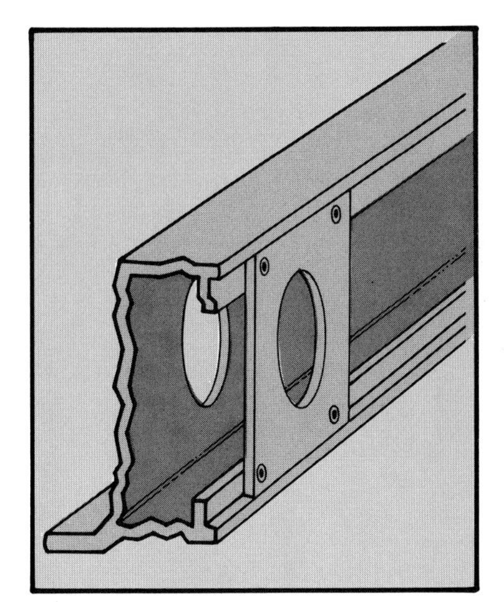

Figure 2.13 Top fly section C-channel beam rail has a tongue on the outside.

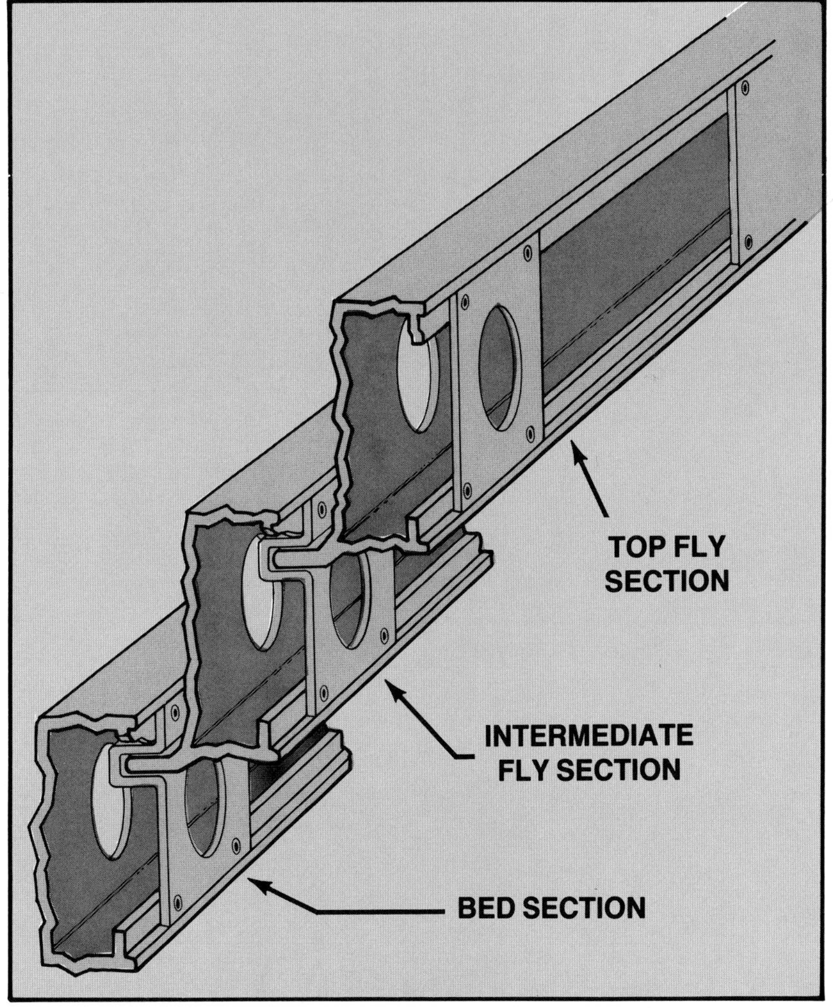

Figure 2.14 C-channel beam rail construction of the tongue and groove type with three sections in place.

same guide design that was originally developed for a wood ladder. A guide is riveted to the fly section beam near the bottom and another guide is riveted to the bed section beam near the top (Figure 2.15).

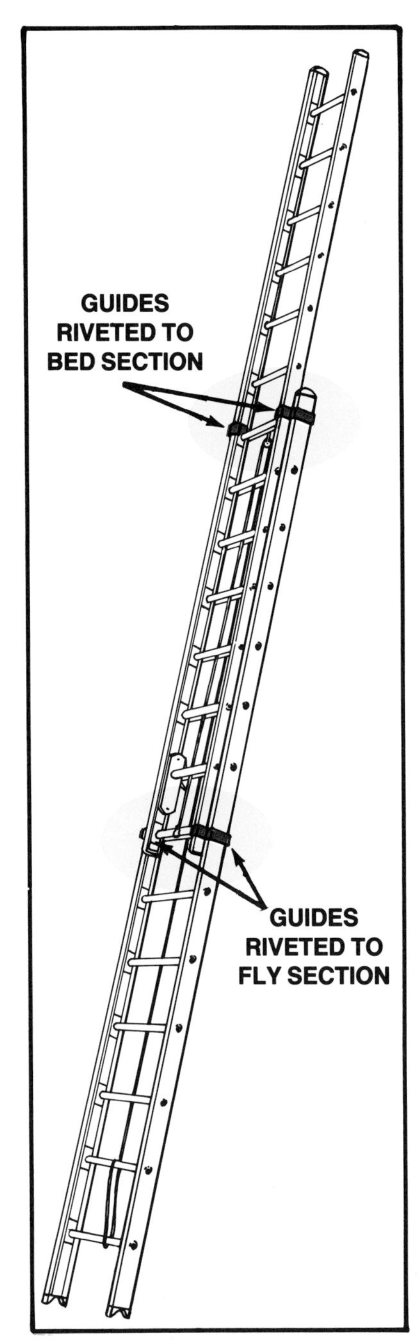

Figure 2.15 Guides riveted to beams.

Another design uses a tubular beam rail made by riveting two extruded structural members together (Figure 2.16). Extension ladders with this type beam rail construction have a plate with a U-channel bolted to the outside of the fly beam near the bottom of the fly section (Figure 2.17). A similar plate and U-channel is bolted to the inside of the bed ladder beam near the top. The top T-rail of the bed section slides through the U-channel of the fly section and the bottom T-rail of the fly section beam rail slides through the U-channel attached to the top of the bed ladder beam rail. Figure 2.18 shows the completed assembly.

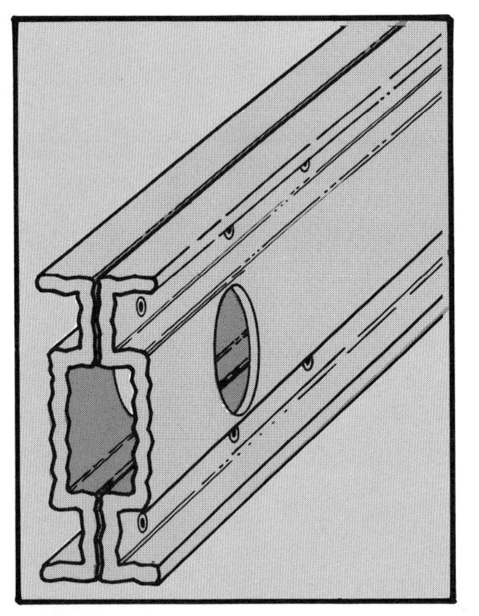

Figure 2.16 Two extruded structural members are riveted together to form this beam rail.

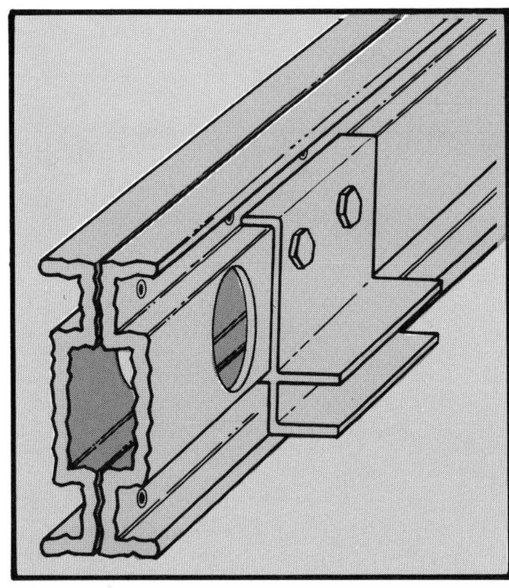

Figure 2.17 U-channel guides bolted to the outside of the fly section beam near its bottom.

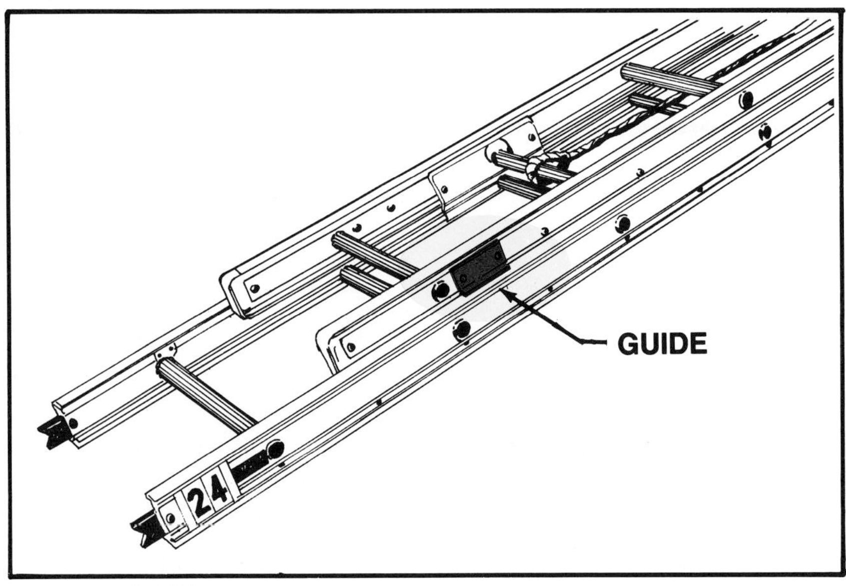

Figure 2.18 Bed and fly section assembled to form an extension ladder.

A one piece extruded tubular beam is manufactured by a Canadian company (Figure 2.19). Extension ladders are manufactured with a C-channel as a part of the top of the bed section beam rail (Figure 2.20). Rungs of the fly section protrude through the outside of the beam rail and slide up and down in the C-channel. The protruding portions of the rungs are fitted with nylon bushings to make them slide easier (Figure 2.21). The completed assembly is shown in Figure 2.22.

2.11 1. Rungs, Beams
2. Truss, Solid

2.12 1. Solid
2. Truss

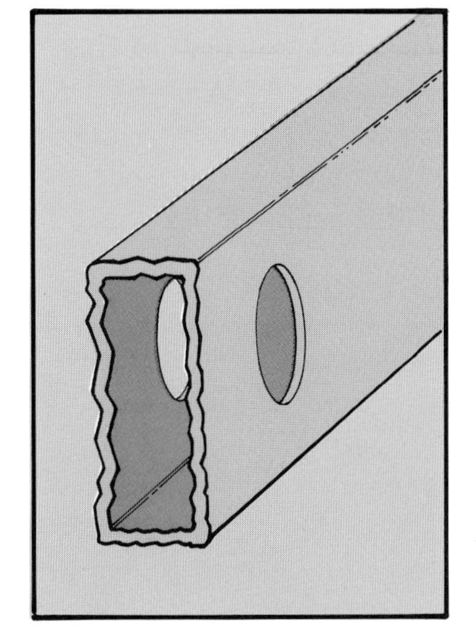

Figure 2.19 One piece of extruded tubular beam rail for a single or roof ladder.

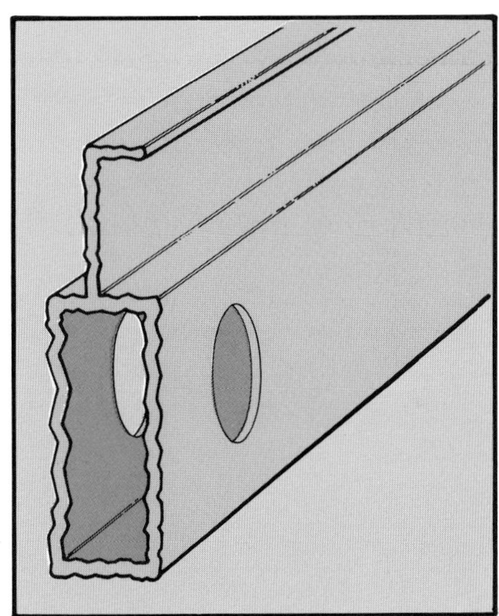

Figure 2.20 One piece extruded tubular beam for the bed section of an extension ladder. Protruding ends of rungs of the fly section travel in the U-shaped channel.

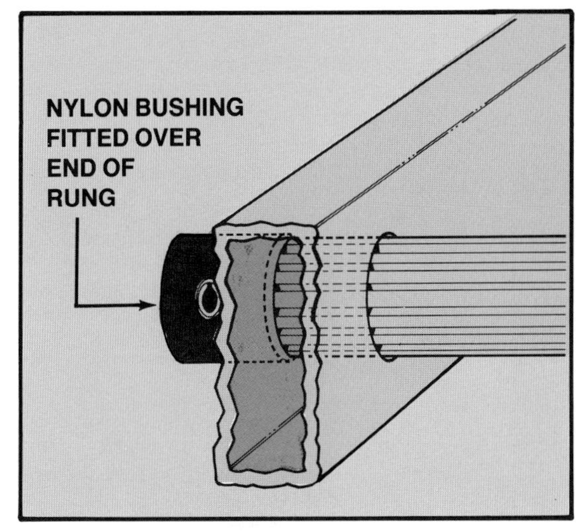

Figure 2.21 The protruding ends of the fly section rungs are fitted with nylon bushings to make them slide easier.

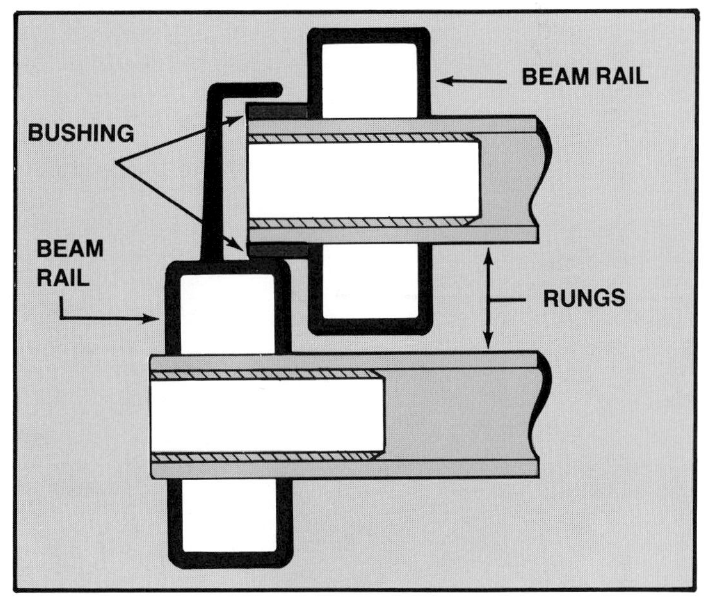

Figure 2.22 Extension ladder assembled.

Truss beam construction also varies. One type uses two rectangular rails for each beam. The rails have a U-shaped channel manufactured on the facing surfaces of each rail. Truss plates are riveted across between the U-channels to hold the two rails together and to form a beam (Figure 2.23).

Extension ladders with this type beam construction are made so that the outside width of the fly is slightly less than the inside width of the bed section. U-channels are attached to the side of the bed section truss plates (Figure 2.24). These channels act as a guide for the rail of the fly section to slide in as it is extended and serve to hold the assembly together. When there is another fly, as in a 35-foot (11 m) three piece extension ladder, the scheme is repeated (Figure 2.25).

A second type utilizes two T-channels joined by riveted truss plates (Figure 2.26). The rungs of this extension ladder protrude from the outside of the beams of the fly section(s) and slide in a U-shaped guide manufactured as a part of the side of the upper T-rail (Figure 2.27).

Figure 2.23 Double rectangular rail truss beam construction.

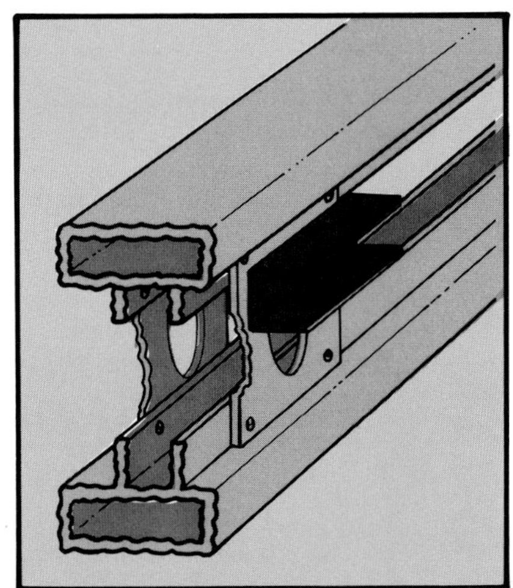

Figure 2.24 U-shaped guide on bed section of an extension ladder.

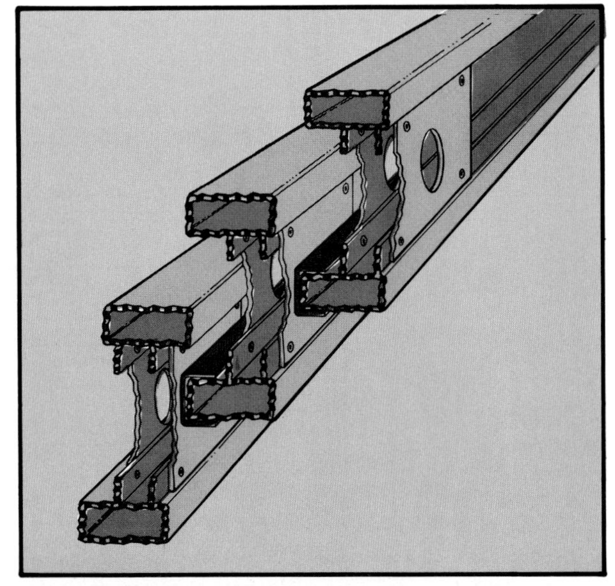

Figure 2.25 Three-section extension ladder assembly.

Figure 2.26 Double T-channel truss beam construction. The bed section of an extension ladder is shown. Ends of rungs protruding from the fly section rail fit into the U-channel on the inside of the top rail.

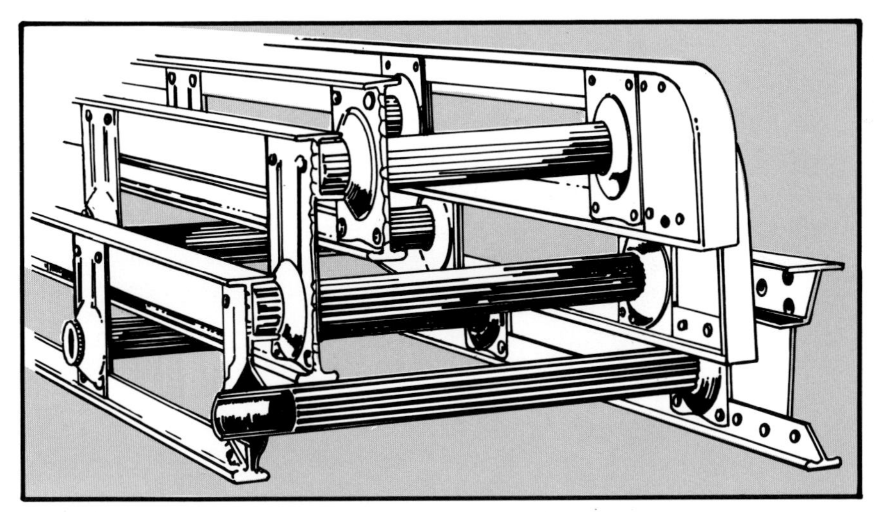

Figure 2.27 Fly section rungs protrude beyond the outside of the beam rail and ride in U-shaped channels which are part of the top T rail.

Metal rungs are used. NFPA 1931 requires rungs, except for those of folding and pompier ladders, to be not less than one and one-fourth inch (32 mm) in diameter, and that they be spaced on 14-inch (356 mm) centers plus or minus one-eighth inch (3 mm). Metal rungs must be constructed of heavy-duty corrugated, serrated, knurled, or dimpled material, or be coated with a skid resistant material.

A variety of methods are used to attach rungs to beams; most use the truss or rung plate as the point of support. In the case of the C-channel "solid" beam construction, a rung plate is riveted across the open side of the C. Holes the size of the rungs are cut through the rung plate and the ladder beam, and the rung is inserted. Then it is expanded on both sides of the rung plate, and the end is welded flush to the outside of the beam (Figure 2.28).

A second design of similar nature is used for truss ladder construction except that the rung end is welded to the outside of the outer truss plate (Figure 2.29).

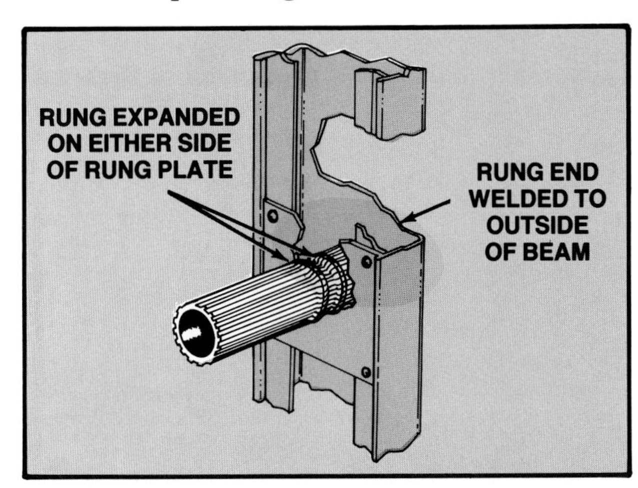

Figure 2.28 With this type of construction the rung is inserted and expanded on either side of the rung plate. The end of the rung is welded to the outside of the beam rail.

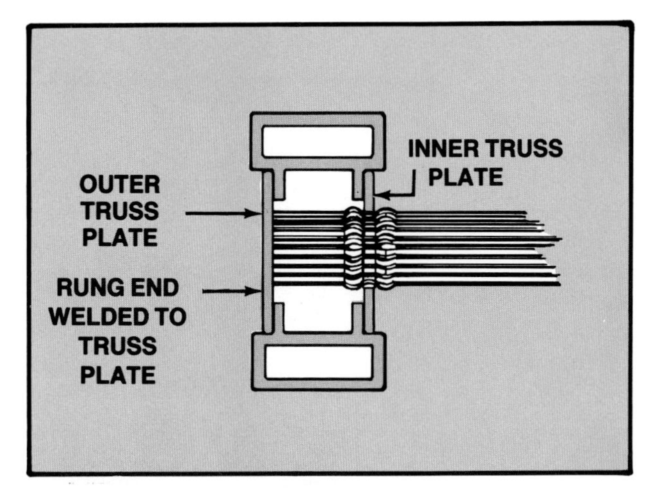

Figure 2.29 This design is similar to that shown in Figure 2.28 except that it is for truss ladders. The rung is expanded on both sides of the inner truss plate and it is welded to the outer truss plate.

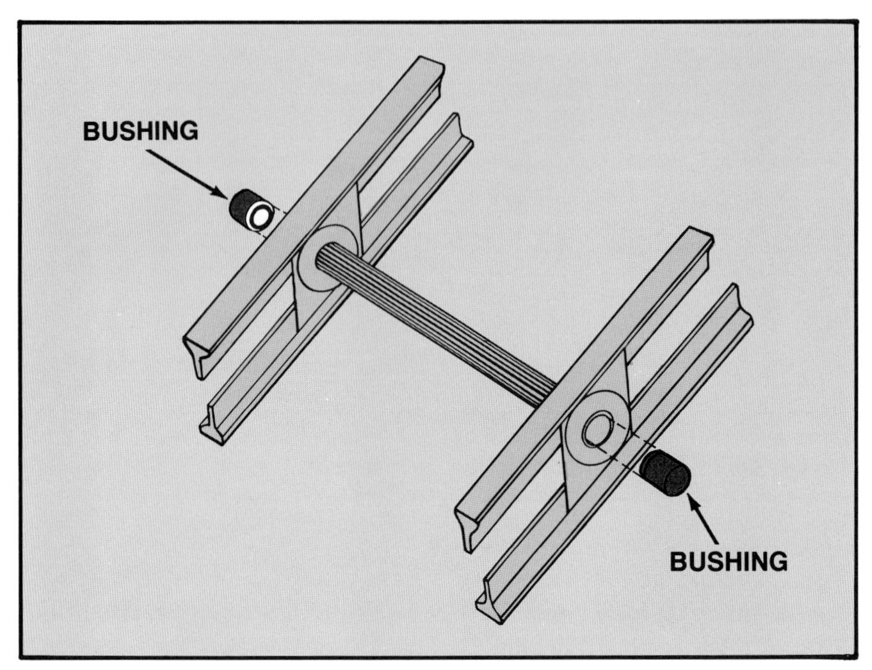

Figure 2.30 This design secures the rung by using bushings which are inserted into the rung ends and then expanded.

A third design uses a bushing that is inserted into the end of the rung and then is expanded into place (Figure 2.30).

NFPA 1931 also requires that all structural components of metal ladders be manufactured of materials that maintain at least 75 percent of their designated design strength at a minimum of 300°F (149°C).

2.13 Check the correct response.

NFPA 1931 requires that all structural components of metal ladders be manufactured of materials that maintain at least ______% of their design strength at a minimum of 300°F (149°C).

- ☐ A. 50%
- ☐ B. 25%
- ☐ C. 75%
- ☐ D. 100%

WOOD LADDERS

Wood ladders are also built in both solid beam and truss beam construction.

Solid Beam Construction

The solid beam of a wood ladder is just what the term implies: a single piece of solid wood, usually Douglas fir. Solid beam extension ladders use a metal guide attached to the beam of the fly section near the bottom and another guide attached to the bed section near the top to hold the sections together (Figure 2.31).

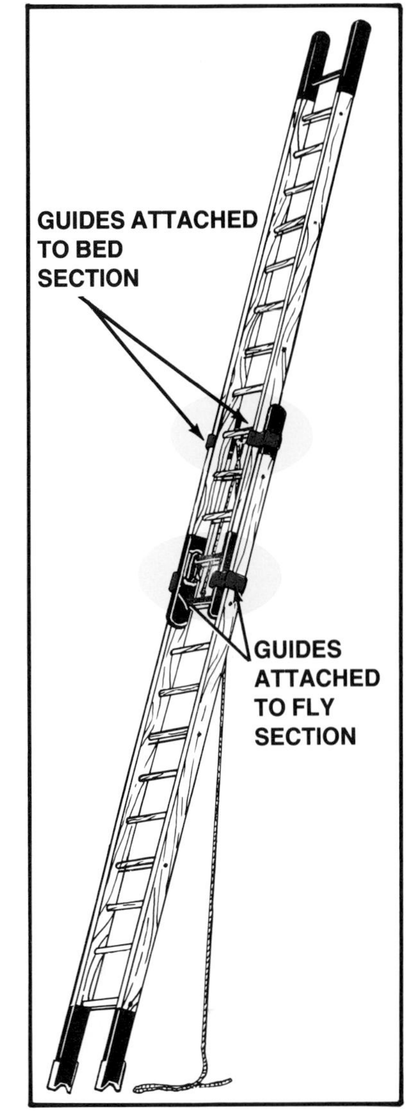

Figure 2.31 Solid beam wood ladder. Note guides on this extension ladder.

Truss Beam Construction

Truss beams have two rails separated by truss blocks and are also usually constructed of Douglas fir. There are three basic designs used.

- Those with rungs mounted in the truss blocks
- Those with rungs mounted in the bottom rail
- Those with rungs mounted in the top rail

NOTE: Extension ladders with rungs mounted in the truss blocks or with rungs mounted in the bottom rail can meet NFPA 1932's requirement for being used with the fly out. However, those with *rungs mounted in the top rail are designed to be used with the fly in.* Wood does not have as much strength in compression as in tension. Reversing the ladder to get the fly out reverses the load stresses in the rails, so the rail designed for tension loading is placed in compression and vice versa.

Rungs Mounted in Truss Blocks

There are two styles: those with one piece rung and truss block construction (Figure 2.32) and those with separate truss blocks. There are three variations of those with separate truss blocks:

- Rung centered in the truss block. Identical size parallel rails (Figure 2.33).

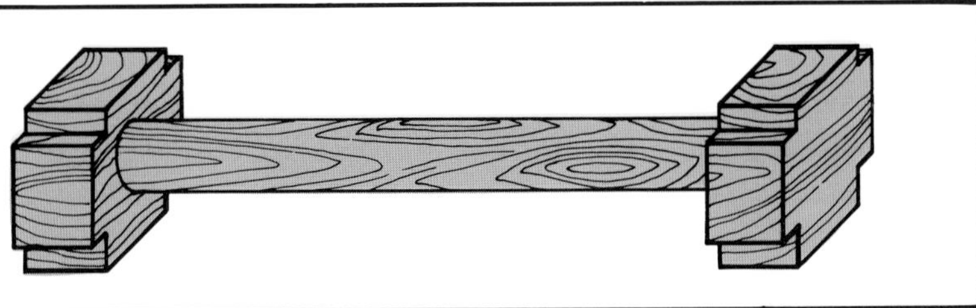

Figure 2.32 One piece rung and truss block construction.

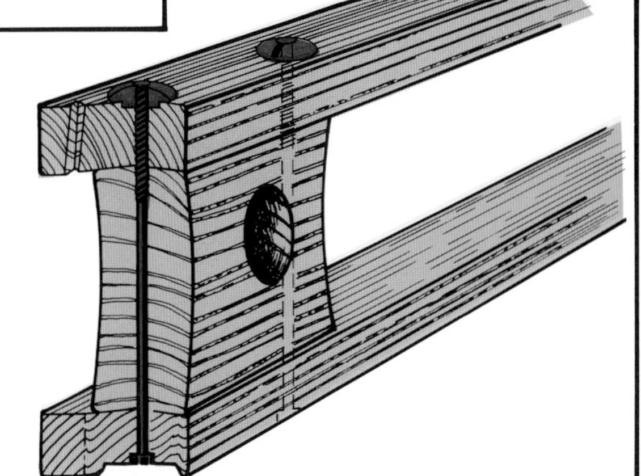

Figure 2.33 This design incorporates rungs mounted in the center of the truss block.

- Rung mounted at bottom center of truss block adjacent to the bottom truss rail. Identical size parallel rails (Figure 2.34).
- Rungs mounted at bottom center of truss block adjacent to the bottom truss rail. Identical size rails. Top rail arches. Bottom rail is straight (Figure 2.35).

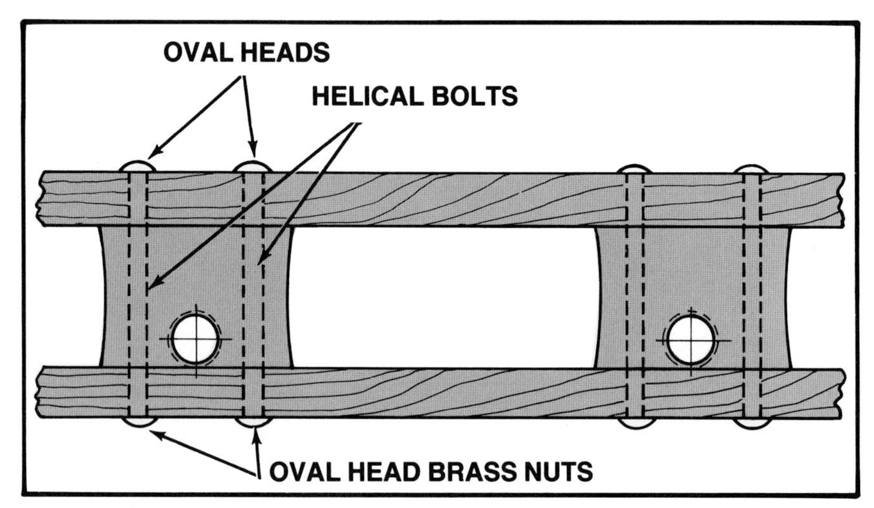

Figure 2.34 Identical parallel beam rails with rungs mounted in the center bottom of truss blocks.

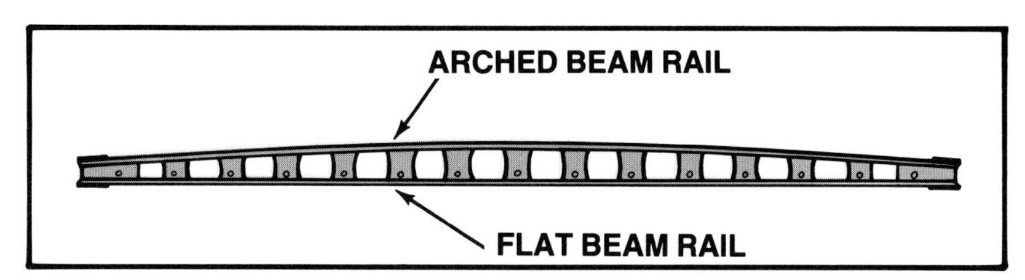

Figure 2.35 Identical size beam rails; one flat, one arched. Rungs mounted at bottom center of the truss block adjacent to bottom truss rail.

Rungs Mounted in Bottom Rail

This design has two parallel rails but the bottom rail is thicker than the top rail so that it can support the rungs (Figure 2.36).

Rungs Mounted in Top Rail

This design has a thicker top rail and except for the roof ladder, both rails arch. The bottom rail of the roof ladder is straight so that it will lie flat on a roof (Figures 2.37 and 2.38).

Figure 2.39 illustrates one method used to hold sections together on wood truss beam extension ladders. The fly beam is constructed with a wood tongue which travels in a U-shaped groove cut into the side of the truss block just above the rungs. Another method employs an oak strip attached to the inside of the beam (Figure 2.40). When rungs are mounted in the bottom beam, the beam rails are made wider than the truss blocks. The resulting groove acts as the guide and holds the assembly together (Figure 2.41).

2.13 C

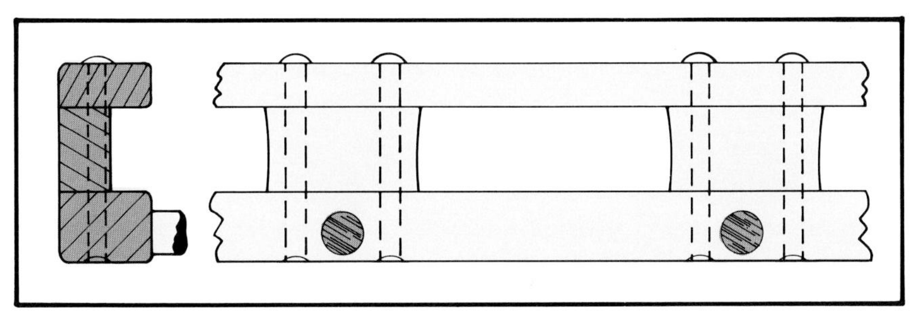

Figure 2.36 In this design the two beam rails are parallel with the bottom rail being thicker so that it can support the rungs.

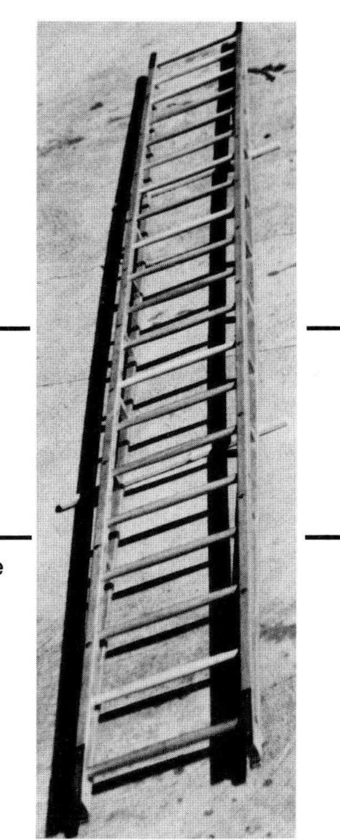

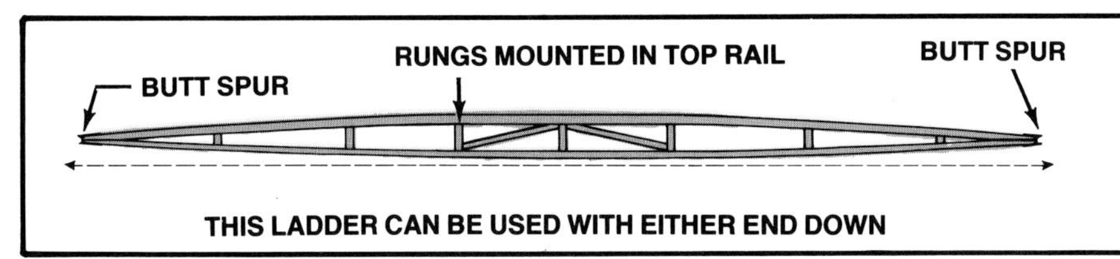

Figure 2.37 This design uses a thicker top rail to support the rungs. Both beam rails arch except for the roof ladder as shown in Figure 2.38. *Photo courtesy of ALACO Ladder Company.*

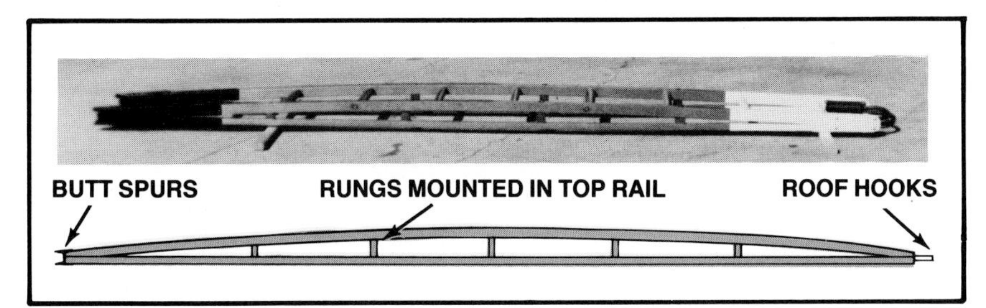

Figure 2.38 The roof ladder has a flat bottom rail so that it will lie on the roof properly. *Courtesy of ALACO Ladder Company.*

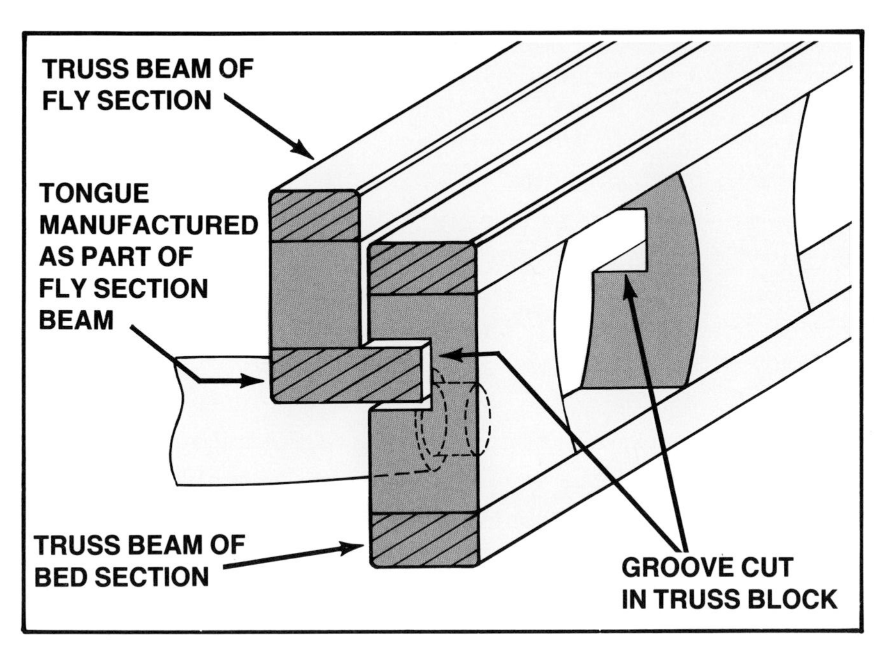

Figure 2.39 A tongue manufactured as part of the bottom truss rail of a fly section travels in a groove cut into the truss blocks of the bed section rail.

Figure 2.40 An oak strip attached to the inside of the bed section beam is used in this design.

Figure 2.41 This design incorporates beam rails that are wider than the truss blocks. The lower half of the upper section beam rail travels in this groove.

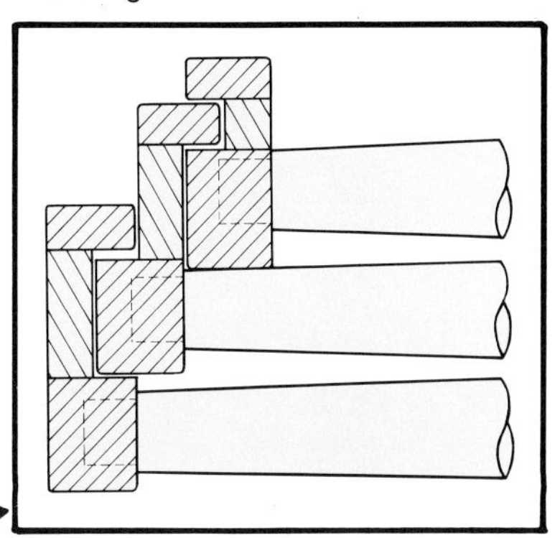

Rungs

Rungs for both solid and truss beam ladders must be a minimum of one and one-fourth inches (32 mm) in diameter to meet requirements of NFPA 1931. Most are the same diameter all the way across but some are thicker in the center. These are called swell center rungs (Figure 2.42). NFPA 1931 now calls for rungs to be spaced 14 inches (356 mm) on centers, plus or minus ⅛ inch (3 mm). Some wood ladders are still being constructed with rungs spaced 12 inches (300 mm) on center and many older ladders were constructed with the shorter spacing. No tread surface is required for wood rungs.

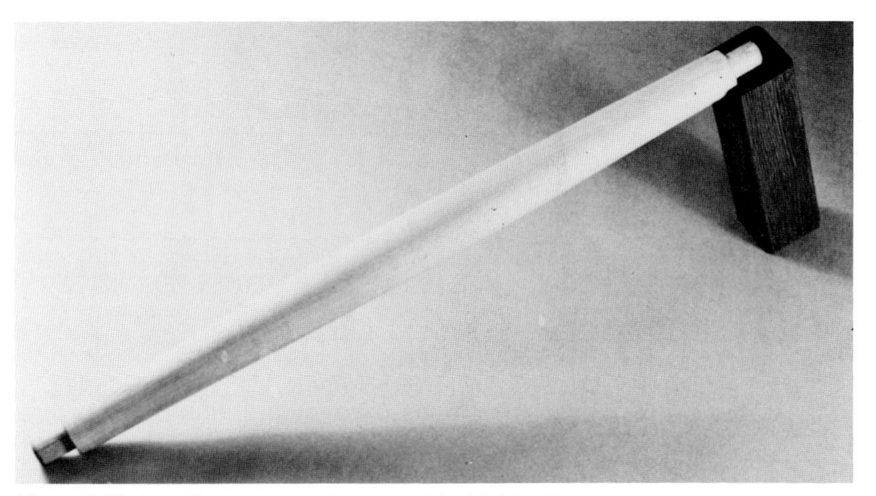

Figure 2.42 A swell center rung. *Courtesy of ALACO Ladder Company.*

Figure 2.43 shows three typical rung installation methods. In both instances the rung is manufactured with a tenon on each end. The drawing on the left has a tenon set in a mortise drilled in the ladder beam rail. When this construction method is used, the ladder tie rods are depended on to keep the beams tight against the rung shoulder. The rung is prevented from turning by a metal key driven into the rung tenon when it is inserted into the ladder beam rail. A nail or wood screw may also be used for this purpose. The construction shown in the drawing on the right has the full diameter of the rung entering the beam rail and it is said to be stronger. Tie rods keep the beams bearing on the rung and a key or screw keeps the rung from turning, but the taper of the rung will usually be sufficient to prevent turning. Some rungs are held in place by gluing. The swell center rung pictured in Figure 2.42 is installed in this manner. Some rungs are both glued and nailed. Rungs set in truss blocks are attached in much the same method except that the tenon passes through the truss block. Its end is flush with the outside surface of the truss block. When the rung and truss block are one piece, the truss block is keyed to both the upper and lower beam rails and a single beam bolt is used to hold it in place (Figure 2.44).

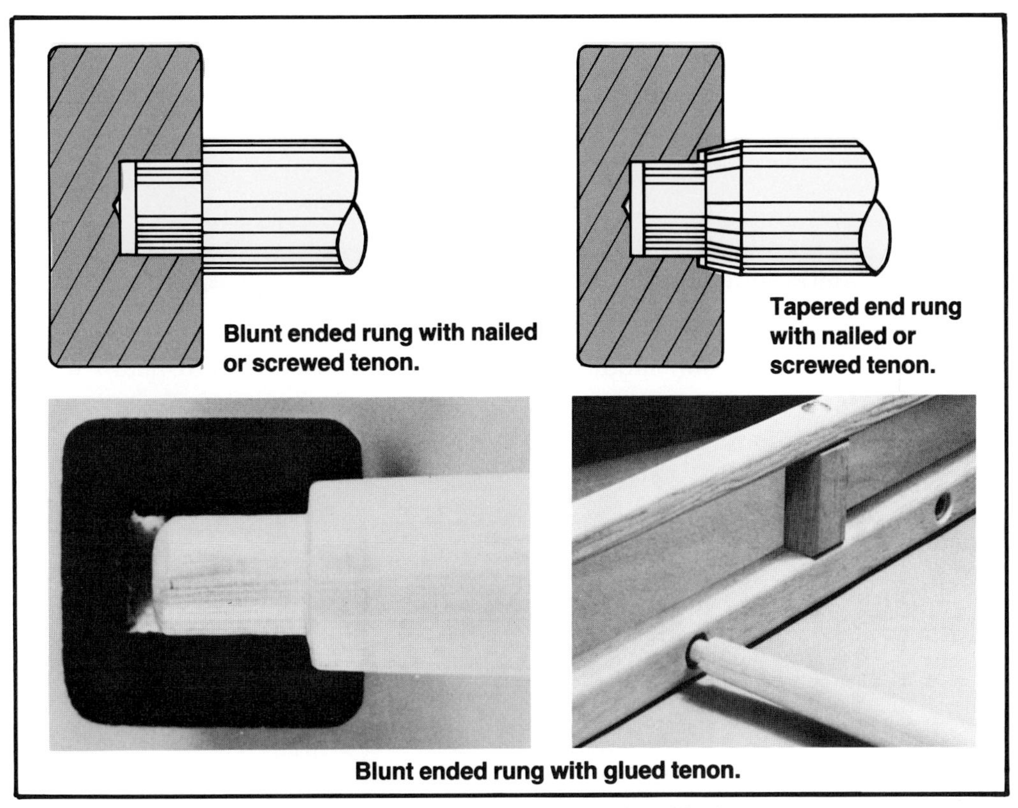

Figure 2.43 Typical rung installation methods. *Courtesy of ALACO Ladder Company.*

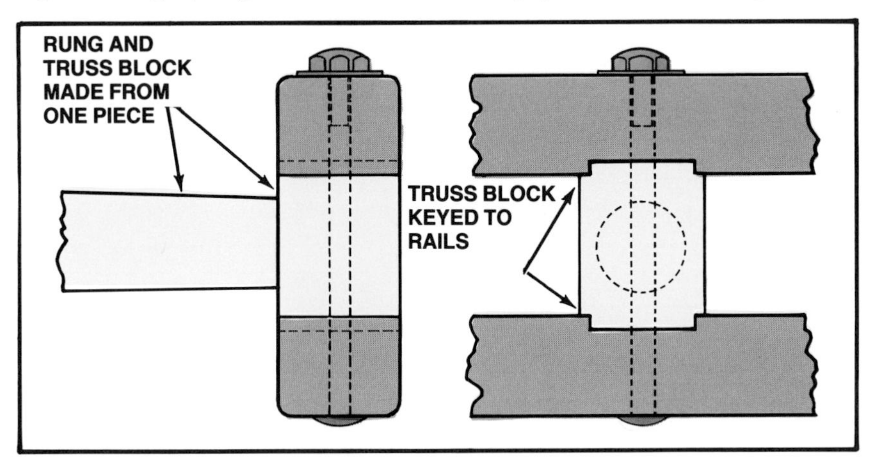

Figure 2.44 When the rung and truss block are one piece, the upper and lower beam rails are keyed and a single beam bolt is used.

2.14 True or False.

	True	False
1. Solid wood beams are not really solid, they are hollow.	☐	☐
2. Wood ladders are constructed with solid or truss beam construction.	☐	☐
3. One method of wood ladder construction uses a one piece rung/truss block.	☐	☐

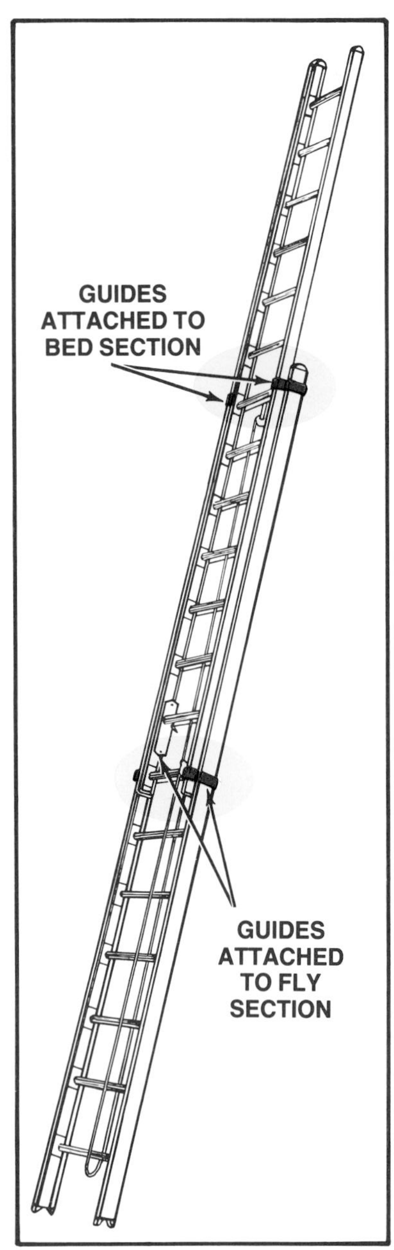

Figure 2.46 Fiber glass ladder uses guides identical to some wood and metal ladders.

FIBER GLASS LADDERS

At present, fiber glass fire service ladders are only manufactured in solid beam construction. The beam is not actually solid but like some metal ladders it employs a C-channel construction. A metal rung plate is riveted across the open side of the C-channel and the rung, which is metal, is attached to this plate just as it is in metal construction. The rungs are constructed the same as for metal ladders and the spacing requirement is the same (Figure 2.45).

Extension ladders use guides identical to solid beam wood ladders. One guide is attached to the fly section beam near the bottom and the other is attached to the bed section beam near the top (Figure 2.46).

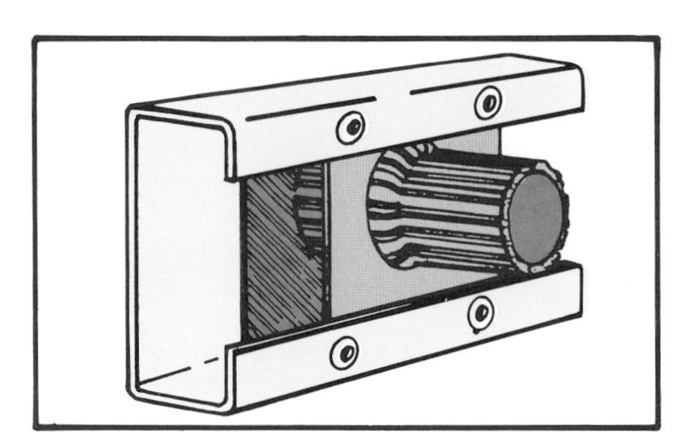

Figure 2.45 Rung attachment for a fiber glass ladder.

COMPOSITE: WOOD AND METAL LADDERS

Composite ladders are manufactured only in truss beam construction. Douglas fir is used for two similar parallel wood rails. Metal truss plates are used to complete the beam construction. They are held together by two beam bolts extending through the beam at each truss plate. Metal rungs of the standard size and spacing are used, and are welded to the outer side of the truss plate. Ribs are expanded on either side of the inner truss plate.

On extension ladders, the sections are held together by interlocking the bottom rail of the fly with the top rail of the bed ladder (Figure 2.47).

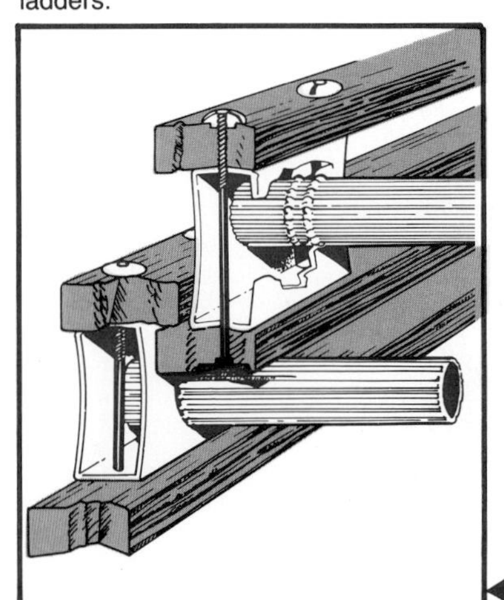

2.15 True or False.

	True	False
1. Fiber glass ladders are manufactured only in solid beam construction.	☐	☐
2. The beams of a fiber glass ladder are solid rectangular shaped fiber glass rails.	☐	☐
3. Composite ladders are manufactured only in truss beam construction.	☐	☐

Figure 2.47 Composite ladders are held together by interlocking the upper truss beam rail of the bed section with the lower truss beam rail of the upper section.

2.14 1. False
2. True
3. True

Folding Ladders

Folding ladders are made from both metal and wood, with metal construction being much more common. NFPA 1931 restricts the length to 14 feet (4 m). The primary differences between these and other fire service ground ladders are

- They are required to have foot pads to prevent slippage.
- The rungs are hinged at both ends so that the ladder can be folded into a compact assembly.
- The rungs are not round and so are not one and one-fourth inches (32 mm) in diameter.

Beams of metal folding ladders are manufactured in three designs. One is a tubular rail with a U-channel manufactured on one side. Each rung consists of two flat metal strips held in place against each side of the U-channel by a single hinge pin (Figure 2.48).

A second design uses a U-channel beam with a smaller U-channel in the center to support the rungs. Rungs are of square tubular construction and are attached to the smaller U-channel by a hinge pin (Figure 2.49).

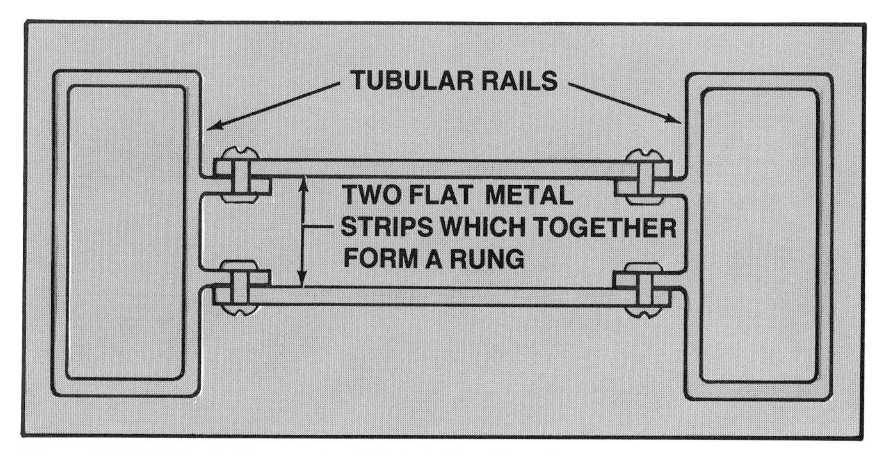

Figure 2.48 Folding ladder constructed of tubular rails and parallel metal strips for rungs.

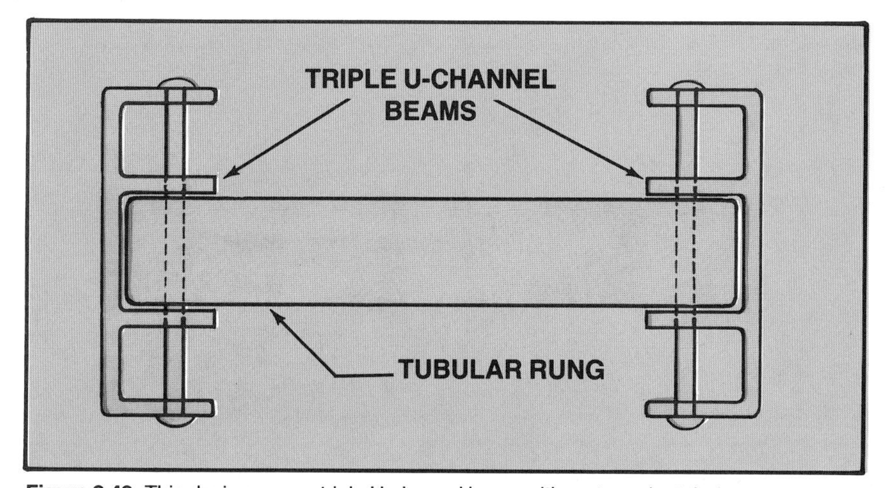

Figure 2.49 This design uses a triple U-channel beam with rectangular tubular rungs.

The third design incorporates a C-channel beam with square tubular rungs attached to the C-channel by hinge pins (Figure 2.50). All three designs use a bracelike device to lock the ladder in the open position (Figure 2.51).

Wood folding ladders utilize two parallel wood beams separated only by the width of the wood rung; no truss block is used. The rung hinge pin holds the assembly together. The rung is square (Figure 2.52).

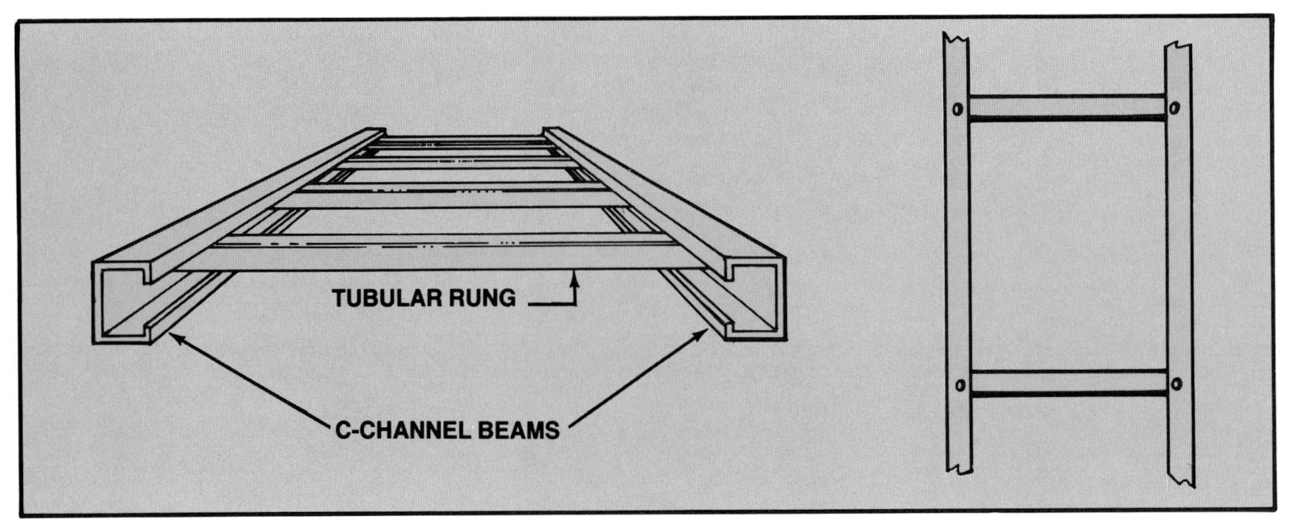

Figure 2.50 Another manufacturer constructs the folding ladder with C-channel beams and rectangular tubular rungs.

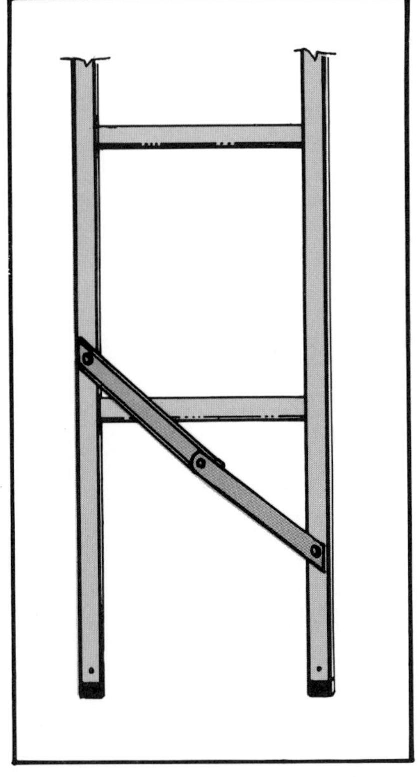

Figure 2.51 Locking brace typical of those used on metal folding ladders.

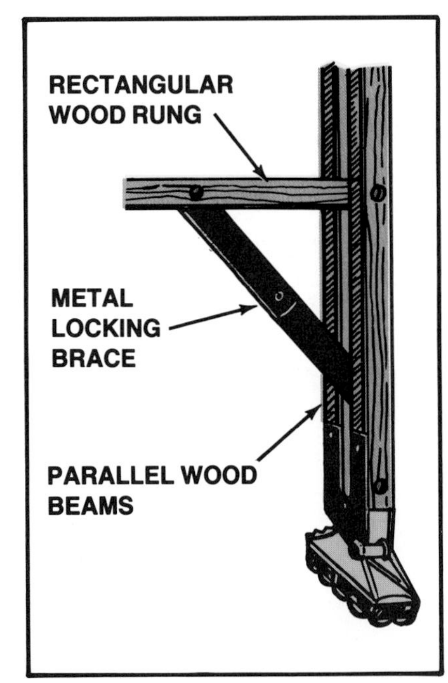

Figure 2.52 Each "beam" of the wood folding ladder is actually two parallel wood beams separated only by the width of the rectangular shaped wood rungs.

2.15 1. True
2. False
3. True

2.16 Fill in the blank.

1. Folding ladders are required to have ______ to prevent slippage.
2. Rungs of folding ladders are ______ at both ends so that the ladder can be folded into a compact assembly.
3. Folding ladder rungs are ______ in shape.

2.17 Check the correct response.

1. Some wood folding ladders utilize two separate beams. The assembly is held together by ______.

☐ A. Metal guides
☐ B. Truss blocks
☐ C. Rung hinge pins

Pompier Ladders

Early pompier ladders were manufactured of wood; however, most of those made today are of metal construction. NFPA 1931 restricts the length to 16 feet (5 m).

Metal pompier ladders have a single beam made of aluminum alloy. The beam is drilled so that the rung, which is also aluminum alloy, passes through it. The rung is attached to the beam on both sides by an L-shaped bracket that is riveted in place. The minimum overall width of rungs is 12 inches (300 mm).

Two metal standoff brackets are attached to the beam (one near the top and one near the bottom) on the side which will be against the building. (The standoff brackets set the ladder away from the building a minimum distance of seven inches (180 mm) so that the climber can get a toehold and a handgrip. A serrated gooseneck-shaped hook made of hardened steel is attached to the top of the beam. Its purpose is to hook over a windowsill or similar surface to keep the ladder from slipping (Figure 2.53).

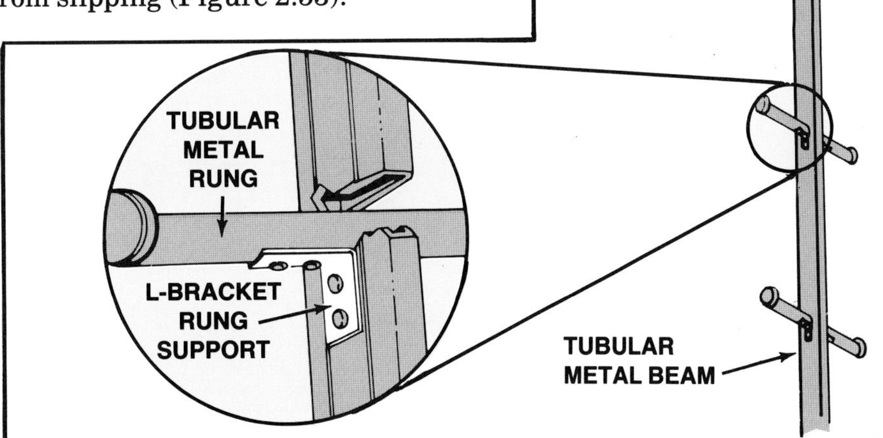

Figure 2.53 Metal pompier ladder construction.

Wooden pompier ladders have a single solid wood beam, usually of hickory. Wood rungs are supported and attached by steel L-brackets. The standoff brackets and gooseneck are much the same as on metal ladders.

2.18 Check the correct response.

1. Standoff brackets on a pompier ladder serve the purpose of ______.

- ☐ A. Setting the ladder off from the building sufficiently for the climber to get a toehold and a handgrip.
- ☐ B. Preventing the ladder from being abraided by coming in contact with the building.
- ☐ C. Providing a means to mount the ladder on the side of a pumper.

Combination Ladders

Most combination ladders are of metal construction, but some are made of wood. Beam and rung construction, unless specifically noted below, is the same as for single, roof, extension, and pole ladders. Four variations of design are found; all provide an A-frame in combination with some other type ladder.

COMBINATION SINGLE LADDER/A-FRAME

The combination single ladder/A-frame consists of two equal length ladder sections joined together at the ends of the beams by steel hinges. The hinges lock open at the 180 degree position to form a single ladder. When used as an A-frame two manually operated metal braces keep the two halves at the proper angle and prevent them from spreading apart when a load is applied (Figure 2.54).

TELESCOPING BEAMS: COMBINATION SINGLE LADDER/A-FRAME

This ladder is constructed of metal only. There are two equal length ladder sections joined together at the ends of the beams by steel hinges. The hinges have two locking positions; one at the 180 degree position is used to form a single ladder. The second lock position, which is used instead of braces between sections, maintains the proper distance between the two halves and prevents them from spreading apart when used in the A-frame configuration.

The lower half of each section has U-channel beams which curve outward toward the butt so that the ladder is wider at the bottom than in the middle and top portions. Square rungs are attached to the top of the beam. This allows the upper half of each section to slide into the U-channel of the lower half. The upper half beams are extruded rectangular channels. Rungs are hollow rectangular units. A latching device is attached to the upper rung of the lower half. It inserts into the hollow end of the upper half rungs to secure the ladder at the desired height (Figure 2.55).

2.16 1. Foot pads
2. Hinged
3. Square or rectangular

2.17 C

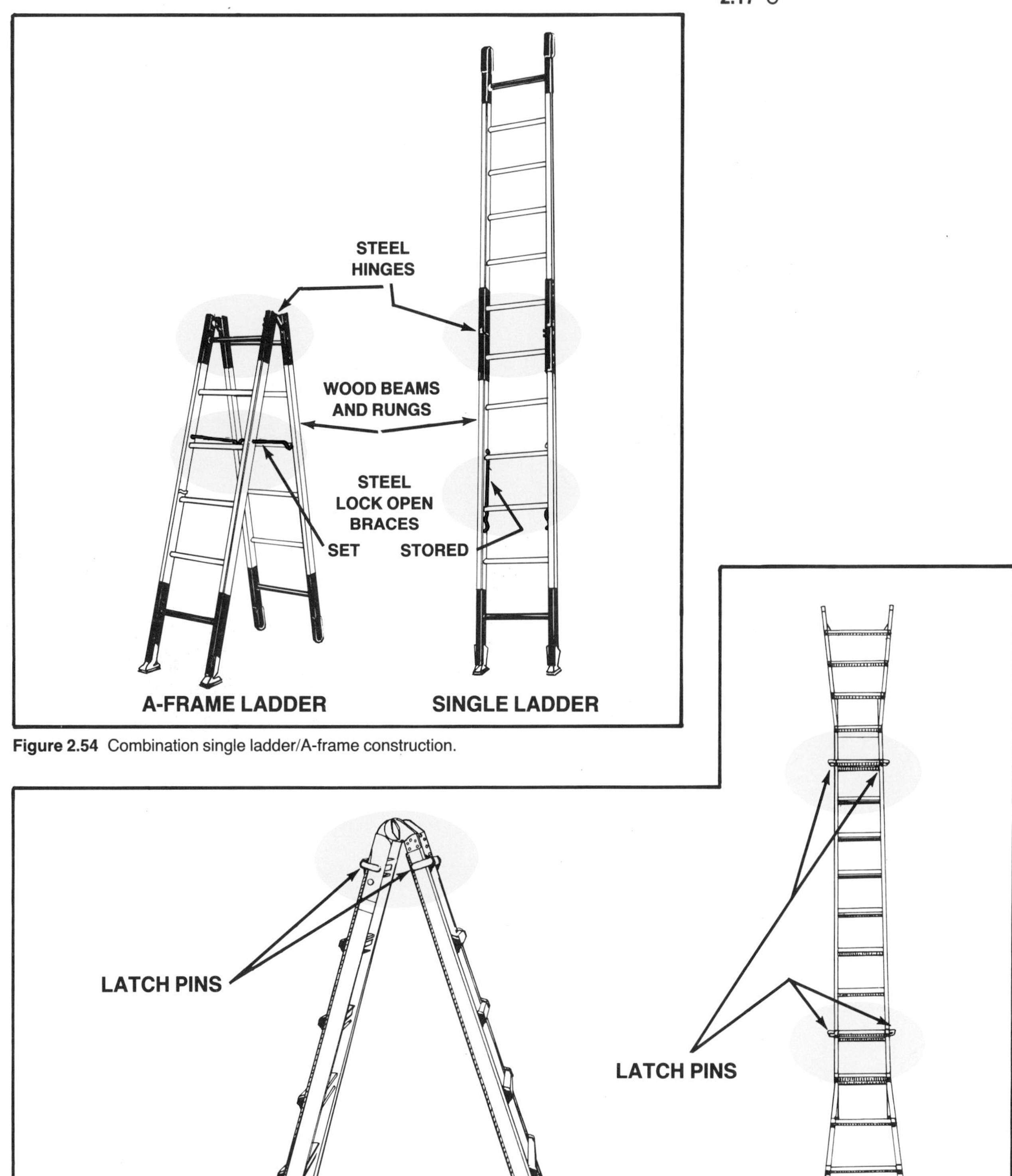

Figure 2.54 Combination single ladder/A-frame construction.

Figure 2.55 Telescoping beams-combination single ladder/A-frame construction.

COMBINATION EXTENSION LADDER/A-FRAME

This ladder is basically a short extension ladder without a halyard that spreads apart or is taken apart to form an A-frame. Some means of joining the two sections together at the top of each section is provided so that an A-frame can be made. One type has a slot in the beam at the tip end of the bed section and a rod that protrudes from the side of the beam at the tip end of the fly section. The rod fits into the slot to make the A-frame (Figure 2.56).

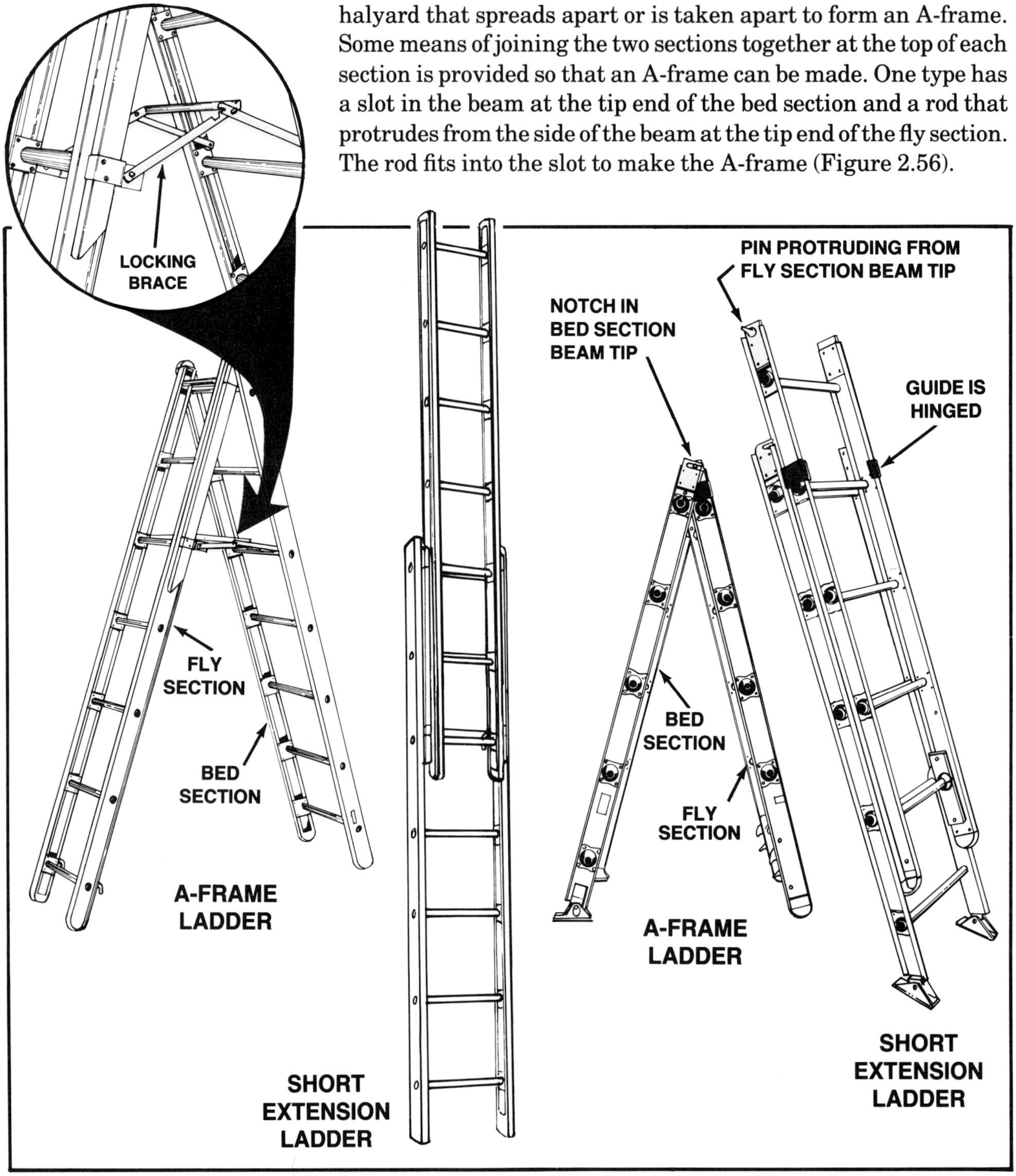

Figure 2.56 Two types of combination extension/A-frame ladders.

EXTENDING A-FRAME

The extending A-frame ladder has a fly section that fits between the two A-frame sections. A steel hinge holds the two sections of the A-frame together and it is locked when the sections are spread open to hold them in the proper position. The hinge is made in such a way that the two A-frame sections are held far enough apart at the top that the fly section can be extended between them. There is no halyard. A modified pawl assembly holds the fly at the desired height (Figure 2.57).

2.19 True or False.

	True	False
1. Combination ladders are designed so that one ladder can serve as a short extension ladder, an A-frame ladder and a single ladder.	☐	☐

2.18 A

Figure 2.57 Extending A-frame ladder construction features.

HARDWARE AND ACCESSORIES

Hardware has to meet the minimum strength requirements of the ladder components; it also must be resistant to corrosion.

Pawls

Pawls (locks) are used on fly sections of extension ladders to hold the fly sections at the desired height. There are two types: automatic and manual latching.

ENCLOSED AUTOMATIC LATCHING TYPE

The enclosed design consists of a metal housing and assembly which is bolted to the inside of the fly section beam, either at the bottom rung or the next rung above it. Two are needed, one on each side. They also support the rung located at this point (Figure 2.58).

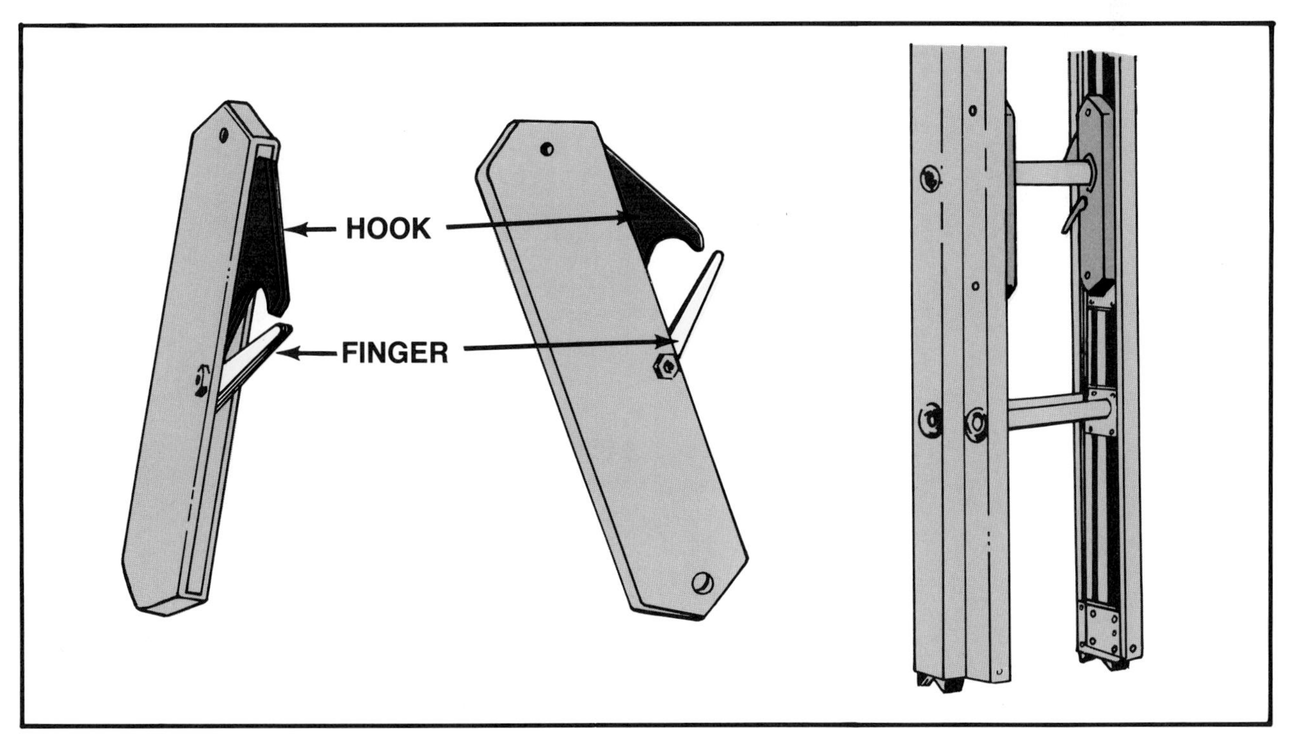

Figure 2.58 Enclosed automatic latching pawls.

There are only three main parts to this pawl: the hook, the finger, and the torsion spring. Figure 2.59 shows the pawl assembly in the locked position. This is the normal position during climbing and nesting. When the fly is raised slightly, as in Figure 2.60, the lower section rung makes contact with the projecting (tapered) end of the finger, depressing it into the assembly housing. When the other (rounded) end of the finger is rotated by this movement, it depresses the hook part way which puts the spring into tension. As soon as the finger clears the rung the spring pushes the hook out again; this movement rotates the rounded end of the finger and brings the tapered end of the finger out again.

When the slanted portion of the top of the hook contacts the bottom of the next higher rung, it again depresses the assembly into the housing. The hook slides by the rung, then immediately springs out; it can be latched at this point. To lower a fly section, the fly is extended until the hooks and the fingers clear the rung; the fly is then lowered until the fingers depress the hooks as in Figures 2.61 and 2.62.

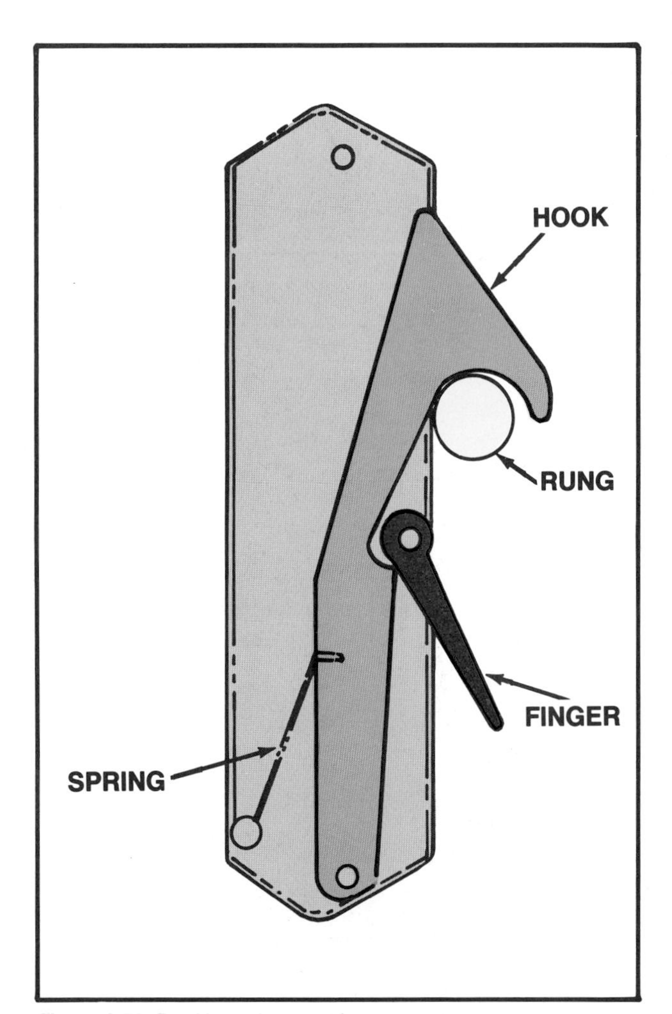

Figure 2.59 Pawl in latched position on a rung.

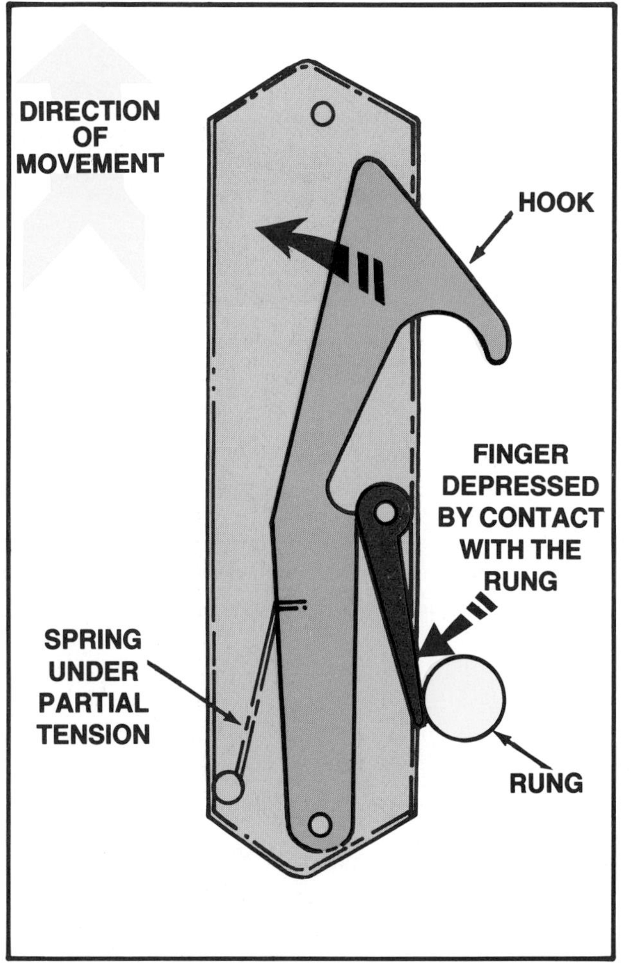

Figure 2.60 As the fly section is raised the hook moves off of the rung. When the finger contacts the rung, the hook is partly depressed.

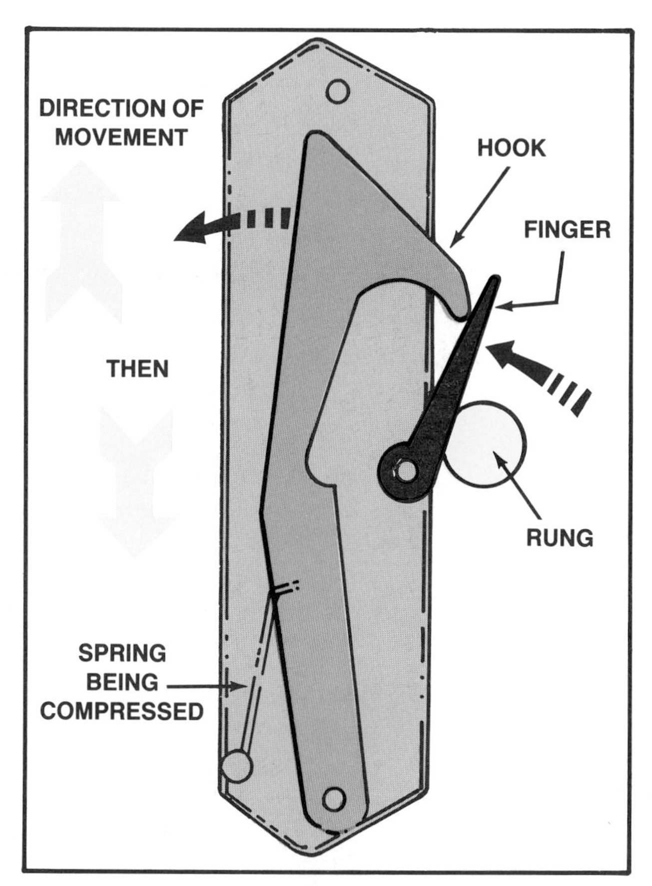

Figure 2.61 When the fly is lowered it must first be raised a short distance to allow the finger to get above the rung. Then, as it is lowered, the finger begins to depress the hook so that it will clear the rung.

2.19 False

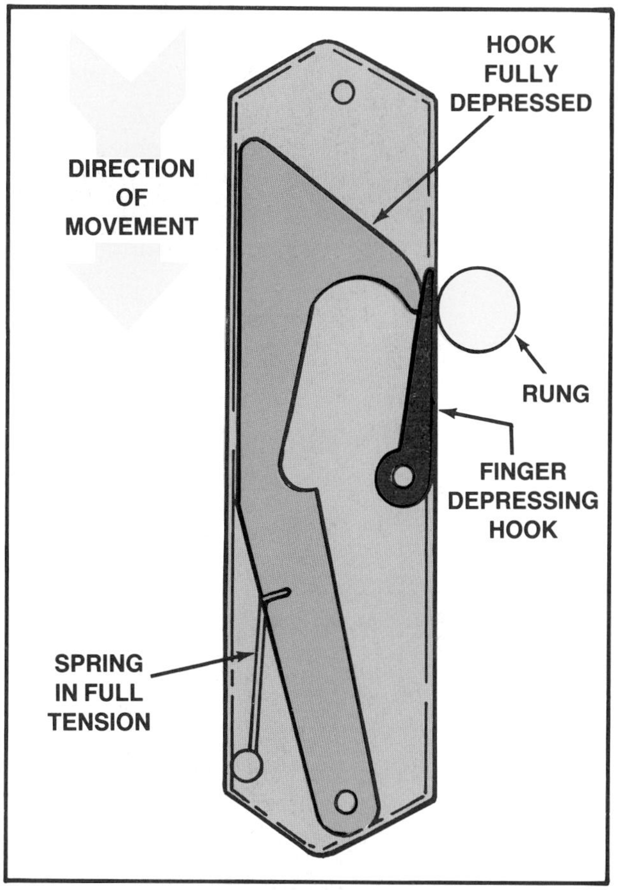

Figure 2.62 As the fly continues downward, the finger depresses the hook into the pawl housing so that it will clear the rung.

OPEN AUTOMATIC LATCHING TYPE

The parts of this pawl assembly are the pendulum, the pendulum spring, the pendulum/main spring support, the pawl, and the main spring. There is also a steel rod which runs across between the beams. The pawls mount and rotate on this rod. The halyard anchor is attached to its center (Figure 2.63).

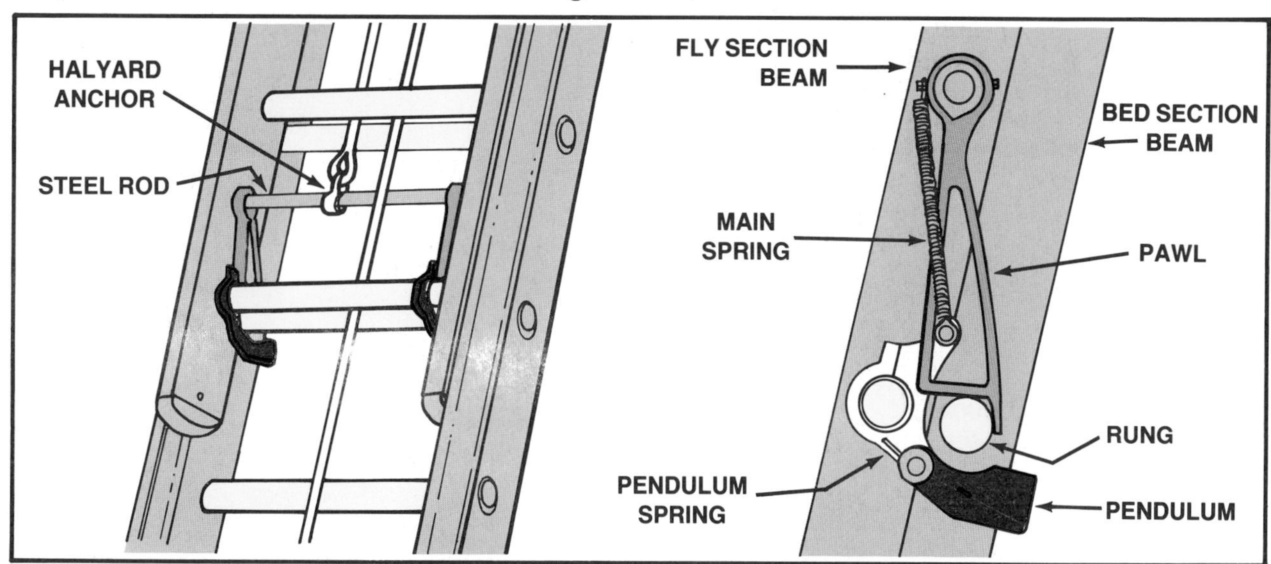

Figure 2.63 Open automatic latching pawl in the latched position on a rung.

When the fly is extended, the pendulum rotates down as it contacts a bed section rung (Figure 2.64). As soon as the pendulum clears the rung, the pendulum spring causes it to swing back into place (Figure 2.65).

When the pawl comes in contact with the next higher bed section rung, it is depressed (rotates on the steel rod) to allow the pawl to clear the rung (Figure 2.66). As soon as the pawl clears the rung, the main spring pulls it back into the latching position.

To lower the fly, it is necessary to raise the fly sufficiently for the pendulum to clear the bed section rung (Figure 2.67). The fly is then lowered. When the pendulum comes in contact with the rung, it rotates upward and depresses the pawl so that both will clear the rung (Figure 2.68).

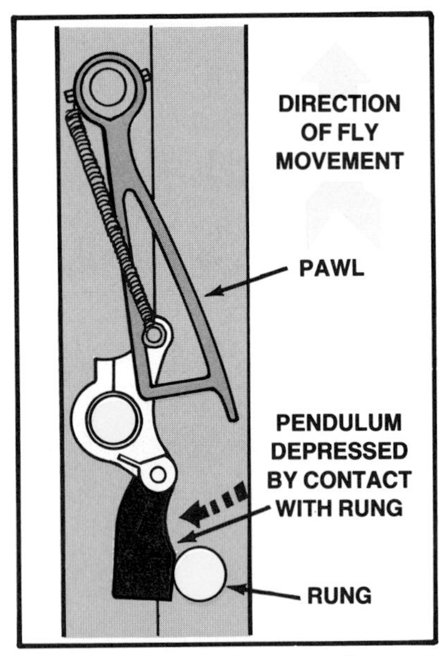

Figure 2.64 As the fly is moved upward, the pendulum is depressed when it contacts the bed section rung.

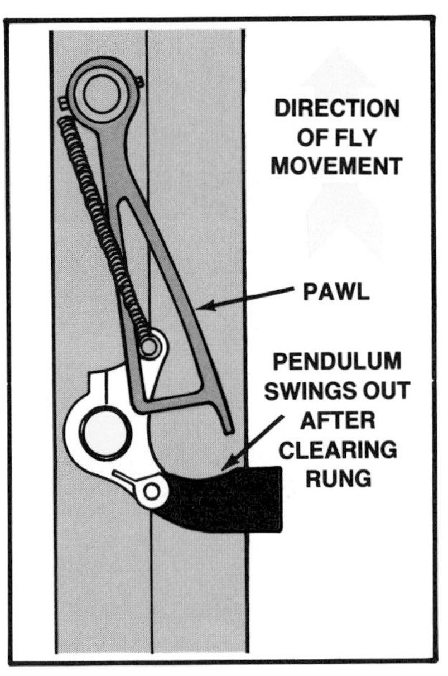

Figure 2.65 When the pendulum clears the rung, the pendulum spring causes it to swing back to normal position.

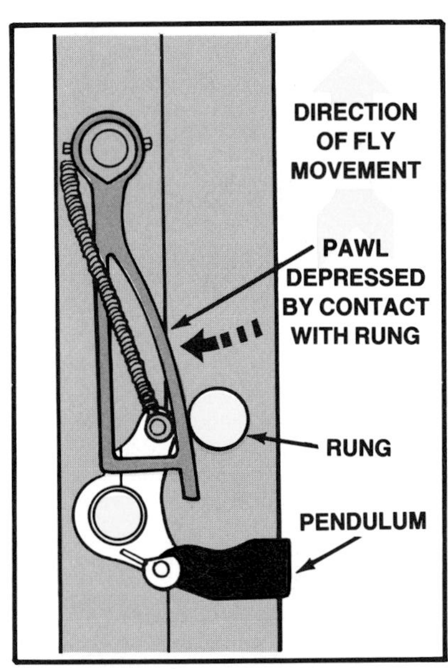

Figure 2.66 As the assembly meets the next highest bed section rung, the pawl contacts that rung and is depressed.

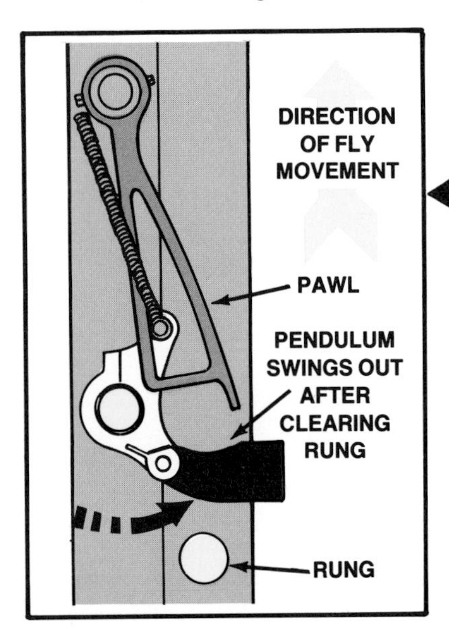

Figure 2.67 To lower the fly, it is first raised until both the pawl and pendulum clear the rung and swing back out.

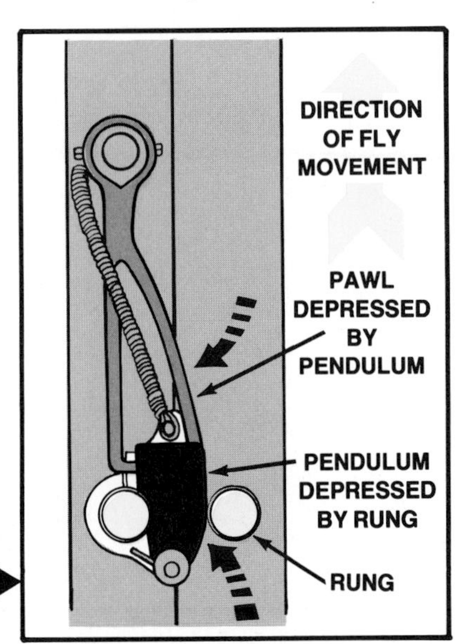

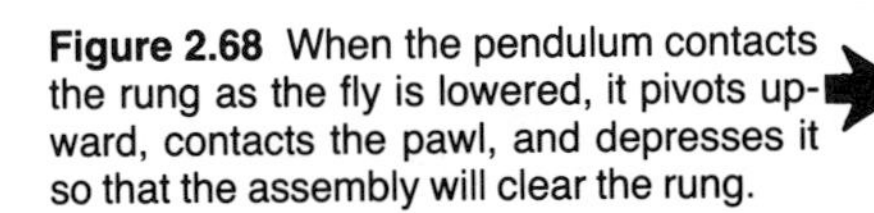

Figure 2.68 When the pendulum contacts the rung as the fly is lowered, it pivots upward, contacts the pawl, and depresses it so that the assembly will clear the rung.

MANUAL LATCHING TYPE

The manual latching type uses two A-shaped steel pawls attached near each end of a steel rod that runs across between the beams approximately four inches (100 mm) above the bottom rung of the fly section (Figure 2.69). This rod is mounted so that it will rotate. Rotation of the rod moves the pawls in and out from between the rungs of the bed section. A continuous halyard is used. Both ends are snapped to an inverted L-shaped halyard anchor which is attached at the center of the steel rod. The end of the halyard that comes down from the pulley is snapped to the short leg of the inverted L part. The end that comes up from the butt end of the ladder is snapped to the long leg of the inverted L part (Figure 2.70).

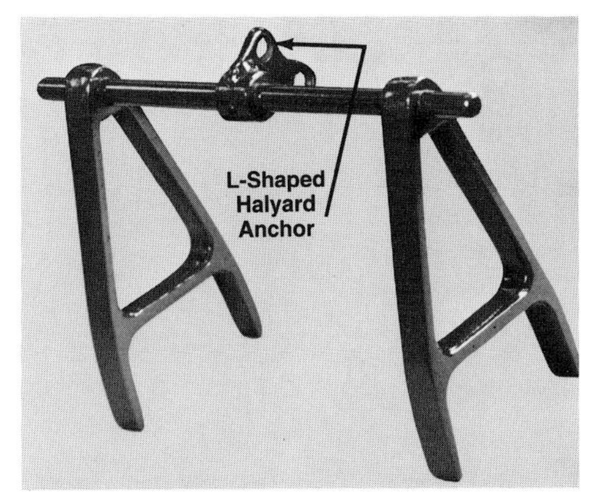

Figure 2.69 Manual Latching pawl. *Courtesy of ALACO Ladder Company.*

Figure 2.70 Manual pawls in "latched" position. *Courtesy of ALACO Ladder Company.*

When the halyard is operated in the normal manner to raise the fly, pressure is applied to the short leg of the L part of the halyard anchor. As soon as the pawls clear the bed section rung, the rod rotates clockwise and the pawls are pulled away from the bed section. This keeps them clear of bed section rungs as the fly travels upward (Figure 2.71).

To latch or set the pawls, the fly is raised until the pawls are slightly above the bed section rung upon which they are to rest. Tension is maintained on the halyard while one hand is used to pull upward on the slack or lower part of the halyard rope (the part that runs down around the bottom rung of the bed section and back up to the long leg of the L part of the halyard anchor). This causes the rod to rotate counterclockwise and the pawls are pulled back up between the rungs of the bed section. The fly section is then eased downward until the pawls engage the bed section rung (Figure 2.72).

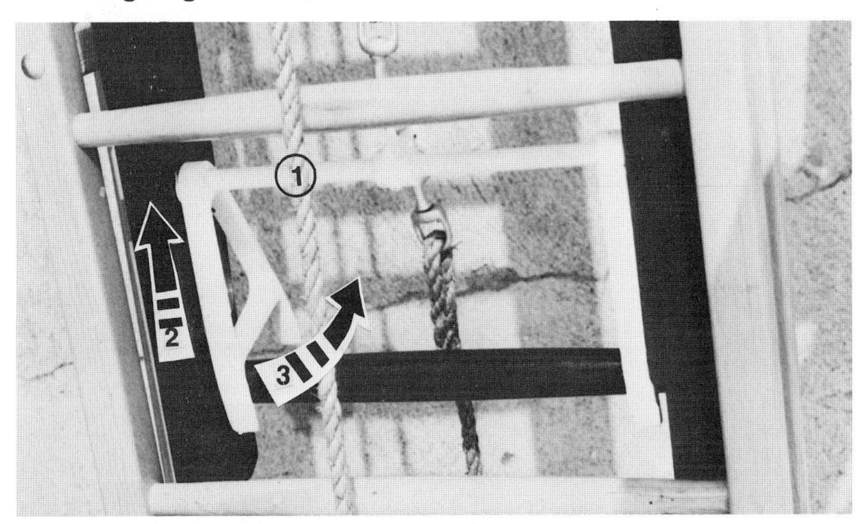

Figure 2.71 Pulling downward on the halyard (1), lifts the fly (2), and allows the pawl to swing back (3). *Courtesy of ALACO Ladder Company.*

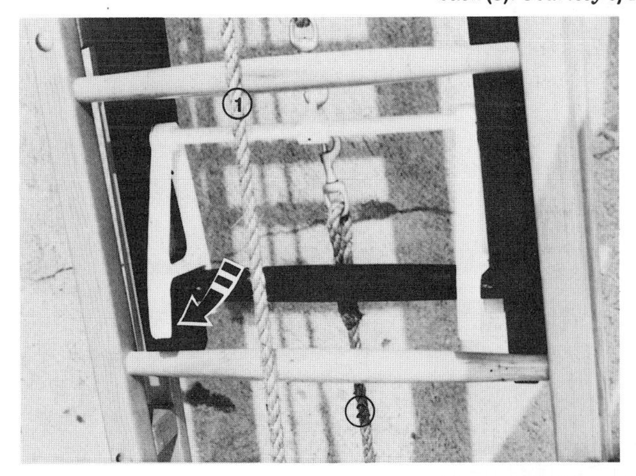

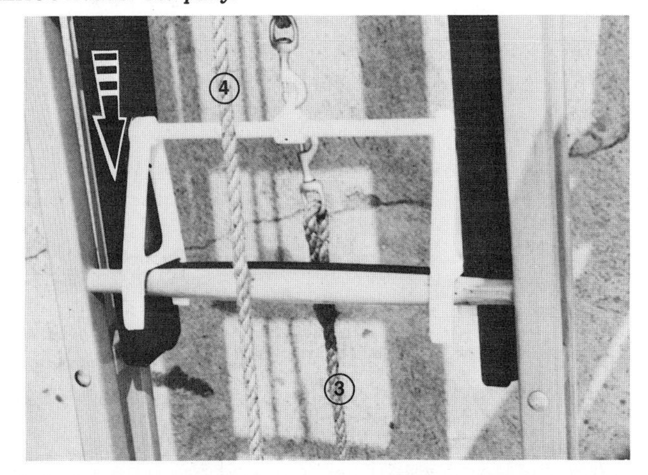

Figure 2.72 When the desired height is attained, tension is maintained on the halyard (1), while the other part of the halyard (2), is pulled with the other hand. This causes the pawl to swing outward above the rung upon which it will rest. Tension is then maintained on the halyard (3), while the first part (4), is slacked off to allow the fly to lower and the pawl to come to rest on the rung. *Courtesy of ALACO Ladder Company.*

Roof Ladder Hooks

Hooks for roof ladders are of solid steel. The end of the hook intended to engage the roof is tapered to reduce slippage. The hook is square where it passes through the top of the spring housing. The top opening in the spring housing is also square. This prevents the hook from rotating unless it is depressed and allows it to lock into two positions, one with the hook nested between the rungs, and the other with the hook parallel to the beam (90° from the rung). A coil spring inside the housing (Figure 2.73) holds the hook in the desired position. To change the position, the hook is manually depressed until the square part clears the hole in the housing; then it is rotated.

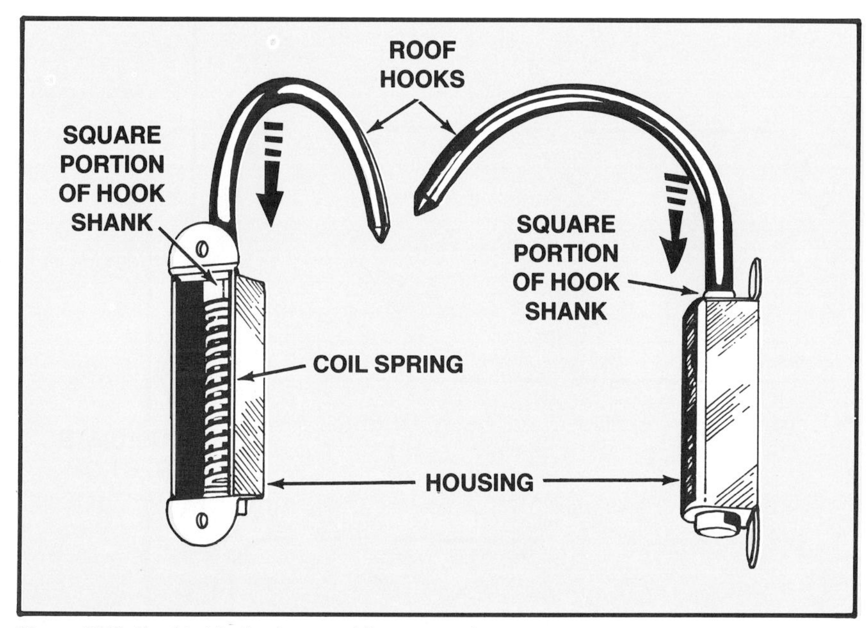

Figure 2.73 Roof ladder hook assembly construction.

Halyard/Halyard Anchor/Pulley

NFPA 1931 specifies that the halyard rope be ⅜-inch (10 mm) in diameter with minimum breaking strength of 825 pounds (374 kg) and be of sufficient length for the purpose intended. On three- and four-section extension ladders the second and third fly sections may be extended by wire rope (cable) in which case the cable must be three-sixteenth inch (5 mm) diameter to meet standards. Where cable is used, a means for adjusting the length of cable has to be provided. No splices are allowed.

The halyard rope is threaded through a pulley attached to the top rung of the bed section. One end is attached to the bottom rung of the fly section. The other end is either free, in which case it is known as a free end halyard, or it runs down the ladder, under the bottom rung and back up and is also attached to the bottom rung of the fly section. This is called a continuous halyard.

Three section extension ladders have a second halyard, usually a cable, that threads through a pulley attached to the top rung of the intermediate fly section. One end is attached to the bottom rung of the top fly section. The other end is attached to the second rung from the tip of the bed section (Figure 2.74).

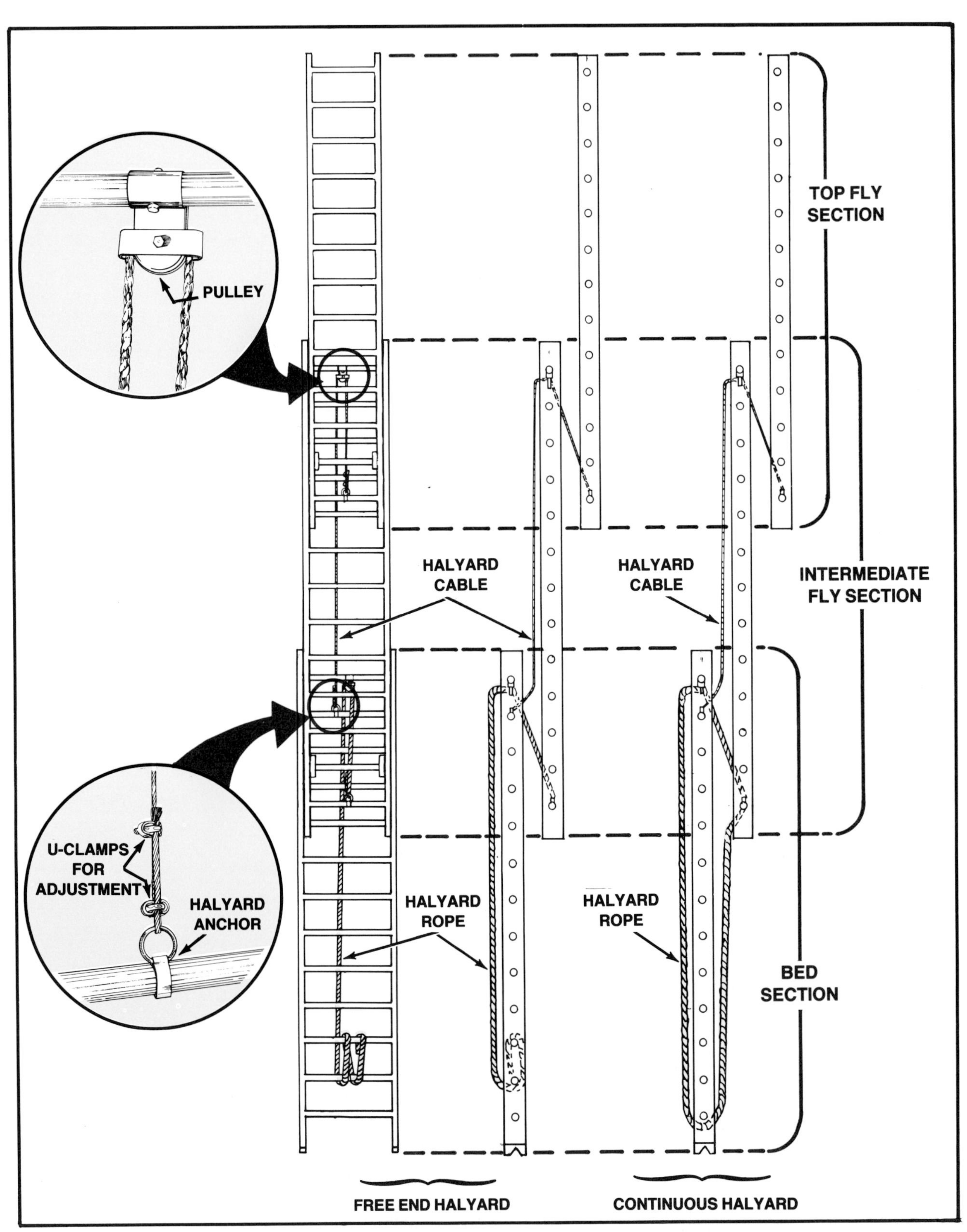

Figure 2.74 Typical halyard arrangement for a three-section extension ladder.

Stops

All extension ladders have stops to keep fly sections from traveling off the top end. One type consists of an L-shaped piece of metal attached to the outside of the fly section beam near its bottom and a similar piece of metal attached to the top of the bed section beam near its top (Figure 2.75). As the fly travels up, the two pieces engage and prevent further travel of the fly(s). One uses blocks set in the guide tracks. Another uses attachments on an upper rung which stop the pawls from traveling beyond that point (Figure 2.76). Still other means are employed which can be seen by examining extension ladders in use.

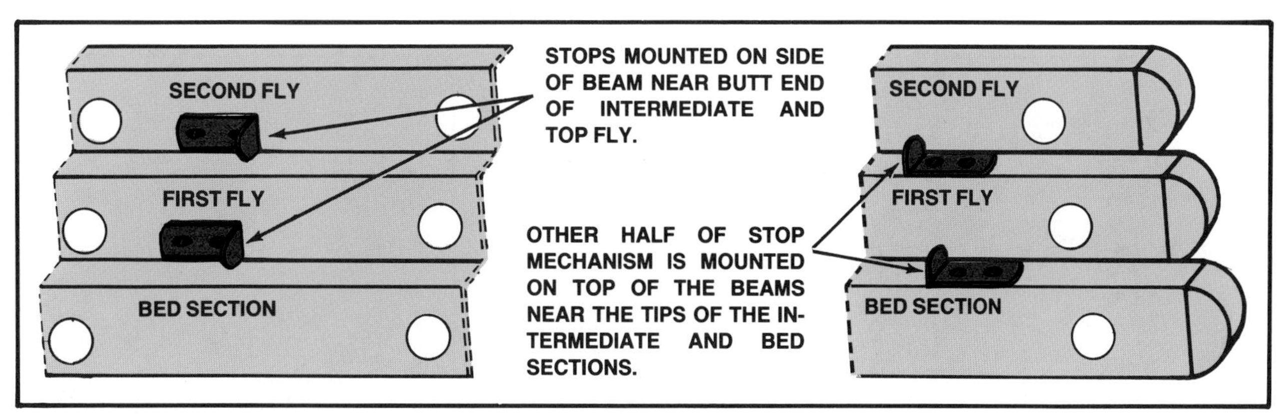

Figure 2.75 Common type of stops used for extension and pole ladders.

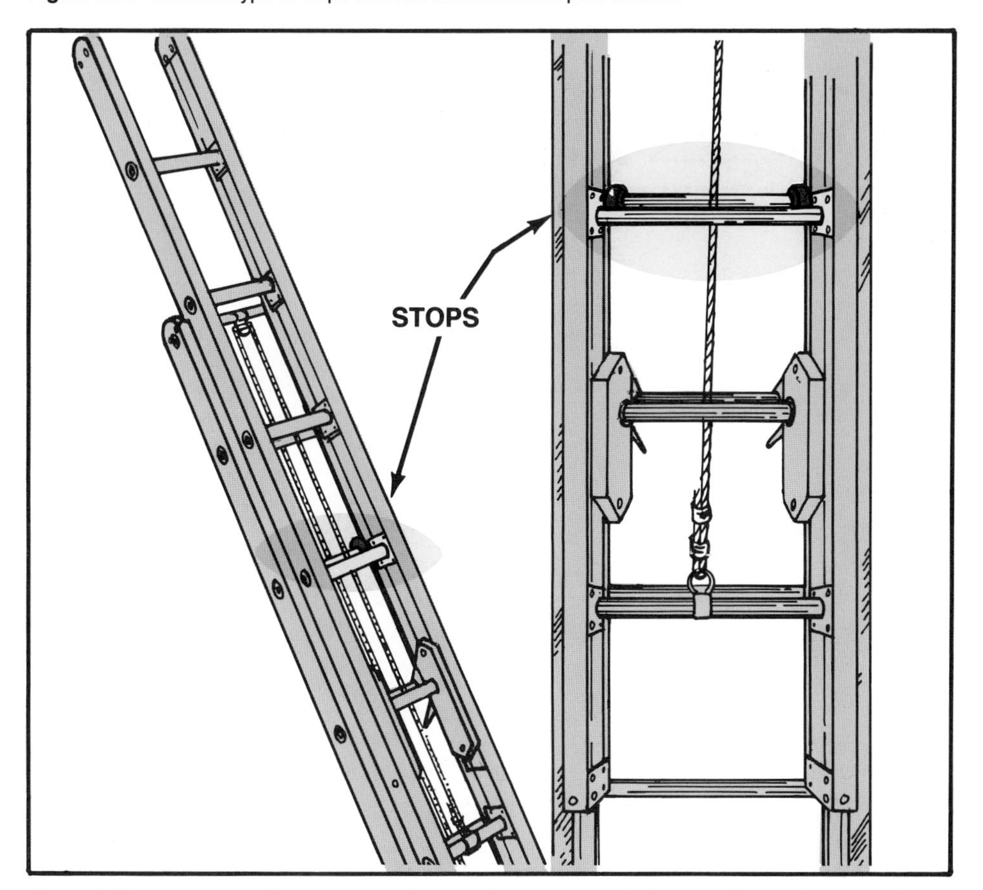

Figure 2.76 Other types of ladders use a ring attached to each end of a rung. The pawl assembly housing will not pass beyond it.

Butt Spurs

As with other features of ladder construction, there are a number of different designs used for the butt spurs. All are intended to prevent the ladder butt from slipping. Three types are shown in Figures 2.77, 2.78, and 2.79. All are metal.

Foot Pads

Foot pads are made of metal with rubber or rubberlike tread. They are bolted or riveted to the butt end of the beam, usually on shorter length ladders, particularly on folding ladders (Figure 2.80). Some are manufactured with a combination rubber tread and steel toe for use either inside or outside (Figure 2.81).

Figure 2.77 Butt spur on a wood ladder.

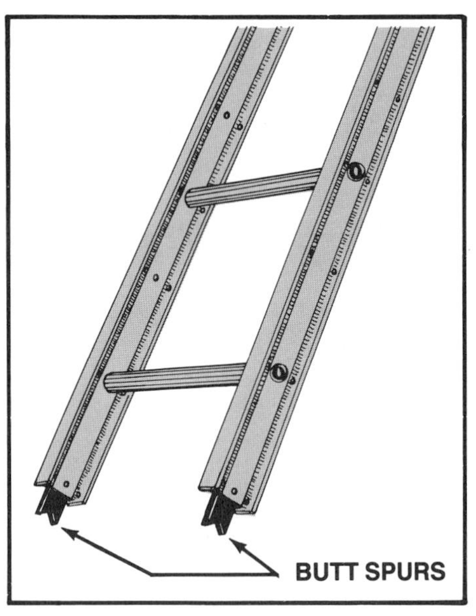

Figure 2.78 Butt spur on a solid beam metal ladder.

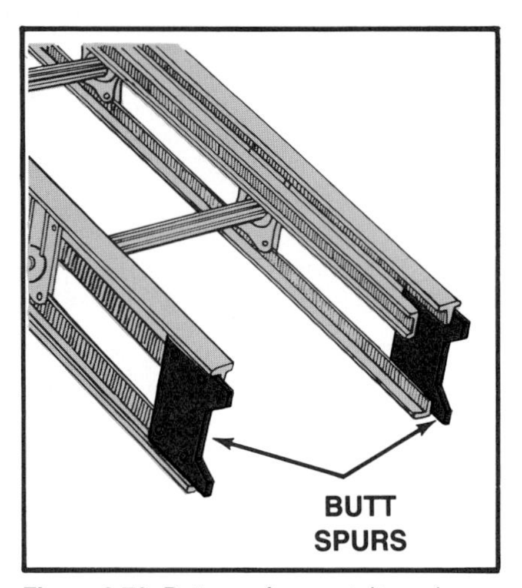

Figure 2.79 Butt spur for a metal truss beam ladder.

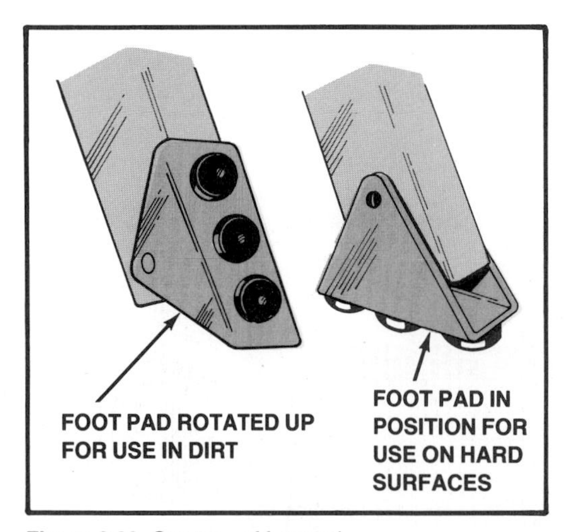

Figure 2.80 One type of foot pad.

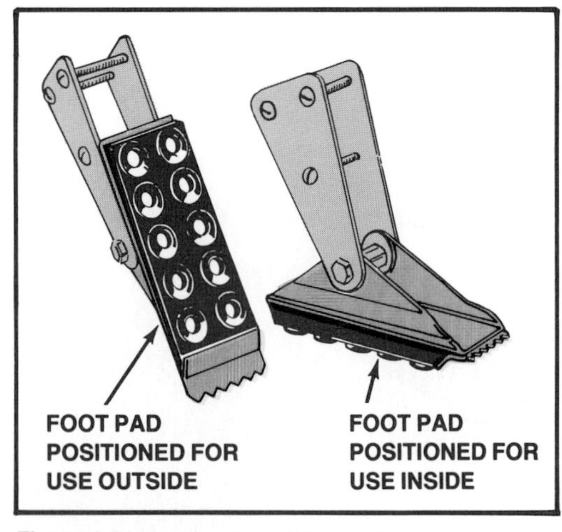

Figure 2.81 Another type of foot pad.

Tie Rods

Wood ladders have steel tie rods to help hold them together (Figure 2.82). Tie rods are usually installed just below every fourth rung. The head and nut are flat and the outside of the beam is countersunk so that they do not protrude (Figure 2.83). The nut is notched so that a two-pronged tool can be used to tighten it.

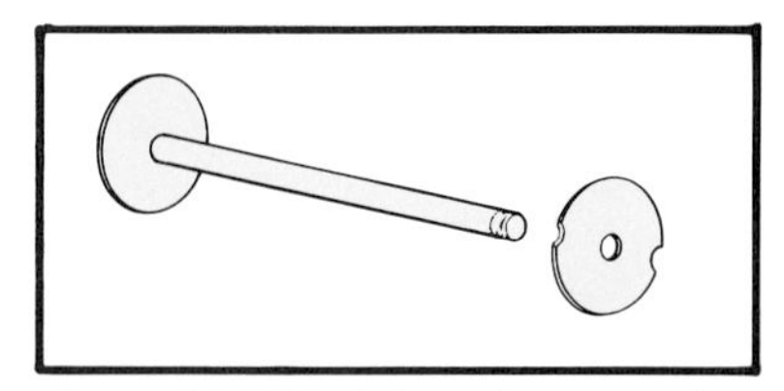
Figure 2.82 Steel tie rods are used to help hold wood ladders together.

Figure 2.83 Tie rods are normally installed just below every fourth rung.

Toe Rod

A toe rod is a steel rod that is installed just above the butt spurs. It provides a place for the firefighter to place the toe of the boot during healing operations (Figure 2.84).

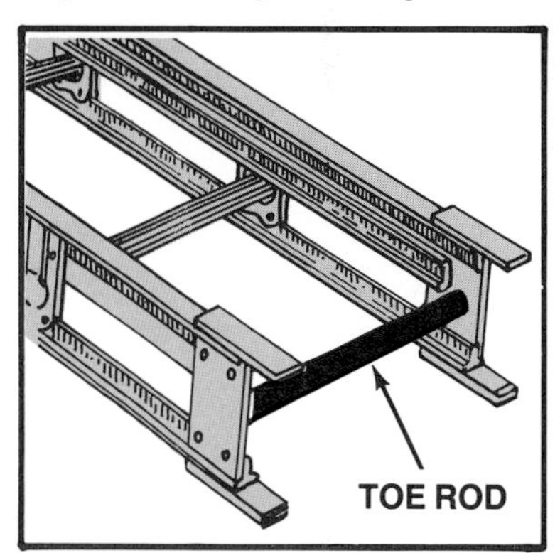

Figure 2.84 Toe rod installed at the butt end of the ladder.

Staypoles, Staypole Spurs, and Toggles

Staypoles are made of both wood and metal. Some are permanently attached (NFPA 1931 now requires permanently attached staypoles) and some latch on. The point of attachment is to the side of the beam near the top of the bed ladder. The attaching device is called a toggle. Metal spurs are provided to prevent slippage when the ladder is in place and the poles have been set (Figure 2.85).

Figure 2.85a Staypole assembly.

Figure 2.85b Staypole spurs.

Mud Guard

A mud guard is an accessory available for metal truss ladders. It is a metal plate attached between the beams at the butt which significantly increases the surface area resting on the ground. Its purpose is to prevent the ladder butt from sinking into soft ground (Figure 2.86).

Protection Plates

Protection plates are strips of metal attached to ladders at chafing points, such as the tip or areas in contact with the apparatus mounting brackets (Figure 2.87).

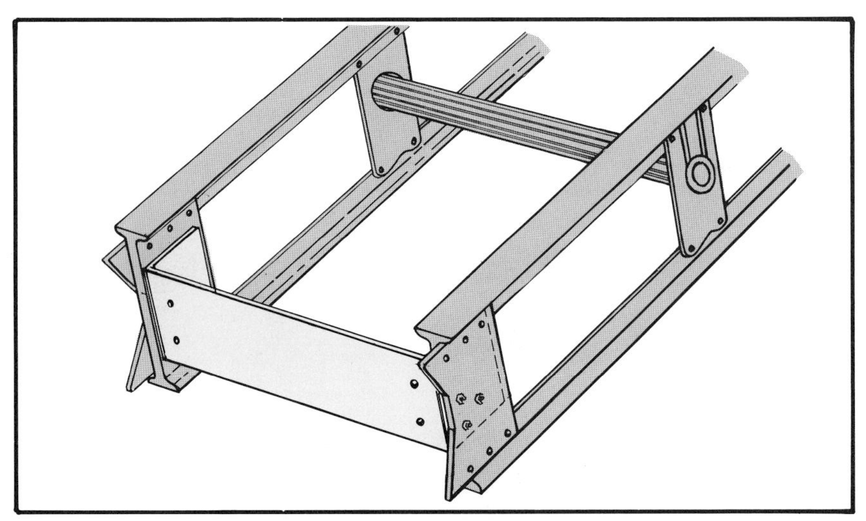

Figure 2.86 A mud guard installed across the butt end of the ladder.

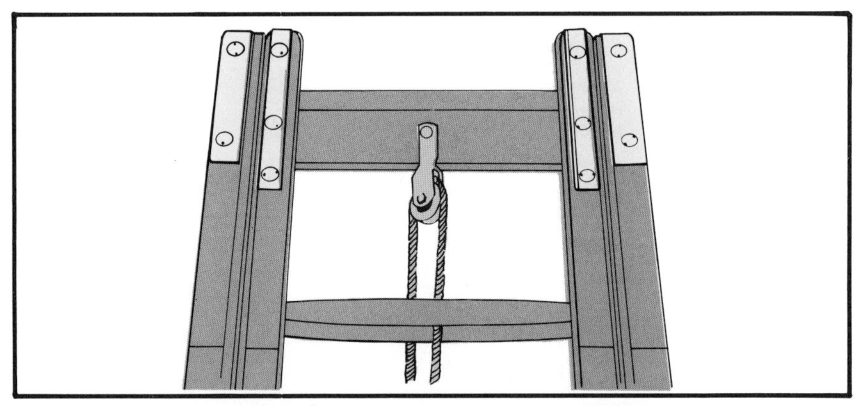

Figure 2.87 Protection plates are used on the tip ends of the beams to prevent scuff damage.

Levelers

Fire department ground ladders (with the exception of roof ladders used on sloped roofs and pompier ladders which hang from windowsills by a gooseneck-shaped hook) are designed to be used from level surfaces. Loads and stresses are thereby distributed evenly between the two beams.

When ground ladders have to be placed on uneven terrain, steps must be taken to shim or otherwise support the beam which is not touching the supporting surface. A device known as a leveler has been developed to overcome this problem. Levelers are used where uneven terrain is regularly encountered, such as in San Francisco, California.

There are two types of levelers available. Figure 2.88 shows one that is permanently attached to the butt end; it automatically adjusts to terrain contours. Figure 2.89 shows a leveler that is carried on the apparatus and attached as necessary. It is manually adjusted.

Figure 2.88 Permanently attached automatic adjusting leveler.

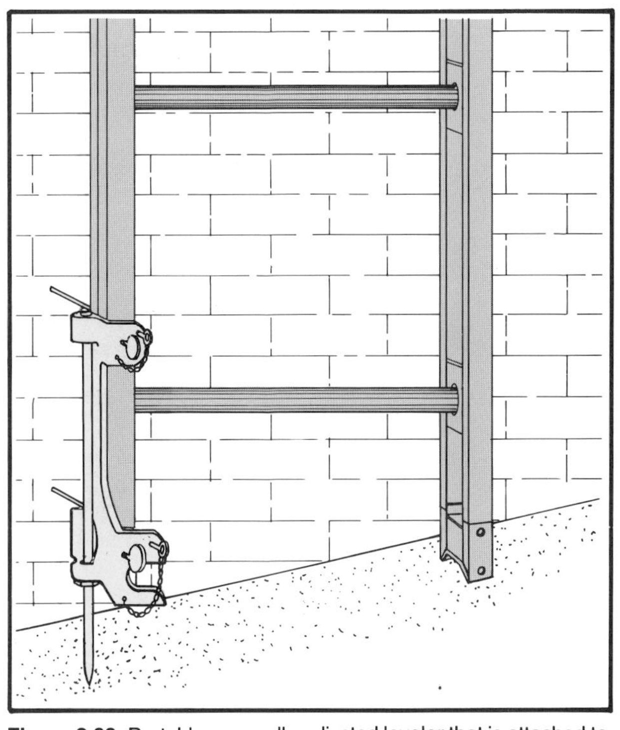

Figure 2.89 Portable, manually-adjusted leveler that is attached to the ladder beam as needed.

Labels

There are a number of labels found on newer fire service ground ladders. NFPA 1931 requires some of them and contains specifications for durability and adhesion of these labels. Labels that are commonly found are as follows:

MANUFACTURER'S IDENTIFICATION

This label gives the manufacturer's name and address and identifies the series or model number of the ladder (Figure 2.90).

Figure 2.90 Label identifying the ladder model or series.

SERIAL NUMBER

NFPA 1931 requires that fire department ground ladders bear a unique individual identification number. It can be embossed, stenciled, branded, or stamped on the ladder, or it may be on a metal plate attached to the ladder (Figure 2.91).

CERTIFICATION LABEL

This label attests that the ladder has been manufactured in accordance with NFPA 1931 and OSHA fire ladder requirements (Figure 2.92).

Figure 2.91 Serial number plate attached to ladder beam rail.

Figure 2.92 Certification label and companion electrical hazard label required by OSHA for metal ladders.

2.20 Match the definition to the ladder accessories.

_______ 1. Pawls
_______ 2. Stops
_______ 3. Butt Spurs
_______ 4. Toe Rods
_______ 5. Mud Guards
_______ 6. Levelers
_______ 7. Protection Plates

A. Provides a pivoting surface for heelman's use.
B. Strips of metal attached to ladders at chafing points.
C. Holds fly section of extension ladder at desired height.
D. Prevents the ladder from slipping.
E. Keeps the fly section from traveling off the top end.
F. Devices used to level the ladder on uneven ground.
G. Prevents the ladder butt from sinking into soft ground.

HEAT SENSOR LABEL

This is a label affixed near the top of each section of the ladder that turns color at a preset temperature. The color change indicates that the ladder has been exposed to a sufficient degree of heat that it should be tested before further use. It is most commonly used on metal ladders, in which case a Hardness Test should be run before further use of the ladder (Figures 2.93 and 2.94).

Figure 2.93 One type of heat sensor label showing appearance before and after being exposed to excessive heat.

Figure 2.94 Another type of heat sensor label. The one for the top fly section indicates exposure to excessive heat.

Design Verification Testing

The 1984 edition of NFPA 1931 requires ladder manufacturers to conduct design verification tests as a part of the initial evaluation of a specific product design and thereafter when there is a change in the design, method of manufacturing, or material. Note that each and every ladder manufactured is NOT tested in this manner only each new model is so tested. Ladders that are used for design verification testing are destroyed after testing is completed.

The design verification tests are summarized here to provide the firefighter with background knowledge of what goes into ladder design and to show fire officials the value of purchasing ladders that meet the criteria of NFPA 1931. Those persons needing further information should consult the standard.

2.21 True or False.

	True	False
1. NFPA 1931 requires design verification tests to be run on every fire service ground ladder.	☐	☐

HORIZONTAL BENDING TEST (NOT APPLICABLE TO POMPIER LADDERS)

The ladder is laid horizontal while supported at each end and a specified weight is applied to the middle area. A load of 300 pounds (136 kg) is applied to folding ladders and 750 pounds (340 kg) is applied to the others. The ladder has to sustain this load without ultimate failure.

DEFLECTION TEST (NOT APPLICABLE TO FOLDING AND POMPIER LADDERS)

The ladder is set at 75½ degrees. A 500 pound (227 kg) test load is applied to the center rung, first at one side and then at the other. The butt spur on the beam opposite the test load has to remain in contact with the ground.

RUNG BENDING STRENGTH TEST (NOT APPLICABLE TO FOLDING AND POMPIER LADDERS)

While the ladder or a three rung test section is lying horizontal and supported at both ends, the center of a selected rung is put under a 1,000 pound (454 kg) load. A formula defines how much permanent deformation of the rung can occur.

RUNG-TO-BEAM SHEAR STRENGTH TEST (NOT APPLICABLE TO FOLDING AND POMPIER LADDERS)

A single section of a ladder or a three rung test section is placed at 75½ degrees. A 1,000 pound (454 kg) test load is applied to a rung as near the beam as possible, first to one side and then the other. No permanent deformation or ultimate failure is allowed.

2.20 1. C
2. E
3. D
4. A
5. G
6. F
7. B

RUNG TORQUE TEST (NOT APPLICABLE TO FOLDING AND POMPIER LADDERS)

A 50 pound (23 kg) weight is suspended from an arm 30 inches (760 mm) long which is attached to the center of a rung. The load is applied first clockwise then counterclockwise 10 times. The standard prescribes how much rung movement may occur.

SIDE SWAY TEST (NOT APPLICABLE TO FOLDING AND POMPIER LADDERS)

In this test each section is tested separately. The test is run with the ladder in the horizontal position turned up on edge, supported at each end at a point in line with the top and bottom rungs. A load of 140 pounds (64 kg) is applied to the center of the span of the bottom beam. This is a test for permanent deformation.

BEAM CANTILEVER BENDING TESTS (NOT APPLICABLE TO FOLDING AND POMPIER LADDERS)

These tests are run on a single section. The ladder is placed in the horizontal position turned up on edge. The end is supported at a point in line with the bottom rung. A load of 850 pounds (386 kg) is suspended first from the extreme bottom end of the upper beam, then from the extreme bottom end of the lower beam. The test is to determine if permanent deformation of the beam in excess of one-half inch (13 mm) occurs.

LADDER SECTION TWIST TEST (NOT APPLICABLE TO FOLDING AND POMPIER LADDERS)

This test is conducted on a ladder base section supported over a seven foot (2 m) span. The ladder is placed in a flat horizontal position, with support for one end fixed. A 1,200 inch-pound (13,680 mm - kg) load is applied, first in a clockwise and then in a counterclockwise direction. The twist from horizontal during either test cannot be more than 14 degrees.

BUTT SPUR SLIP TEST

This is a test for skid resistance of butt spurs. The test is run with a fully extended 16 foot (5 m) extension ladder set at 75½ degrees. Both the top and the butt rest against the "A" surface of A-C plywood. A 500 pound (227 kg) test load is attached to the third rung from the tip. A horizontal force of 50 pounds (23 kg) is applied to the bottom one inch (25 mm) above the test surface. The ladder may not move more than one-fourth inch (6 mm) across the test surface.

ROOF HOOK STRENGTH TEST: ROOF LADDERS ONLY

The roof ladder is hung solely by the hooks. A test load of 2,000 pounds (907 kg) is placed over as many rungs as needed. There can be no damage from this test and deformation to the hooks may not exceed five degrees.

BEAM AND HARDWARE TEST: EXTENSION, POLE, AND COMBINATION LADDERS ONLY

The fly section is extended one rung. The ladder is placed at 75½ degrees. A 2,000 pound (907 kg) test load is applied to the top rung of the fly section. There can be no visible weakening of the beams and hardware.

SINGLE PAWL LOAD TEST: EXTENSION, POLE, AND COMBINATION LADDERS ONLY

A special fixture is used in lieu of the bed ladder. A fly section is positioned so that only one pawl is engaged to a 1¼-inch (32 mm) steel rod that is a part of the test fixture. The assembly is positioned at 75½ degrees. A 2,000 pound (907 kg) test load is then applied to the tip end of the beam. The pawl may not disengage or come loose from the beam.

PAWL TIP LOAD TEST: EXTENSION, POLE, AND COMBINATION LADDERS ONLY

The ladder or test section is positioned at 75½ degrees. Both pawls are partly engaged so that the tips rest on special plates attached to a bed section rung. The pawls are prevented from pivoting and the butt of the ladder is chocked to prevent slippage. A downward distributed 2,000 pound (907 kg) load is applied to the tip of the ladder. The test load has to be sustained without ultimate failure.

CYCLIC RUNG-PAWL TEST: EXTENSION AND POLE LADDERS

A machine is used to engage and disengage the pawl 6,000 times. The assembly has to perform with no failure.

MULTISECTION EXTENDING FORCE TEST: EXTENSION AND POLE LADDERS

This test measures how much force it takes to raise the fly. Three tests are run. The average force taken to raise the fly cannot exceed two times the weight of one ladder fly section.

COMPRESSION TEST FOR COMBINATION LADDERS

The combination ladder is placed in the A-frame position and a test load of 2,000 pounds (907 kg) is applied uniformly to the top rungs. The ladder has to sustain this load without ultimate failure.

DESIGN VERIFICATION TEST FOR POMPIER LADDERS

The pompier ladder is tested in the vertical hanging position supported only by the hook and the standoff brackets. A weight of 2,000 pounds (907 kg) is applied to the two lower rungs. The ladder has to withstand this load without ultimate failure.

2.21 False

Table 2.4 Summarizes the Design Verification Testing.

TABLE 2.4
Summary of Design Verification Testing

DESIGN VERIFICATION TEST	TYPE OF LADDER: SINGLE	ROOF	EXTENSION	POLE	COMBINATION	FOLDING	POMPIER
750 pound (340 kg) HORIZONTAL BENDING TEST	●	●	●	●	●		
300 pound (136 kg) HORIZONTAL BENDING TEST						●	
500 pound (227 kg) DEFLECTION TEST	●	●	●	●	●		
1000 pound (454 kg) RUNG BENDING STRENGTH TEST	●	●	●	●	●		
1000 pound (454 kg) RUNG TO BEAM SHEAR TEST	●	●	●	●	●		
1500 inch pounds (165 N/m) RUNG TORQUE TEST	●	●	●	●	●		
140 pounds (64 kg) SIDE SWAY TEST	●	●	●	●	●		
850 pound (386 kg) BEAM CANTILEVER BENDING TEST	●	●	●	●	●		
1200 inch pounds (132 N/m) LADDER SECTION TWIST TEST	●	●	●	●	●		
500 pound (227 kg) LOAD/50 pound (27.5 kg) PULL BUTT SPUR SLIP TEST	●	●	●	●	●		
2000 pound (907 kg) ROOF HOOK STRENGTH TEST		●					
2000 pound (907 kg) BEAM AND HARDWARE TEST			●	●	●		
2000 pound (907 kg) SINGLE PAWL LOAD TEST			●	●	●		
2000 pound (907 kg) PAWL TIP LOAD TEST			●	●	●		
6000 CYCLES RUNG PAWL TEST			●	●	●		
MULTISECTION EXTENDING TEST NOT TO EXCEED 2X WEIGHT OF ONE FLY SECTION			●	●	●		
2000 pound (907 kg) COMPRESSION TEST: IN A-FRAME POSITION					●		
2000 pound (907 kg) HANGING STRENGTH TEST							●

NOTE: Design Verification Testing is *not performed* by fire service personnel.

INSPECTION AND MAINTENANCE

Like any equipment used in the fire service, ground ladders require periodic and thorough inspection, regular cleaning, and lubrication to ensure that they are safe and 100 percent operational. This degree of reliability does not occur by accident. A fire department must follow a systematic program to accomplish this.

The 1984 edition of NFPA 1932 provides an excellent basis for a fire department ground ladder inspection and maintenance program. Much of the information on the following pages is excerpted from this standard.

Inspection

The NFPA standard calls for an inspection monthly and after each use. These inspections include but are not limited to checking:

- The heat sensor label on metal and fiber glass ladders, and on wood ladders when provided, for change indicating heat exposure. Ladders without a heat sensor label may also show signs of heat exposure: bubbled or blackened varnish on wood ladders, discoloration of fiber glass ladders, heavy soot deposits, or bubbled paint on tips of any kind of ladder.

  CAUTION: The strength of metal ladders depends on the hardness of the metal. In the case of metal ladders, the aluminum alloy's hardness was obtained by a heat treating process. Metal ladders which have been exposed to heat *should be placed out of service* until a hardness test can be performed.
- All rungs for snugness and tightness.
- All bolts and rivets for tightness (bolts on wood ladders should not be so tight that they crush the wood).
- Welds, for any cracks or apparent defects.
- Beams and rungs for cracks, splintering, breaks, gouges, checks, wavy conditions, or deformation.

If any of the conditions described in the preceding items are found, the ladder should be removed from service until it can be repaired and tested or it should be destroyed.

2.22 Fill in the blanks.

1. NFPA 1932 calls for ground ladders to be inspected ______ and ______.
2. Ladders discovered with color changes in their heat sensor labels should be taken out of service until a ______ test has been performed.

Other items to be checked are dictated by the type of ladder. This additional inspection includes but is not limited to checking:

WOOD LADDERS/WOOD COMPONENTS OF COMPOSITE LADDERS

- For areas where the varnish finish has been chafed or scraped off
- For darkening of the varnish (indicating age deterioration of the varnish)

- For dark streaks in the wood (indicating deterioration of the wood)

  NOTE: Any indication of deterioration of the wood should be cause for the ladder to be removed from service until it can be service tested.

ROOF LADDERS

- For proper operation of roof hook assemblies. In addition, the assembly should not show signs of rust, the hooks should not be deformed, and parts should be firmly attached with no sign of looseness.

  NOTE: Serious problems found should result in removal from service pending service testing.

EXTENSION AND POLE LADDERS

- For proper operation of pawl assemblies. The hook and finger should move in and out freely.
- For fraying or kinking of the halyard. If condition is found the halyard should be replaced.
- For snugness of any halyard cable when the ladder is in the bedded position (to insure proper synchronization of upper sections during operation).
- For free-turning pulleys.
- For the condition of the guides and for free movement of the fly sections.

  NOTE: Any indication of failure or weakness of the guides should be cause for removing the ladder from service pending service testing.

- For free operation of the staypole toggles and their condition. Detachable staypoles are provided with a latching mechanism at the toggle. This should be checked to be sure that it is latching properly.

2.23 True or False.

	True	False
1. When wire cable is used as a halyard for the second and third fly sections, it does not require inspection.	☐	☐
2. When extension ladder fly sections are fully retracted the halyard cables should have a slight tension on them.	☐	☐

Maintenance

Maintenance items detailed in NFPA 1932 are

- Keeping ground ladders free of moisture

- Not storing them in an area where they are exposed to the elements
- Not painting them except for the top and bottom 12 inches (300 mm) of the beams for purposes of identification or visibility

METAL AND FIBER GLASS LADDERS ONLY

- An occasional application of a good automotive paste wax to restore the surface finish

WOOD LADDERS ONLY

- Store away from steam pipes, radiators, and out of direct sunlight or areas where the humidity is artificially reduced.

  NOTE: Continued exposure to a heating source or direct sunlight will cause the wood to dry out and lose its strength.

- Protection of the wood by at least two coats of a good quality, clear spar varnish, or polyurethane varnish. The varnish finish should be redone at least annually.

  NOTE: It is important that the varnish coating be maintained as it preserves the wood, which is a factor in maintaining its strength.

The standard provides the following procedure for repairing small areas of damage to the varnish coating.

Step 1: Remove peeling areas by scraping and sanding with sandpaper to remove all the loose or damaged finish.

Step 2: Spot prime bare sanded spots with varnish.

Step 3: Resand when dry and coat with at least two coats of a good quality, clear spar varnish, or polyurethane varnish.

FIBER GLASS LADDERS ONLY

- Store out of direct sunlight. Continued exposure to ultraviolet rays will erode the fiber glass surface and cause the fibers to be exposed, thereby weakening the ladder.

ROOF LADDERS ONLY

- When roof ladder hooks are rusted in or around the spring assembly housing, they should be disassembled, cleaned, and lubricated.

EXTENSION LADDERS ONLY

- Pawl assemblies should be kept clean and lubricated in accordance with manufacturer's instructions.

- Pawl torsion springs should be replaced every five years or sooner if pawl operation appears weak.

  NOTE: When reinstalling pawl assemblies use caution to prevent overtightening of pawl assembly fasteners, as this will cause binding of pawl assembly parts.

- Ladder slide areas (including guides) should be kept lubricated in accordance with manufacturer's instructions. If instructions are not available, the lubricant should be cleaned off and reapplied. Dry or spray lubricant is used on metal and fiber glass ladders. Candle wax should be applied to wood extension and pole ladder guides at least once every six months. A wax thinner than candle wax does not sufficiently lubricate the surfaces, leaving the fly sections unable to travel freely.

  CAUTION: A *safety solvent* should be used to remove old lubricant.

- When replacing halyards on a three- or four-section extension ladder, it may be wise to sketch the cable arrangement before removing the old halyard. The sketch can then be used to ensure that the replacement is threaded properly.
- Halyard pulleys with ball bearing centers require a small amount of lubricant periodically to operate smoothly.

2.24 True or False.

	True	False
1. Painting fire department ground ladders is not recommended; however, the top and bottom 12 inches (300 mm) may be painted for identification or visibility.	☐	☐
2. Wood ladders in storage should be stored in areas where the humidity is artificially reduced.	☐	☐
3. Continued exposure of fiber glass ladder beams to sunlight will cause ultraviolet erosion of the surface of the fiber glass.	☐	☐
4. Automatic latching pawl torsion springs are normally replaced yearly, usually at the time of the annual service test.	☐	☐
5. Ladder slide areas should be kept lubricated with petroleum grease. Dry or spray lubricant should not be used.	☐	☐
6. Kerosene should be used to remove old lubricant from ladders.	☐	☐

2.22 1. Monthly, After Each Use
2. Hardness

2.23 1. False
2. True

Cleaning

Regular and proper cleaning of ladders is more than a matter of appearance. Unremoved dirt or debris may collect and harden to the point where ladder sections are no longer operable. Therefore it is recommended that ladders be cleaned after every use.

A brush and running water are used as in Figure 2.95. Tar, or oily or greasy residues are removed with safety solvents. After the ladder is rinsed, or anytime a ladder is wet, it should be wiped dry. During each cleaning firefighters should be alert for defects.

Figure 2.95 Proper cleaning is an important part of ladder maintenance.

2.25 Fill in the blanks.

1. To remove dirt from ladders, use a ______________________ and ________________ water.
2. To remove tar residues from a ladder, ______________________ is used.

Repairing Ladders

During the inspection of ladders, the inspector should mark all defects with chalk or some other suitable marker (Figure 2.96). Legible marks help keep defects from being overlooked when repairs are made. Special tools not readily available to most fire departments are usually needed for major repairs to metal ladders. Repair to ladders must be made in accordance with manufacturer's recommendations.

Figure 2.96 Defects marked with chalk during an inspection.

2.24 1. True
2. False
3. True
4. False
5. False
6. False

Special precautions must be observed when repairing wood ladders. Small splinters in wood can be removed by cutting the wood across the grain with a sharp knife at the large end or base of the splinter. The splinter can then be removed and the part sanded smooth. When tightening wood ladder beam bolts and tie rods, a special spanner should be used and the nut should only be tightened snugly to prevent crushing the wood cells. Minor repairs to wood ladders can usually be made in fire department shops. The finish of the wood is very important in maintaining the maximum useful life of the ladder. As a wood finish decomposes or oxidizes, the moisture, resins, and oils in the wood escape and leave the wood cells exposed to air and elements. Varnish may be depended upon to preserve the wood since it seals in the natural oils and resins and keeps moisture out. Varnish also prevents dry rot and fungus growth. When the varnish finish becomes worn or scratched, it should be redone without delay. Paint is not recommended on fire department ladders except to identify the ladder ends, balance point, racking points, or length. Paint used as an outer cover of a wood ladder makes it practically impossible to detect fungus growth, dry rot, or cracks during inspection.

The preparation of the wood to be refinished and the refinishing procedure are of prime importance. First, remove as much of the old finish as possible by scraping with a scraper tool, but do not scrape against the grain of the wood. Then sand all surfaces well with medium sandpaper or garnet paper; follow with fine sandpaper. An appropriate liquid remover may be used when conditions warrant. To preserve the wood, apply two coats of sealer and allow sufficient drying time between coats. After this, two coats of varnish are applied, allowing sufficient drying time between coats. All repairs should be recorded on the correct ladder report form.

2.26 Fill in the blanks.

1. During an inspection of a ladder, all defects should be marked by ______________ or some other ______________.

2.27 True or False.

	True	False
1. Paint used on a wood ladder preserves the wood but makes it practically impossible to detect fungus growth, dry rot, or cracking.	☐	☐
2. Varnish is used to protect wood ladders because it allows the wood to breathe, and moisture can pass in and out.	☐	☐

SERVICE TESTING GROUND LADDERS

If one reviews all the possible events that may require use of fire service ground ladders, especially events beyond the control of fire service personnel, a conclusion can easily be made that fire service ladders are subject to abuse and overloading.

No manufacturer or fire department official can guarantee that a ladder will not fail under such circumstances but the chances can be minimized by testing them in accordance with NFPA 1932, 1984 edition. This standard recommends that *only the tests specified be conducted* either by the fire department or an approved testing organization.

This standard further recommends that caution be used when performing service tests on ground ladders to prevent damage to the ladder or injury to personnel.

NOTE: Personnel need to be aware that there is a possibility of sudden failure of the ground ladder undergoing service testing. All safety precautions possible should be taken to avoid injury.

Service tests of ground ladders may require the purchase of measuring instruments and testing should be done only by personnel trained in service test procedures and operation of the service test equipment.

2.28 True or False.

	True	False
1. If all the instructions are followed, there is no possibility of ladder failure during service testing of ground ladders.	☐	☐

2.29 Check the correct response.

Service testing of ground ladders is detailed in what NFPA standard?

☐ A. NFPA 1901 ☐ C. NFPA 1932 ☐ E. NFPA 1962
☐ B. NFPA 1931 ☐ D. NFPA 1961

2.25 1. Brush, Running
2. Safety Solvent

2.26 Chalk, Suitable Marker

When Should Ladders be Service Tested?

NFPA 1932 contains the following requirements pertaining to frequency of service testing.

- At least annually
- Any time a ladder is suspected of being unsafe
- After the ladder has been subjected to overloading
- After the ladder has been subjected to impact loading or unusual conditions of use
- After heat exposure

  NOTE: Metal ground ladders being tested because of exposure to heat may be subjected to either the Strength Service Test or the Hardness Service Test; both are not required. However, if the Hardness Service Test is used and the ladder fails, a Strength Service Test shall be conducted.

- After any deficiencies have been repaired, unless the only repair was replacing the halyard

2.30 Fill in the blank.
1. NFPA 1932 requires service testing of ground ladders at least ______________.

2.31 True or False.

	True	False
1. NFPA 1932 states that metal ground ladders being tested because of exposure to heat are subjected to *either* the Strength Service Test or the Hardness Test, both are not required.	☐	☐
2. A Strength Service Test is not required after ladder repairs.	☐	☐

What Constitutes Failure?

Any signs of failure during service testing shall be sufficient cause for the ground ladder to be removed from service. It should either be repaired and retested or be destroyed.

How to Service Test Ground Ladders

There are two sets of criteria: one for all ladders except pompier ladders and another for just pompier ladders.

STRENGTH SERVICE TESTING REQUIREMENTS FOR ALL LADDERS EXCEPT POMPIER LADDERS

Horizontal Bending Test

The ladder shall be positioned for testing and tested as shown in [Figure 2.97]. The ladder shall be placed in a flat horizontal position supported under the first rung from each end of the ladder. Extension, [pole] and combination ladders shall be extended to their maximum extended length, with pawls engaged, for this test. The test load shall be applied equally to a center span covering 16 in. [410 mm] each side of the center inclusive. The test load shall be applied to a flat test surface resting on the beams in the center area. The test load shall consist of weight increments consistent with safety and ease of handling. All test loads shall include the weight of the test surface.

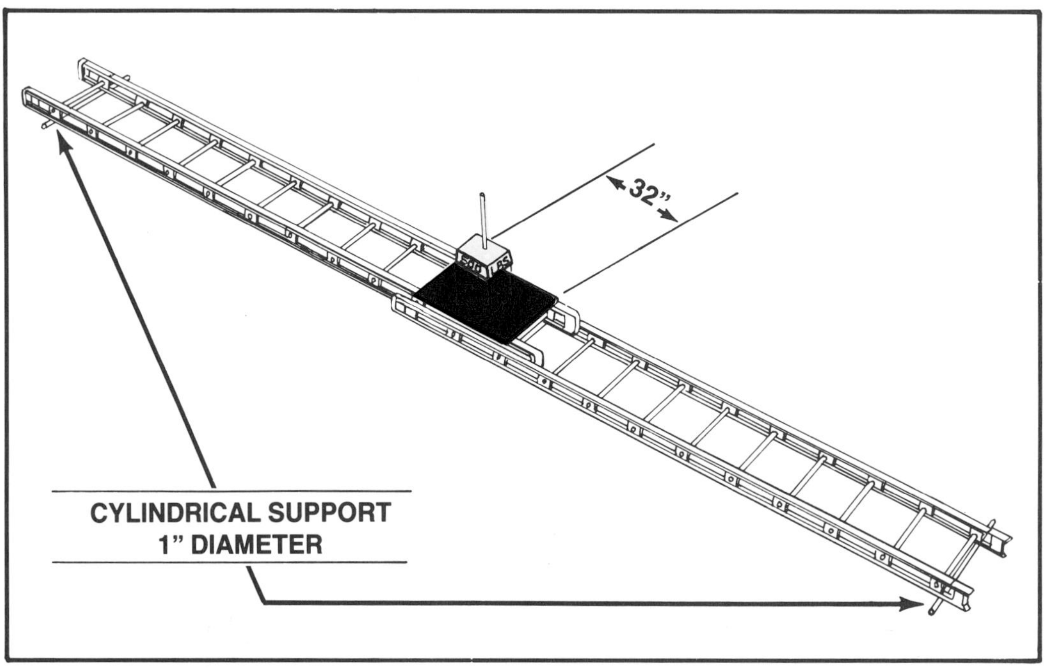

Figure 2.97 Position of ladder for the in-service horizontal bending test.

Test Procedures for Metal and Fiber Glass Ground Ladders Only

[Step 1] The ladder shall be loaded with a preload of 350 pounds [136 kg] applied equally to the center span covering 16 in. [410 mm] each side of the center inclusive. Caution shall be exercised whenever applying or removing the weights to minimize any impact loading. The load shall be allowed to remain for at least one minute, to "set" the ladder prior to completing the rest of the test.

[Step 2] After removing the preload, the distance between the bottom edge of each side rail and the surface upon which the ladder supports are placed shall be

measured. All measurements shall be taken at a consistent location as near as practical to the center of the ladder.

[Step 3] The ladder shall be loaded with a test load of 500 pounds [227 kg] applied equally to the center span covering 16 in. [410 mm] each side of the center inclusive. The test load shall remain in place for five minutes.

[Step 4] The test load shall then be removed and the distance between the bottom of each side rail and surface upon which the ladder supports are placed shall be measured. Five minutes shall elaspse before conducting this measurement after removing the test loads.

Differences between measurements taken in [Step 2 and Step 4] shall not exceed ½-in. [13 mm] for ladders 25 ft. [8 m] in length and under. This difference shall not exceed 1-in. [25 mm] for ladders over 25 ft. [8 m] in length. Any ladder that does not meet these criteria shall be removed from fire service use and destroyed. There shall be no visible permanent change or failure of any hardware.

TEST PROCEDURES FOR WOOD GROUND LADDERS ONLY

The ladder shall be loaded with a test load of 500 pounds [227 kg] applied equally to a center span covering 16 in. [410 mm] each side of the center inclusive. The test load shall remain in place for five minutes and then removed.

To pass the test, the ladder and all components shall not show ultimate failure. Any ladder that fails the test shall be removed from fire service use and destroyed.

2.32 Fill in the blank.

1. Strength Service Tests are required for all fire department ladders except _______ ladders.
2. The Horizontal Bending Test employs a test load of _______.

Additional Requirements for Roof Ladders Only - Roof Hook Test

The ladder shall be positioned for testing and tested as shown in [Figure 2.98]. The ladder shall be hung solely by the roof hooks, with the hooks supported only by the points of the hooks, in a vertical position from a fixture capable of supporting the entire test load and weight of the ladder. The ladder shall be secured in the test position to prevent injury to test personnel if the hooks fail.

2.27 1. True
2. False

2.28 False

2.29 C

2.30 Annually

2.31 1. True
2. False

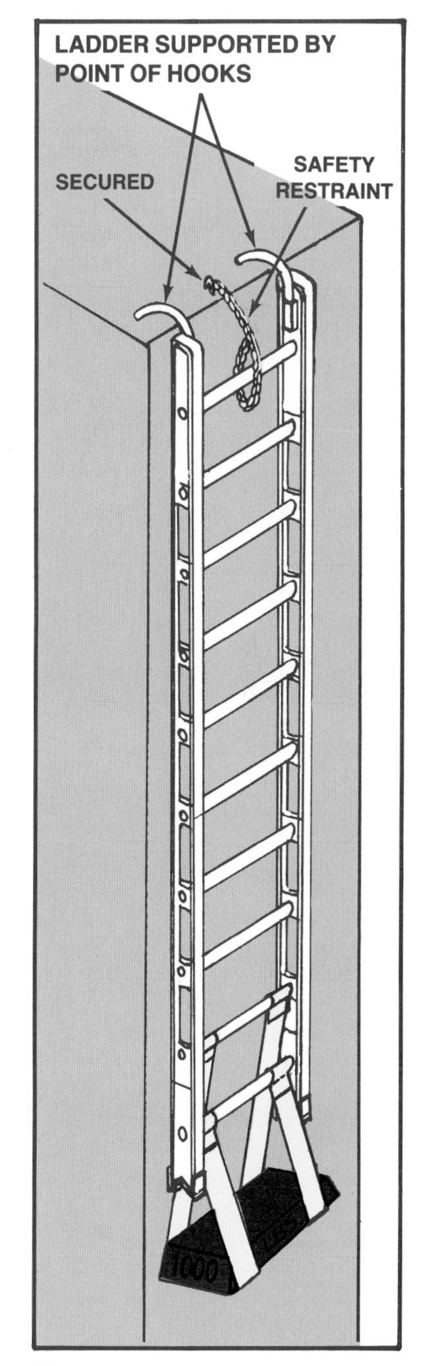

Figure 2.98 Equipment and ladder placement for in-service roof ladder hook test.

A test load of 1,000 pounds [454 kg] shall be placed over as many rungs as needed. The test load shall consist of weight increments consistent with safety and handling ease.

Test load shall be applied for a minimum of one minute.

Ladders and roof hook assemblies shall sustain this test load with no damage to the structure, and any deformation to the hooks shall not exceed 10 degrees, as shown in [Figure 2.99].

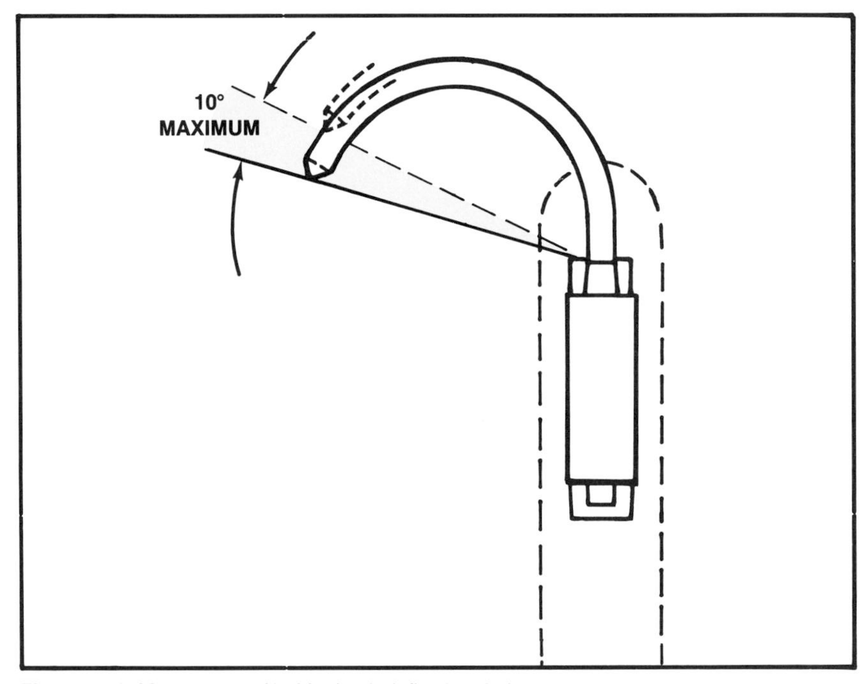

Figure 2.99 Maximum roof ladder hook deflection during testing.

2.33 Fill in the blanks.

1. Roof ladders are subject to a horizontal bending test employing a weight of ______________ and a roof hook test employing a weight of ______________ .

Additional Requirements for Extension [and Pole] Ladders Only-Hardware Test

The ladder shall be positioned for testing and tested as shown in [Figure 2.100]. The ladder shall be extended a minimum of one rung beyond the bedded position.

A test load of 1,000 pounds [454 kg] shall be placed on the rungs of the fly section. The test load shall consist of weight increments consistent with safety and ease of handling. The test load shall be applied for a minimum of one minute.

Ladders shall sustain this test load with no permanent deformation or other visible weakening of the structure.

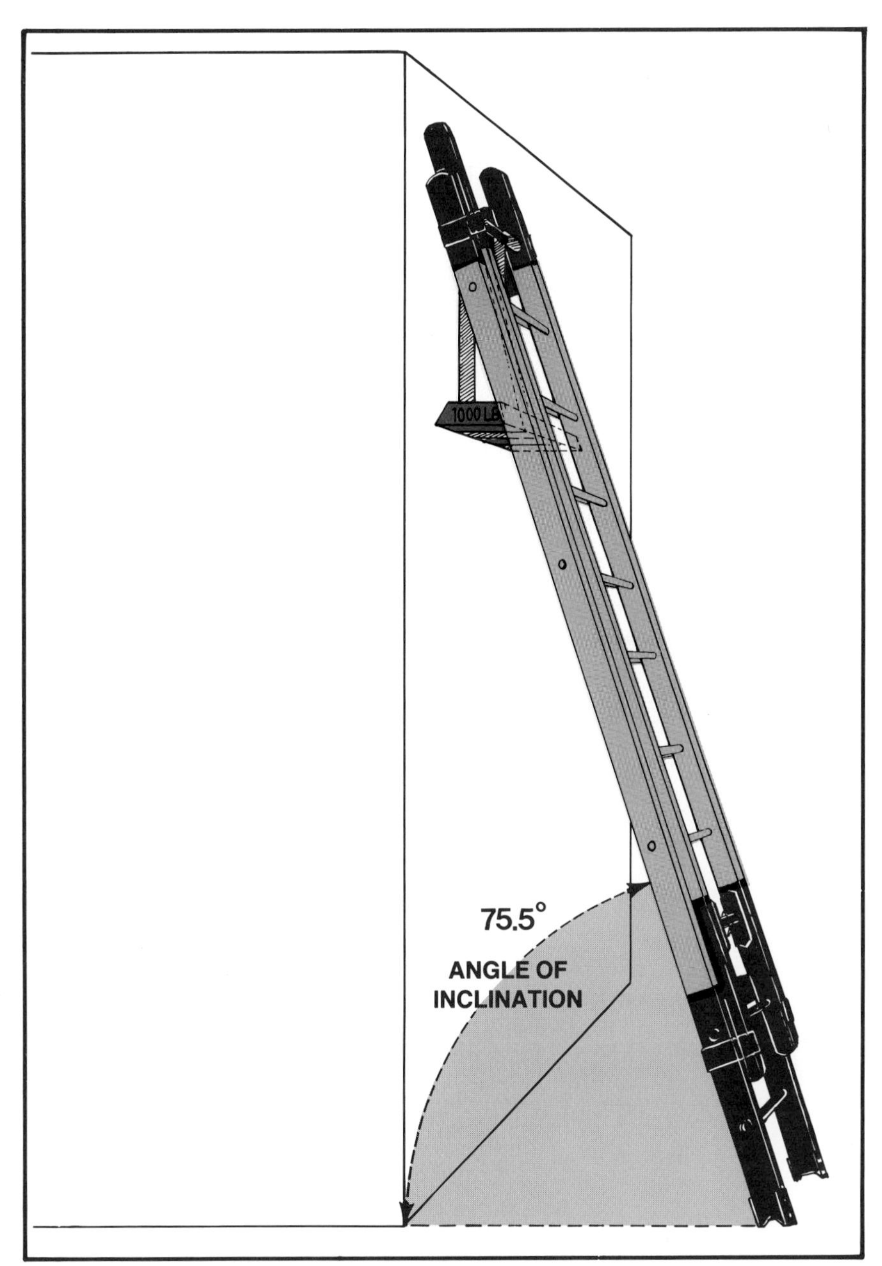

Figure 2.100 In-service extension and pole ladder hardware test.

2.32 1. Pompier
2. 500 (227 kg)

2.34 Fill in the blank.

1. Pawls on extension, pole, and some combination ladders are subject to a test that employs a weight of ______.

Hardness Service Testing Requirements for Metal Ground Ladders Only

The testing criteria specified in this section shall ONLY apply to metal ground ladders constructed from 6061-TB aluminum alloy. For other aluminum alloys or for other metals, the ladder manufacturer shall supply the hardness testing criteria.

The Hardness Service Test shall be performed at a test point located between every rung on both beams. For beams of truss construction, the test point shall be located on both the top chord and the bottom chord of the truss between every rung on both beams. One reading shall be taken at each test point [Figures 2.101 and 2.102].

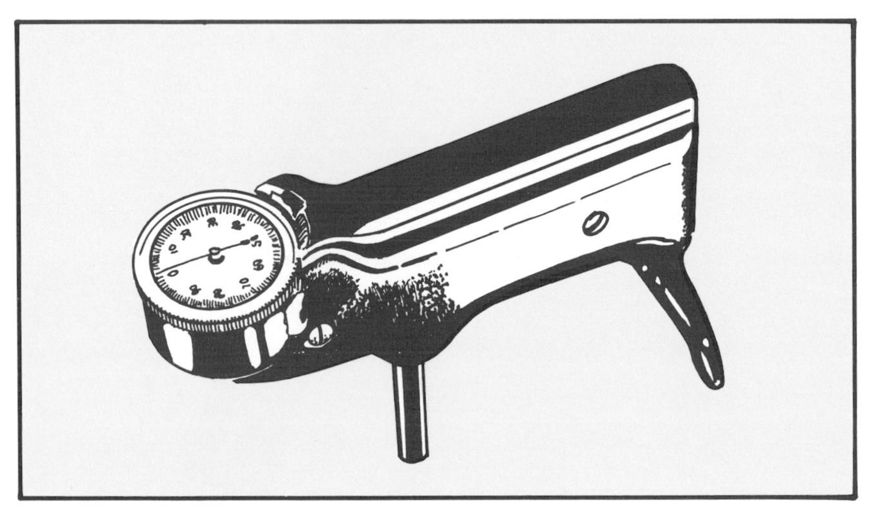

Figure 2.101 Hardness tester.

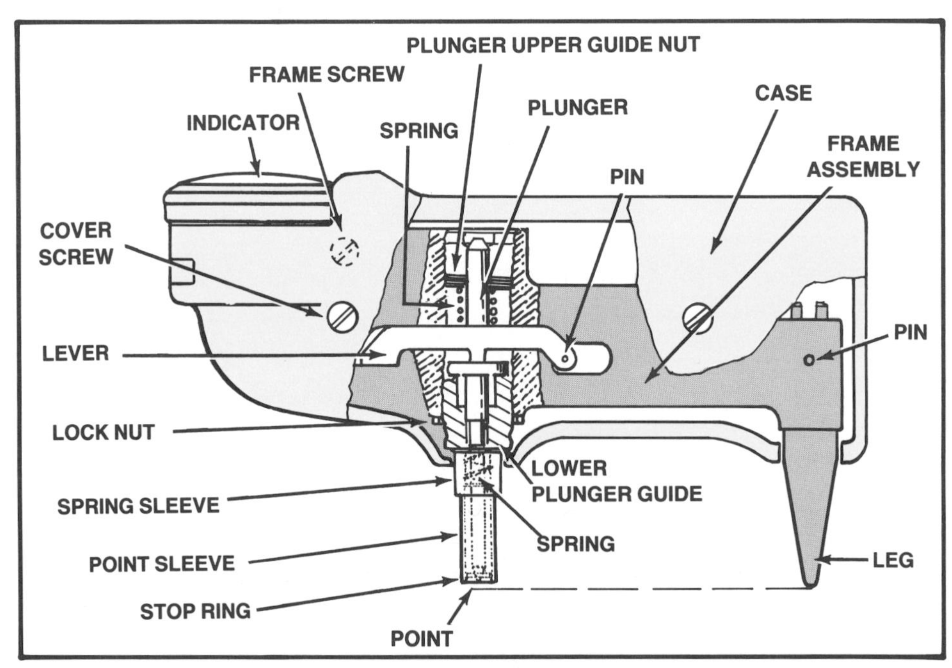

Figure 2.102 Schematic of a hardness tester showing major components.

The hardness testing device shall be calibrated immediately before testing and calibration verified immediately after testing in accordance with the manufacturer's specifications.

The reading obtained at each test point shall not be less than the value given for any of the hardness measuring scales specified in [Table 2.5].

TABLE 2.5

Hardness Test Minimum Readings

Hardness Testing Scale	*Minimum Reading*
Barber Coleman	76
Brinell	80
Rockwell B	48
Rockwell E	84
Rockwell F	84
Rockwell H	103
Vickers	88

If a reading at a test point is less than the value given in Table 2.5 for the respective hardness testing scale, three readings shall be taken at that test point. The average of the three readings shall not be less than the value given in Table 2.6. No one reading of these three shall be less than the minimum value given in Table 2.6 for the respective hardness testing scale.

TABLE 2.6

Additional Hardness Test Readings

Hardness Testing Scale	*Average of 3 Readings Not Less Than*	*No One Reading At Or Less Than*
Barber Coleman	76	73
Brinell	80	71
Rockwell B	48	33
Rockwell E	84	79
Rockwell F	84	79
Rockwell H	103	100
Vickers	88	76

If the ladder does not meet the hardness service test requirements, the ladder is then required to be subjected to a Strength Service Test.

Strength Service Testing Requirements for Pompier Ladders Only

The ladder shall be positioned for testing and tested as shown in [Figure 2.103 on next page]. The ladder shall be tested in the vertical hanging position supported only by its hook from a fixture capable of supporting the entire test load and weight of the ladder. The ladder shall be secured in the test position to prevent injury to test personnel if the hook fails.

ADDITIONAL TESTING

When a fire department desires additional nondestructive testing of metal ground ladders constructed from 6061-T6 aluminum alloy, it is usually contracted to an approved

2.33 500 pounds (227 kg), 1,000 pounds (454 kg)

2.34 1,000 pounds (454 kg)

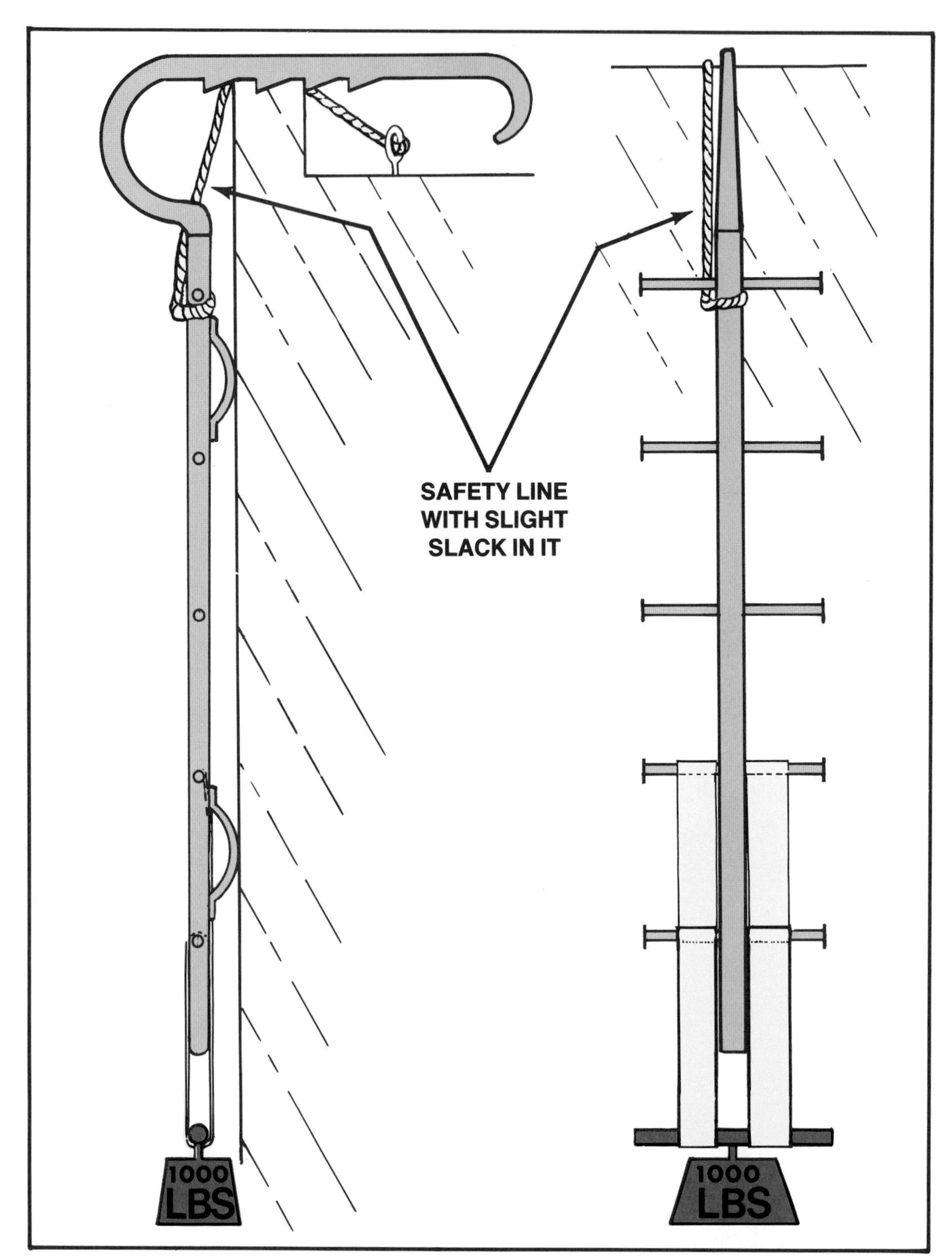

Figure 2.103 Positioning and equipment needed for in-service pompier ladder tests.

testing organization because interpretation of testing results is critical and must be performed only by certified personnel. There are two additional tests performed: Eddy Current Percent Electrical Conductivity of International Annealed Copper Standard (IACS) and *Liquid Penetrant Testing.*

*Reprinted by permission from NFPA Standard No. 1932. *Standard on Use, Maintenance, and Service Testing of Fire Department Ground Ladders*. Copyright 1984, National Fire Protection Association.

Eddy Current Percent Electrical Conductivity Test

The purpose of the eddy current test is to find areas of aluminum ladders which have become annealed by exposure to heat. For a fuller explanation of this test see paragraph A-5-4.71 of the appendix to NFPA 1931.

2.35 True or False.

	True	False
1. The Eddy Current Percent Electrical Conductivity Test is so simple that anyone with the equipment can perform the test.	☐	☐

Liquid Penetrant Testing

The purpose of liquid penetrant testing is to detect otherwise invisible cracks in the metal. It is particularly valuable to check welds. For a fuller explanation of this test see paragraph A-5-4.71 of the appendix to NFPA 1932.

RECORDS

It is important that records be kept of strength service tests, repairs, and retesting. These records will be valuable to

- Provide a means of showing that testing was performed.
- Provide the information necessary (dates) to assure that testing is done annually.
- Identify the service status (in or out of service).
- Identify the ladder location.
- Identify ladders that have been subjected to abuse and show that they were tested after the incidents.
- Provide information for investigation of ground ladder failures.
- Provide information for use in legal actions resulting from ladder failures.
- Provide records for use in ISO inspections.
- Provide information for evaluations of one model or brand as compared with another model or brand when drawing specifications for purchasing or evaluating purchasing bids.

Appendix A provides a sample testing and repair record sheet for fire department ground ladders.

Review

Answers on page 386

Fill in the blanks.

1. NFPA 1931 requires all manufacturers to provide certification attesting that the particular ladder was constructed in accordance with the standard's requirements. This certification is in the form of a ______________ which is ______________ ______________.

2. If a metal ladder loses its strength during exposure to heat, it ______________ (will, will not) regain its original strength when it cools.

3. A metal ladder hot enough to cause water to sizzle upon contact should be subjected to a ______________ test before further use. If it fails this test, it should be given a ______________.

Essay

4. What indication, other than a change in the heat sensor label, may be present if a fiber glass ladder is exposed to excessive heat?

____

____

Fill in the blanks.

5. The maximum ladder load weight for folding and pompier ladders is ______________, for all other ladders it is ______________.

6. The three main parts of the enclosed automatic latching pawl assembly are: (1) ______________, (2) ______________ and (3) ______________, which should be replaced every _____ years.

True or False.

		True	False
7.	Solid beam ladders in lengths of 24 feet (7 m) and under are lighter than comparable truss beam ladders.	☐	☐
8.	NFPA 1932 allows halyards to be spliced if the splice is made by a trained person and the rope is weight tested before being put back in service.	☐	☐
9.	When an extension ladder using wire rope (cable) for the second and third flys is in the fully retracted position the cable should be slack.	☐	☐
10.	NFPA 1931 states that design verification testing can be done periodically by fire departments if they desire, are properly trained, and have the necessary equipment.	☐	☐

	True	False
11. Manufacturers perform a series of design verification tests on every ladder built to specifications which require compliance with NFPA 1931.	☐	☐

2.35 False

Fill in the blank.

12. ________________ should be applied to metal and fiber glass ladders to enhance their finish.

Essay

13. Why is maintenance of ground ladders critical?

____

____

True or False.

	True	False
14. The weight of a charged hoseline being used off of a ground ladder is NOT counted when figuring maximum ladder loading.	☐	☐
15. The new NFPA 1931 now requires all ladders except pompier ladders to be the same minimum width of 16 inches (410 mm).	☐	☐
16. Hardware on ground ladders has to meet the same minimum strength requirements as the ladder components.	☐	☐

Fill in the blanks.

17. Halyard rope has to be a minimum of ______ in diameter with a minimum breaking strength of ____________________.

18. ____________ or ____________ lubricant is used on metal and fiber glass ladder slide areas when recommendations of NFPA 1932 are followed.

19. NFPA 1932 recommends that ________________ be used to remove old lubricant from ladders.

True or False.

	True	False
20. When a splinter is found on a wood ladder the proper procedure is to use a sharp knife to cut across the grain at the base of the splinter to remove it and then sand and refinish it.	☐	☐
21. Liquid varnish remover is an approved method of removing old varnish from wood ladders.	☐	☐

	True	False
22. NFPA 1932 requires that after any repair is made to a ladder, except replacement of a halyard, a strength service test be conducted.	☐	☐
23. Three firefighters are climbing a properly positioned 20 foot (6 m) single ladder. All are on ladder at one time. One weighs 150 pounds (68 kg), another 175 pounds (79 kg), and the other weighs 190 pounds (86 kg). Each is wearing protective clothing weighing 40 pounds (18 kg) and each has on an SCBA weighing 30 pounds (14 kg). The ladder is NOT OVER-LOADED according to maximum weight limitations specified in NFPA 1932.	☐	☐
24. The purpose of the Eddy Current Percent Electrical Conductivity Test is to find areas of fiber glass ladders that have been damaged by exposure to heat.	☐	☐

Fill in the blanks.

25. Strength Service Tests should be performed at least annually, at anytime a ladder is suspected of being unsafe, after a ladder has been subjected to ________________, after the ladder has been subject to ______________ loading or ______________________ ____________________ conditions of use, and after exposure to ______________________.

LADDERS

Chapter 3

Handling Ladders

NFPA STANDARD 1001
Fire Fighter I

3-9.2 The fire fighter, operating as an individual and as a member of team, shall demonstrate the following ladder carries:

(a) One person carry
(b) Two person carry
(c) Three person carry
(d) Four person carry
(e) Five person carry
(f) Six person carry

Chapter 3
Handling Ladders

The subject, Handling Ladders, as used in this text, details the movement of ground ladders from the apparatus rack or another temporary position to placement at the point of use. The topics discussed include what ladders are carried, where they are mounted, and how they are removed, carried, and placed. It is important that these tasks be accomplished in a safe and efficient manner that will not damage the ladder or other property. Movements need to be smooth and instinctive because speed is essential in many instances. Since more than one firefighter is frequently required, development of teamwork is another important factor. Therefore, proficiency in handling ladders will only be realized with repeated practical training.

3.1 Fill in the blanks.
It is important that tasks associated with handling ladders be accomplished in a ______________________ and ______________________ manner.

3.2 Fill in the blank.
______________________ is an additional important factor in handling ladders because more than one firefighter is usually involved.

LOCATION AND METHODS OF MOUNTING ON APPARATUS

There are no established standards for the location or method of mounting ground ladders on fire apparatus. This varies according to each manufacturer's policy, type of mounting bracket or racking used, type of apparatus, body design, individual fire department requirements, and the type of ladder. These differences make it necessary for each fire department to develop and administer its own training procedures for removing and using ladders. General information is provided as follows:

Ladders Carried on Pumpers

NFPA 1901 sets the minimum lengths and types of ladders to be carried: one 14-foot (4 m) roof ladder, and one 24-foot (7 m) extension ladder. A 35-foot (11 m) three-section extension ladder is frequently carried instead of the 24-foot (7 m) ladder. These minimums affect the location and mounting arrangement. Ladders on pumpers are usually mounted vertically (on the beam) on the right side of the apparatus (Figure 3.1). This racking location usually results in one end of the ladder overhanging the rear of the apparatus. A padded guard is sometimes placed over the protruding ends to prevent injury to persons if they should accidentally walk into them (Figure 3.2).

Figure 3.1 The right side vertical ladder racking with side removal shown here is the most prevalent used for carrying ladders on a pumper.

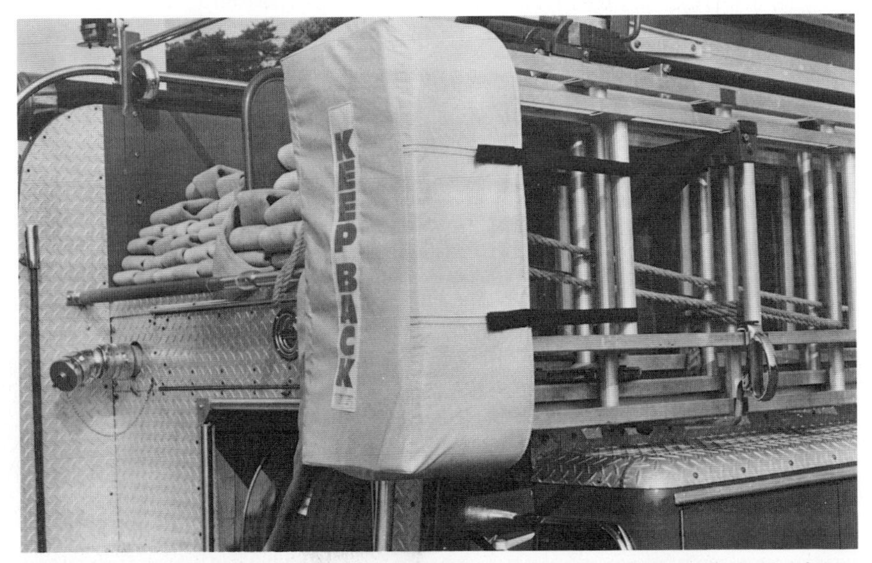

Figure 3.2 This illustration shows a commercially manufactured padded guard placed over protruding ends of ladders racked on the side of a pumper.

When a pumper is constructed with high side compartments on both sides, ladders can be mounted in an overhead rack (Figure 3.3). An overhead rack is also used when it is necessary to carry a ladder longer than 35 feet (11 m) on a pumper (Figure 3.4). Other pumpers have a rack atop or above the side compartments. Some of these are on hydraulic arms that swing the ladders down to a convenient height for removal (Figure 3.5).

3.1 Safe, Efficient

3.2 Teamwork

Figure 3.3 A normal pumper ladder complement being carried in an overhead rack. This arrangement may make loading hose more difficult.

Figure 3.4 A pole ladder carried on a pumper in an overhead rack. These ladders are usually in addition to the normal complement carried on the side.

Figure 3.5 An alternate racking arrangement for pumpers with high side compartments is shown in this series of illustrations. The upper left illustration shows the ladders nested flat in an overhead rack located above the compartments. This arrangement leaves the hose body unobstructed. When the ladders are needed, hydraulically operated arms pivot the rack downward and turn it towards vertical as shown in the upper right illustration. Note that this rack carries a 14-foot (4 m) roof ladder, a 24-foot (7 m) extension ladder and a 35-foot (11 m) extension ladder. The "ready to use" position shown in the illustration at right has the ladders vertical at a convenient height for removal. The apparatus shown was designed so that side compartment doors can be opened while the rack is in the down position. Note the folding ladder which has been mounted to the inside of the rack. *Courtesy of Larry Arnold, San Gabriel, CA.*

A 10-foot (3 m) folding ladder is usually also carried even though it is not required by the NFPA Standard (Figure 3.6).

Figure 3.6 Ten-foot (3 m) folding ladders are frequently carried on pumpers. This illustrates just one of many locations where they are mounted.

3.3 Check the correct response(s).
NFPA 1901 sets the minimum types and lengths of ladders carried on a pumper as ______________.
- ☐ A. One 10-foot (3 m) folding ladder.
- ☐ B. One 14-foot (4 m) roof ladder.
- ☐ C. One 24-foot (7 m) extension ladder.
- ☐ D. One 28-foot (9 m) extension ladder.
- ☐ E. One 35-foot (11 m) extension ladder.

Ladders Carried on Aerial Apparatus

Aerial apparatus carry a larger complement of ground ladders, usually ranging in length from a 10-foot (3 m) folding ladder to a 40-foot (12 m) or longer pole ladder. NFPA 1901 provides a minimum requirement for numbers, types, and lengths of ground ladders to be carried on aerial apparatus:

One 10-foot (3 m) folding ladder
One 14-foot (4 m) extension ladder*
One 16-foot (5 m) roof ladder
One 20-foot (6 m) roof ladder
One 28-foot (9 m) extension ladder
One 35-foot (11 m) extension ladder
One 40-foot (12 m) pole ladder

The racking arrangement of ground ladders on aerial apparatus is frequently influenced by the amount of space needed for the mounting of the aerial device. Because of this a variety of

*Many departments carry a combination ladder instead of the 14-foot (4 m) extension ladder.

racking schemes are used to carry the required number, types, and lengths of ground ladders. All are variations of two basic options:

- Flat or vertical racking
- Rear or side removal

Figures 3.7-3.10 illustrate some of the combinations being used. A roof ladder is sometimes carried mounted near the top of

Figure 3.7 Two examples of vertical, side removal racking. The first for a pumper and the second for an aerial apparatus.

Figure 3.8 Vertical racking, rear removal on an aerial apparatus. This particular apparatus also has a ladder mounted vertical on the side that removes from the side (1), and ladders racked flat that removes from the rear (2).

Figure 3.9 Flat racking, side removal on an aerial apparatus is shown in the upper photograph. Flat racking, side removal with optional rear removal on an aerial apparatus is shown in the lower photograph.

Figure 3.10 Flat racking, rear removal on an aerial apparatus.

aerial devices (Figure 3.11). Folding ladders, because of their compactness, are mounted wherever it is convenient.

3.3 B, C

Figure 3.11 Roof ladder mounted inside the top fly side rail of an aerial-ladder tower apparatus. Note the latching device indicated by the arrow.

3.4 True or False.

	True	False
1. A 10-foot (3 m) folding ladder is required by NFPA 1901 for aerial apparatus.	☐	☐
2. The space needed for the mounting structure of the aerial device is commonly modified to suit the number and length of ground ladders carried.	☐	☐

Ladders Carried on Other Apparatus

NFPA 1901 requires one 16-foot (5 m) metal extension ladder be carried on tankers and two 10-foot (3 m) metal roof ladders on salvage trucks. While not required, ground ladders are frequently carried on minipumpers and rescue trucks (Figure 3.12).

Figure 3.12 An A-frame ladder being carried inside a squad truck.

Before firefighters are drilled in removing ground ladders from apparatus, each firefighter should be able to answer the following questions:

- What ladders (types and lengths) are carried and where are they carried on the apparatus?
- Are the ladders racked with the butt toward the front or rear of the apparatus?
- Where ladders are nested together, can one ladder be removed leaving the other(s) secured in place? (In particular can the roof ladder be removed from the side of the pumper and leave the extension ladder secured in place?)
- In what order do the ladders that nest together rack? (Pumper extension ladder goes on first, roof ladder second, or vice versa?)
- Is the top fly of the extension ladder on the inside or outside when the ladder is racked on the side of the apparatus?
- How are the ladders secured in place?
- When ladders are mounted vertically on the side of apparatus which rungs go in the brackets? (Many departments find it a good practice to mark ladders to indicate which rungs go in the brackets, as shown in Figure 3.13.

Figure 3.13 Stripes painted on ladder beams indicate where it fits into the bracket.

> **3.5** Fill in the blanks.
> Before removing ground ladders from apparatus firefighters need to know ________________ ladders are carried and ________________ they are carried.

SELECTING THE CORRECT LADDER FOR THE JOB

3.4 1. True

A key factor in selecting the correct ladder for the job is knowledge, or at least the ability to make a good estimate of the length of ladder required to reach various levels of windows or roofs. Generally, a residential story will average 8 feet (2 m) to 10 feet (3 m) per story, and the distance from the floor to the windowsill is about three feet (1 m). A commercial story will average 12 feet (4 m) from floor to floor, and the distance from the floor to windowsill is about four feet (1 m).

Working rules for ladder length include the following:

- The ladders should extend a few feet (three rungs) beyond the windowsill or roof edge when it is being used to gain access. This extension of the ladder provides handholds when stepping on or off the ladder (Figure 3.14).

Figure 3.14 Tips of ladders placed for access to roofs and narrow windows should be long enough to extend a few feet (meters) above the roof edge or windowsill, as illustrated in the pictures above and at right.

- When ladders are used to rescue from a window opening, the tip of the ladder is placed at or just below the window-sill (Figure 3.15).

Figure 3.15 Tips of ladders placed for rescue from a narrow window should be below the windowsill.

The next step is to determine how far various ladders will reach. Knowledge of the designated length of a ladder can be used to answer this question. Remember that the designated length (this figure is normally displayed on the ladder) is derived from a measurement of the maximum extended length. This is NOT THE LADDER'S REACH because ladders are set at approximately 75 degrees. Reach will therefore be LESS than the designated length. One other thing needs to be considered: single, roof, and folding ladders meeting NFPA 1931 are required to have a measured length equal to the designated length. However, the maximum extended length of extension and pole ladders may be as much as six inches (150 mm) LESS than the designated length.

Table 3.1 provides information on the reach of various ground ladders when placed at the proper climbing angle.

Note that

- For lengths of 35 feet (11 m) or less, reach is approximately one foot (300 mm) less than the designated length.
- For lengths over 35 feet (11 m), reach is approximately two feet (600 mm) less than the designated length.

3.5 What, Where

TABLE 3.1
Maximum Working Heights for Ladders Set at Proper Climbing Angle

Designated Length of Ladder		Maximum Reach	
10 foot	(3.0 m)*	9 feet	(2.7 m)*
14 foot	(4.3 m)	13 feet	(4.0 m)
16 foot	(4.9 m)	15 feet	(4.6 m)
20 foot	(6.1 m)	19 feet	(5.8 m)
24 foot	(7.3 m)	23 feet	(7.0 m)
28 foot	(8.5 m)	27 feet	(8.2 m)
35 foot	(10.6 m)	34 feet	(10.4 m)
40 foot	(12.3 m)	38 feet	(11.6 m)
45 foot	(13.7 m)	43 feet	(13.1 m)
50 foot	(15.2 m)	48 feet	(14.6 m)

RULE OF THE THUMB

Ladders 35 feet (11 m) and under reach 1 foot (.3 m) less than the designated lengths.

Ladders over 35 feet (11 m) reach 2 feet (.6 m) less than the designated length.

**Elsewhere in this text metrics have been rounded off to the nearest whole number. In this instance metrics are rounded off to the nearest tenth to more accurately show the difference.*

For purposes of figuring the length of ladder needed, the differences between the designated length and reach will be known as the REACH FACTOR. Some examples will help in understanding the practical application of this information.

EXAMPLE 1: What ladder length is needed to reach the eaves of a two-story residence, assuming eight feet (2.4 m) per story?

		METRICS
Feet Per Story........	8 feet	2.4 m
Number of Stories: 2..	x2	x2
	16 feet	4.8 m
Length Needed For Handhold............	4 feet	1.2 m
Total	20 feet	6 m
Reach Factor..........	1 foot	.3 m
LENGTH OF LADDER NEEDED.............	21 feet	6.3 m

EXAMPLE 2: What ladder length is needed to gain access to a third-story window of a boarding house, assuming 10 feet (3 m) per story?

		METRICS
Feet Per Story.........	10 feet	3 m
Number of Stories: 2..	x2	x2
	20 feet	6 m
Floor to Windowsill....	3 feet	1 m
Length Needed For Handhold	4 feet	1.2 m
Total	27 feet	8.2 m
Reach Factor..........	1 foot	.3 m
LENGTH OF LADDER NEEDED............	28 feet	8.5 m

EXAMPLE 3: What ladder length is needed to perform rescue from a fourth-story window of a commercial structure?

		METRICS
Feet Per Story.........	12 feet	3.6 m
Number of Stories: 3..	x3	x3
	36 feet	10.8 m
Floor to Windowsill....	4 feet	1.2 m
Total	40 feet	12.0 m
Reach Factor..........	2 feet	.6 m
LENGTH OF LADDER NEEDED.............	42 feet	12.6 m

When all factors are considered, the 35-foot (11 m) extension ladder emerges as the most versatile of all extension ladders. This is because it is suitable for one- to three-story residential, one- and two-story commercial, and one-story industrial structures; it can be carried and raised by two firefighters; and it is compact enough to be carried on the side of a pumper.

Ladders carried on aerial apparatus are sometimes preferred to pumper ladders because they are frequently wider, making them easier to work from; and in the 35-foot (11 m) length, they are usually lighter because they have only two sections.

3.6 Fill in the blank.
For lengths of 35 feet (11 m) or less, reach is approximately ______________________ less than the designated length.

3.7 Fill in the blank.
For lengths over 35 feet (11 m), reach is approximately ______________________________ less than the designated length.

REMOVING LADDERS FROM APPARATUS

Ladders are secured in racks by various means, and the method of releasing the locking mechanism is unique for each manufacturer (Figure 3.16). It is necessary for each firefighter to determine how those of the particular apparatus in the department operate. Once the securing device is released there are other considerations:

Figure 3.16a This latching device, common to pumpers, pulls outward and turns 90 degrees.

Figure 3.16b This latching device flips up to release an eye on the end of a cable. The yoke then pivots downward to horizontal.

Figure 3.16c In this instance a pin is pulled upward to allow the ladder to slide out of the rack.

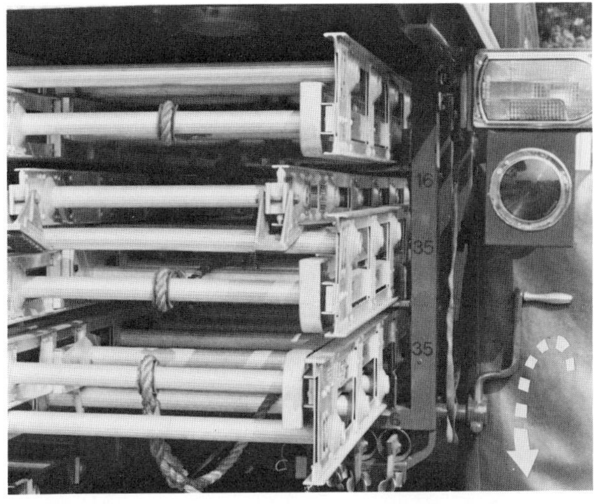

Figure 3.16d Ladders racked in this manner usually are released by rotating a handle, located off to the side, downward.

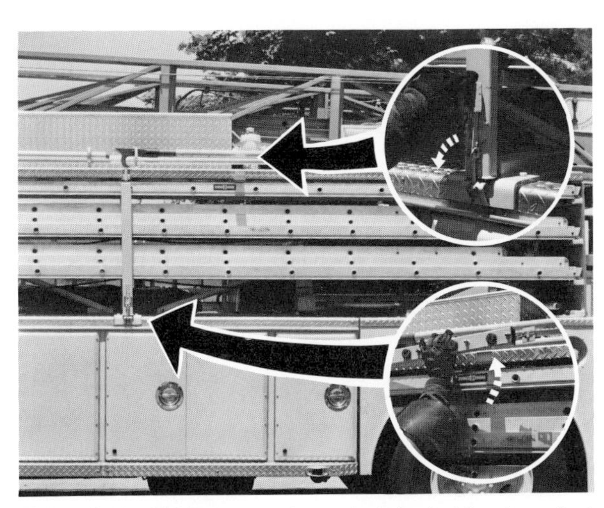

Figure 3.16e This apparatus has a single bar holding the rack of ladders in place. A lever at the bottom of the bar is pulled downward to release it. The bar is then rotated upward 90 degrees into a bracket which holds it out of the way.

Figure 3.16f A typical bracket for a folding ladder is shown here. A pin rotates 90 degrees and is then pulled outward.

Figure 3.16 A sampling of latching arrangements are illustrated in photographs a-f above.

- Ladders are usually carried nested, with more than one in a racking, so it may be necessary to remove one or more to get to the ladder needed.
- Removal of any ladder from the racking frequently leaves the securing device nonfunctional. The remaining ladder may fall from the apparatus, particularly if the vehicle is moved. Apparatus with ladders mounted vertically on the side are prone to this problem. Some have a separate holding device for each ladder to prevent this.
- Unused ladders should be put back on the apparatus and secured or stored in a safe and available location where individuals will not trip over them or where vehicles will not run over them (Figure 3.17). Departments should adopt a standard policy governing this so that all personnel will be doing the same thing.

Figure 3.17 Two examples of proper placement of unneeded ladders are shown.

3.8 True or False.

	True	False
1. Removal of a ladder from racking frequently leaves the securing device nonfunctional.	☐	☐
2. Unused ladders that have to be removed to get to a needed ladder should be laid against the side of the rear wheels.	☐	☐

PROPER LIFTING AND LOWERING METHODS

Each year many firefighters are injured when using improper lifting and lowering techniques. These injuries are preventable. The following procedures are recommended:

- Have adequate manpower for the task.
- Bend knees, keeping back as straight as possible, and lift with the legs NOT THE BACK OR ARMS (Figure 3.18).
- When two or more firefighters are lifting a ladder, lifting should be done on command of one of the firefighters at the rear who can see the whole operation (Figure 3.19). If any of the firefighters is not ready, that person should make it known immediately so that the operation will be halted. Lifting should be done in unison.

3.6 1 Foot (.3 m)

3.7 2 Feet (.6 m)

Figure 3.18 Kneel, then lift with the leg muscles.

Figure 3.19 The firefighter at the rear is responsible for giving the command to lift the ladder.

- When it is necessary to place a ladder on the ground before raising it, the reverse of the procedure of lifting is used. Lower the ladder with the leg muscles. Also, be sure to keep the body and toes parallel so when the ladder is placed it does not injure the toes (Figure 3.20).

Figure 3.20 When lowering a ladder to the ground, keep the toes parallel to the ladder.

3.9 True or False.

	True	False
When lifting a ladder, it is important to lift in such a manner that the stress is placed equally on the back and legs.	☐	☐

CARRYING LADDERS

The procedures for initiating ladder carries for ladders lying on the ground differ from those for ladders that are carried on the apparatus. Different racking schemes require different procedures that must be adapted to the individual situation.

NOTE: Whenever the following text describes a carry initiated from the racking on apparatus, it presumes that the securing device has been released.

One-Firefighter Carries

There are three methods by which one firefighter may carry a ladder: the low-shoulder method, the high-shoulder method, and the arm's length method.

LOW-SHOULDER METHOD

The one-firefighter low-shoulder carry is shown in Figure 3.21. **NOTE:** The forward end of the ladder is carried slightly lowered. Lowering the forward portion provides better balance when carrying, improves visibility by allowing the firefighter to view the way ahead, and if the ladder should strike another person the butt spurs will make contact with the body area instead of the head (Figure 3.22).

3.8 1. True 2. False

Figure 3.21 The one-firefighter low-shoulder carry.

Figure 3.22 This series of photographs illustrates the reason for carrying the forward end of the ladder lowered. Photographs a-c are of an enactment of a collision occuring when the ladder is carried improperly. Photograph d shows the same collision when the ladder is carried properly. A less vital part of the body is involved.

The steps for performing this carry from various racking locations and from flat on the ground are as follows:

One-Firefighter Low-Shoulder Carry

From Vertical Racking/Side Removal

Step 1: The firefighter faces the side of the apparatus near the center (lengthwise) of the ladder. This position will be on the balance point for the carry.

NOTE: Some fire departments mark this point on the ladder for convenience in initiating one-firefighter carries.

Figure 3.23 The firefighter grasps two rungs.

Step 2: Convenient rungs of the ladder are grasped (Figure 3.23).

Step 3: The firefighter lifts the ladder free of the rack (Figure 3.24), at the same time takes a step backward (Figure 3.25).

Step 4: The rung towards the tip is released, the firefighter then pivots toward the butt, simultaneously inserting the free arm between the rungs and bringing the upper beam onto the shoulder.

Figure 3.24 The ladder is lifted from the rack. Note that the latching device may flop in the way; in which case, it must be held out of the way while the ladder is removed.

Figure 3.25 Once the ladder has cleared the rack the firefighter takes a step backward with it.

One-Firefighter Low-Shoulder Carry

From Vertical Racking/Rear Removal

Step 1: Firefighter grasps bottom rung and pulls the ladder straight back (Figure 3.26), halting when the ladder is just far enough in the rack that it will remain there without support from the firefighter.

Step 2: Firefighter releases grip on the rung and proceeds to the ladder midpoint, turns toward the butt end (assuming the butt end is out), and inserts the near arm between the rungs, grasps a convenient rung, and proceeds forward with the ladder. When the ladder clears the rack sufficiently, it is settled upon the shoulder (Figure 3.27).

Figure 3.26 The ladder is pulled straight back.

Figure 3.27 The firefighter proceeds to the midpoint of the ladder, inserts an arm between two rungs, and brings the upper beam onto the shoulder.

3.9 1. False

One-Firefighter Low-Shoulder Carry

From Flat Racking/Side Removal

Step 1: Firefighter takes a position alongside the apparatus facing the midpoint of the ladder, grasps the ladder by the near beam, and pulls it outward.

Step 2: Before the ladder clears the rack, the firefighter's grip is shifted to two convenient rungs (Figure 3.28).

Step 3: The ladder is pulled clear of the rack, and the far beam is allowed to swing downward until the ladder is vertical.

Step 4: The rung toward the tip is released. The firefighter then pivots toward the butt end, simultaneously inserting the free arm between the rungs, and brings the upper beam onto the shoulder.

Figure 3.28 As the ladder is pulled part of the way from the rack, the firefighter's grip is shifted to two convenient rungs.

One-Firefighter Low-Shoulder Carry

From Flat Racking/Rear Removal

Step 1: Firefighter pulls the ladder nearly out of the rack and lowers the butt end to the ground.

NOTE: Ladders racked overhead may require lowering the tip from the edge of the rack to the tailboard or an intermediate step.

Step 2: Firefighter takes a position at midpoint facing the ladder and grasps two convenient rungs (Figure 3.29).

Step 3: The far beam is then brought upward as the body is pivoted toward the butt, the arm nearest the tip end is inserted between two rungs, and the ladder is brought onto the shoulder.

Figure 3.29 The firefighter shifts to the ladder's midpoint and grasps two convenient rungs.

One-Firefighter Low-Shoulder Carry

From Flat on the Ground

Step 1: Firefighter kneels beside the ladder facing the tip. The middle rung is grasped with the near hand, palm facing forward (Figure 3.30).

Step 2: The ladder is then lifted. As the ladder rises, the firefighter pivots into the ladder, placing the free arm between two rungs so that the upper beam comes to rest on the shoulder (Figure 3.31).

Figure 3.30 The center rung is grasped, palm forward.

Figure 3.31 The firefighter pivots into the ladder as it is lifted.

3.10 Check the correct response(s).
The one-firefighter low-shoulder carry is done with the forward end of the ladder lowered because ________.
☐ A. It provides better balance.
☐ B. Improves visibility.
☐ C. Keeps the butt spur from striking someone in the head.

Special Procedures for Carrying a Roof Ladder

When a roof ladder is going to be used on a sloping roof supported by the hooks and one firefighter is to climb another ladder with it, the ladder is carried by the low-shoulder method with the *tip (hooks) forward* (Figure 3.32). Procedures previously described are for ladders that are carried butt forward.

When the tip is carried forward *the pivot is made toward the tip*. When the ladder is to be picked up from the ground, the firefighter will have to kneel facing the butt instead of the tip.

Normally the roof ladder is carried to the foot of the second ladder with the hooks closed and a second firefighter opens the hooks while the first firefighter maintains the carry. When no second firefighter is present, the first firefighter sets the ladder down, moves to the tip, picks up the tip, opens the hooks, lays the tip down, returns to the midpoint, picks up the ladder again, and resumes the carry.

There may be occasions when there is no second firefighter to open the hooks, time is critical, and there is no crowd of people through which the ladder must be carried. In this case the hooks may be opened at the apparatus before the carry is begun; they are turned outward in relation to the firefighter carrying the ladder.

Figure 3.32 When the roof hooks are going to be used, the ladder is carried tip end forward.

3.11 Fill in the blank.
Roof ladders that are going to be carried up another ladder and used with the hooks over the peak of a sloped roof are carried with the ______________________ end forward.

HIGH-SHOULDER METHOD

The one-firefighter high-shoulder carry is shown in Figure 3.33. This carry is particularly well suited for making beam raises; however, it is not used for carrying a roof ladder when it is to be taken up another ladder.

Figure 3.33 The one-firefighter high-shoulder carry is shown in both of the above illustrations. The first shows the standard way of gripping the ladder, the second an alternate hand grip that is more convenient from some of the ladder racking arrangements.

One-Firefighter High-Shoulder Carry

From Vertical Racking/Side Removal

Step 1: Firefighter faces the ladder at midpoint. The hand nearest the butt is used to grip the top beam. The ladder is lifted slightly so that the other hand can be placed palm up under the bottom beam (Figure 3.34).

Step 2: The ladder is lifted free of the rack. The firefighter simultaneously pivots toward the butt and brings the bottom beam to rest on the shoulder (Figure 3.35).

Step 3: The hand holding the bottom beam is shifted to grasp the outside of that beam.

Figure 3.34 The hand nearest the butt grasps the top beam. The ladder is lifted slightly and the bottom beam is grasped with the other hand, palm up.

Figure 3.35 As the ladder is lifted from the rack, the firefighter pivots toward the butt and puts the bottom beam on the shoulder.

One-Firefighter High-Shoulder Carry

From Vertical Racking/Rear Removal

3.10 A, B, & C

3.11 Tip

Step 1: Ladder is pulled most of the way out of the rack. Firefighter is positioned at midpoint (Figure 3.36).

Step 2: Ladder is lifted, firefighter pivots toward the butt and places the bottom beam on the shoulder (Figure 3.37).

Step 3: Handholds are shifted so that one hand is grasping the top beam and the other the bottom beam or a convenient rung at a point next to the bottom beam.

Figure 3.36 With the ladder nearly out of the rack, the firefighter is positioned at its midpoint.

Figure 3.37 The ladder is lifted and the firefighter pivots toward the butt and places the bottom beam on the shoulder.

One-Firefighter High-Shoulder Carry

From Flat Racking/Side Removal

Step 1: Firefighter takes a position alongside the apparatus facing the midpoint of the ladder, grasps the ladder by the near beam, and pulls it outward.

Step 2: As the ladder reaches the edge of the rack, the outside beam is tilted downward onto the shoulder.

Step 3: The ladder is then tilted up until it is vertical and the hands are adjusted so that the near hand is grasping the outside of the lower beam and the far hand is grasping the top beam or high up on a rung (Figure 3.38).

Figure 3.38 The ladder is grasped at midpoint and pulled out as in the upper illustration. The lower left illustration shows the ladder beam being placed on the shoulder after the firefighter has pivoted toward the butt. The far beam is then tilted up to vertical and the hand grip is adjusted as in the lower right illustration.

One-Firefighter High-Shoulder Carry

From Flat Racking/Rear Removal

Step 1: Firefighter pulls the ladder nearly out of the rack and lowers the butt end to the ground (Figure 3.39).

Step 2: Firefighter takes a position at midpoint facing the ladder, grasps a convenient rung and the near beam (Figure 3.40).

Step 3: The ladder is lifted and simultaneously turned to vertical as the firefighter pivots under the bottom beam and places it on the shoulder (Figure 3.41).

Figure 3.39 The ladder is pulled out until the butt can be placed on the ground.

Figure 3.40 The firefighter shifts to the ladder's midpoint, faces the tip, and grasps a rung and the near beam.

Figure 3.41 The ladder is tilted to vertical and lifted as the firefighter pivots under the bottom beam and places it on the shoulder.

One-Firefighter High-Shoulder Carry

From Flat on the Ground

Step 1: Firefighter kneels beside the ladder facing the tip. The middle rung is grasped with the near hand, palm facing forward (Figure 3.42).

Step 2: Firefighter then lifts the ladder, simultaneously turning it to vertical and pivoting under the bottom beam, which is placed on the shoulder (Figure 3.43).

Figure 3.42 The firefighter kneels at midpoint facing the tip and grasps the middle rung with the near hand.

Figure 3.43 As soon as the ladder is lifted and tilted to vertical, the firefighter begins to pivot towards the butt as shown in the first photograph. The pivot is completed simultaneously with placing the lower beam on the shoulder as in the second photograph. The hand grip may be as shown or reversed.

ARM'S LENGTH METHOD

The one-firefighter arm's length carry is shown in Figure 3.44. It is best suited for a lighter weight ladder because all of the ladder's weight is carried in one hand. Control of the ladder's side-to-side movement is minimal with this carry and there is a tendency for the ladder to bump against the leg.

Figure 3.44 One-firefighter arm's length carry.

One-Firefighter Arm's Length Carry

From Vertical Racking/Side Removal

Step 1: Firefighter faces the ladder at midpoint, grasps two convenient rungs, lifts the ladder out of the rack, and places it on the ground resting on one beam (Figure 3.45).

Figure 3.45 The ladder is grasped at midpoint, lifted clear of the rack, and placed on the ground resting on one beam.

Step 2: Firefighter pivots toward the butt, crouches slightly, and grasps the upper beam with one hand (Figure 3.46).

Step 3: Ladder is lifted off the ground using the leg muscles.

Figure 3.46 The firefighter crouches facing the butt and grasps the upper beam.

One-Firefighter Arm's Length Carry

From Vertical Racking/Rear Removal

Step 1: Ladder is pulled most of the way out of the rack. Firefighter is positioned at midpoint.

Step 2: Two convenient rungs are grasped, the ladder is pulled clear of the rack, and placed on the ground in the vertical position.

Step 3: Firefighter pivots toward the butt, crouches slightly, and grasps the upper beam with one hand (For an illustration refer back to Figure 3.46).

Step 4: Ladder is lifted off the ground using the leg muscles.

One-Firefighter Arm's Length Carry

From Flat Racking/Side Removal

Step 1: Firefighter takes a position alongside the apparatus facing the midpoint of the ladder, grasps the ladder by the near beam, and pulls it outward.

Step 2: Before the ladder clears the rack, the hand nearest the butt is shifted to grasp a rung (Figure 3.47).

Step 3: The ladder is pulled clear of the rack and is brought downward, simultaneously being tilted to vertical (Figure 3.48).

Step 4: Grip on the rung is released as the firefighter pivots toward the butt.

One-Firefighter Arm's Length Carry

From Flat Racking/Rear Removal

Step 1: Firefighter pulls the ladder from the rack (Figure 3.49).

Step 2: When the ladder clears the rack it is lowered and brought to vertical (Figure 3.50).

Step 3: Firefighter then releases grip of the hand nearest the butt and pivots toward the butt.

Figure 3.47 When the ladder is part of the way out, the hand grip is shifted to two convenient rungs.

Figure 3.48 The ladder is pulled clear of the rack and is turned to vertical as it is lowered.

Figure 3.49 The firefighter slides the ladder out of the rack.

Figure 3.50 After the ladder clears the rack, it is lowered and brought to vertical.

One-Firefighter Arm's Length Carry

From Flat on the Ground

Step 1: Firefighter crouches facing the ladder at midpoint and tilts the ladder up onto one beam (Figure 3.51).

Step 2: Firefighter grasps the upper beam and stands, at the same time pivoting toward the butt. The hand nearest the butt is released to complete the evolution.

Figure 3.51 The firefighter crouches at midpoint and tilts the ladder up onto one beam.

> **3.12** Fill in the blank.
> The one-firefighter ______________________ carry is particularly well suited for making beam raises.

Two-Firefighter Carries

The three methods most commonly used when two firefighters carry a ladder are the low-shoulder method, the hip or underarm method, and the arm's length on-edge method.

LOW-SHOULDER METHOD

The two-firefighter low-shoulder carry is shown in Figure 3.52. The forward firefighter places the free hand over the upper butt spur. This is done to prevent injury in case there is a collision with someone while the ladder is being carried.

Two-Firefighter Low-Shoulder Carry

From Vertical Racking/Side Removal

Step 1: Both firefighters stand facing the ladder, one near the tip and the other near the butt; each firefighter uses both hands to grasp the ladder and remove it from the rack (Figure 3.53).

Step 2: As soon as the ladder clears the rack, the firefighters continue to grasp the ladder with the hand nearest the butt end while they pivot, place the other arm between two rungs, and bring the upper beam onto the shoulder.

Figure 3.52 The two-firefighter low-shoulder carry. Note that the forward firefighter's hand is placed over the upper butt spur.

Figure 3.53 The first illustration depicts two firefighters lifting an extension ladder from the rack. The hand nearest the tip is released and inserted between two rungs as the firefighters pivot toward the butt and bring the upper beam onto their shoulders as in the second illustration.

Two-Firefighter Low-Shoulder Carry

From Vertical Racking/Rear Removal

Step 1: One firefighter grasps the bottom rung and pulls the ladder from the rack. The second firefighter stands to the side of the ladder adjacent to the rear of the apparatus and assists with removal (Figure 3.54).

Step 2: When the ladder is almost out of the rack the first firefighter shifts to a position adjacent to the butt on the same side as the second firefighter (Figure 3.55).

Step 3: Both firefighters turn to face the butt end, simultaneously placing the arm on the side towards the apparatus between the two bottom and two top rungs respectively. The ladder is pulled clear of the rack and brought onto the shoulder.

Figure 3.54 Both firefighters pull the ladder out, one stands at its end, the other on the tailboard side.

Figure 3.55 When the ladder is almost clear of the rack, the firefighter at its end shifts to the same side as the other firefighter.

Two-Firefighter Low-Shoulder Carry

From Flat Racking/Side Removal

Step 1: Firefighters position themselves facing the ladder, one at each end. Both grasp the beam and pull the ladder outward. As soon as the ladder is part way out, at least one hand is shifted to grip a rung.

Step 2: The ladder is pulled clear of the rack and brought toward vertical by swinging the far beam downward toward the firefighters. Both firefighters pivot toward the butt, simultaneously placing the arm on the side toward the tip between two rungs and bringing the ladder onto the shoulder (Figure 3.56).

3.12 High Shoulder

Figure 3.56 In the first illustration the firefighters have pulled the ladder about halfway out of the rack and they are in the process of shifting the hand grip to the rungs. The ladder is then pulled clear, tilted to vertical simultaneous with the firefighters pivoting toward the butt and placing the top beam on their shoulders as in the second illustration.

Two-Firefighter Low-Shoulder Carry

From Flat Racking/Rear Removal

Step 1: Ladder is pulled from the rack far enough that the two firefighters can position themselves one at each end on the same side facing the ladder. Each firefighter grasps two rungs from underneath the ladder (Figure 3.57).

Step 2: The ladder is pulled clear of the rack. The outside beam is lowered as the inside beam is raised to the shoulder. At the same time, both firefighters pivot toward the butt and place the near arm between two rungs (Figure 3.58).

Figure 3.57 The ladder is pulled almost from the rack. The firefighters face the ladder from the side near each end and grasp two rungs from underneath.

Figure 3.58 When the ladder clears the rack, the far beam is lowered as the near beam is brought onto the shoulder.

Two-Firefighter Low-Shoulder Carry

From Flat on the Ground

Step 1: The two firefighters position themselves on the same side of the ladder, one near the butt and the other near the tip. They then kneel next to the ladder facing the tip and grasp a convenient rung with one hand, palm forward (Figure 3.59).

Step 2: The firefighter at the butt gives the command to "shoulder the ladder." Both firefighters stand up, using their leg muscles to lift the ladder.

Step 3: As the ladder and the firefighters rise the far beam is tilted upward. The firefighters pivot and place the free arm between two rungs. The upper beam is placed on the shoulders with the firefighters facing the butt end.

NOTE: The lift should be smooth and continuous.

Figure 3.59 Firefighters kneel beside the ladder near each end and grasp a convenient rung, palms forward.

> **3.13** Fill in the blank.
> When using the two-firefighter low-shoulder carry, the forward firefighter places the free hand ______________________.

Special Procedures for Carrying Roof Ladders

When two firefighters carry a roof ladder that is going to be taken up another ladder and placed on a sloping roof with the hooks supporting it, the ladder is carried using the low-shoulder method with the *tip end forward* (Figure 3.60). When this is the

Figure 3.60 When a roof ladder is going to be used on a sloped roof, it is carried with the tip end forward.

case, instructions for the firefighters to face the tip and pivot to face the butt, or to face the side of the ladder and pivot to face the butt should be reversed; the firefighters face the butt and pivot to face the tip. When a two-firefighter carry is used, the ladder is carried as far as the foot of the second ladder with the hooks closed. The firefighter at the tip opens the hooks before beginning to climb the second ladder.

HIP OR UNDERARM METHOD

The two-firefighter hip or underarm carry is shown in Figure 3.61. **NOTE:** This carry is not suitable for three-section ladders due to their bulk.

Figure 3.61 Two-firefighter hip or underarm carry. Note that the forward firefighter places a hand over the upper butt spur.

Two-Firefighter Hip or Underarm Carry

From Vertical Racking/Side Removal

Step 1: The firefighters position themselves near the tip and butt ends and face the ladder.

Step 2: The ladder is grasped with both hands and lifted off the rack. Then it is lowered until the top beam is approximately chest high (Figure 3.62).

Figure 3.62 The ladder is lifted off of the rack and lowered until the top beam is chest high.

3.13 Over The Upper Butt Spur

Step 3: Each firefighter then releases the hand farthest from the butt end, reaches across the top beam while pivoting the body toward the butt end, and grasps a rung (Figure 3.63).

Step 4: The top beam is placed against the body just under the armpit. The firefighters should now be facing the butt end.

Figure 3.63 Each firefighter reaches across the top beam with the hand farthest from the butt, pivots toward the butt, and grasps a rung.

Two-Firefighter Hip or Underarm Carry

From Vertical Racking/Rear Removal

Step 1: One firefighter grasps the bottom rung and pulls the ladder from the rack. The second firefighter stands to the side of the ladder adjacent to the back of the apparatus and assists with the removal as in Figure 3.54 on page 132.

Step 2: When the ladder is almost out of the rack, the first firefighter shifts to a position adjacent to the butt on the same side as the second firefighter as shown in Figure 3.55 on page 132.

Step 3: Rungs are grasped with both hands and the ladder is lifted clear of the rack. Then it is lowered until the top beam is approximately chest high.

Step 4: Each firefighter then releases the hand farthest from the butt end, reaches across the top beam while pivoting the body toward the butt end, and grasps a rung.

Step 5: The top beam is placed against the body just under the armpit. The firefighters should now be facing the butt end.

NOTE: Steps 3, 4, and 5 above are identical to steps 2, 3, and 4 of the section on vertical racking/side removal and the same illustrations apply.

Figure 3.64 As the ladder is pulled from the rack, each firefighter reaches across the ladder and grasps a rung near the opposite beam.

Two-Firefighter Hip or Underarm Carry

From Flat Racking/Side Removal

Step 1: Firefighters position themselves facing the ladder, one at each end. Both grasp the beam and pull the ladder outward. As soon as the ladder is part way out, at least one hand is shifted to grip a rung, as in Figure 3.56 on page 133.

Step 2: The ladder is pulled clear of the rack and brought toward vertical by swinging the far beam downward toward the firefighters.

Step 3: Then it is lowered until the top beam is approximately chest high. Firefighters then proceed as in steps 4 and 5 of the previous section.

Two-Firefighter Hip or Underarm Carry

From Flat Racking/Rear Removal

Step 1: Both firefighters position themselves on the same side of the ladder after pulling it almost from the rack. As the ladder clears the rack, each reaches across the ladder and grasps a convenient rung near the opposite beam (Figure 3.64).

Step 2: The firefighters then pivot toward the butt, simultaneously allowing the far beam to swing downward while the near beam swings up against the body just under the armpit.

Two-Firefighter Hip or Underarm Carry

From Flat on the Ground

Step 1: Both firefighters position themselves on the same side of the ladder facing the butt end. One is positioned near the tip end, the other near the butt end.

Step 2: The firefighters kneel beside the ladder and each reaches across and grasps a rung near the far beam (Figure 3.65).

Step 3: The ladder is tilted up so that it is resting on what was the near beam.

Step 4: Each firefighter then reaches across the upper beam and grasps a convenient rung (Figure 3.66).

Step 5: The firefighters then stand and simultaneously lift the ladder against the body with the upper beam located just under the armpit.

CAUTION: It is important to lift with the leg muscles, keeping the back straight.

Figure 3.65 While kneeling beside the ladder near either end, the firefighters each reach across and grasp a rung at a point near the far beam.

Figure 3.66 With the ladder resting on one beam each firefighter reaches across the upper beam and grasps a convenient rung.

ARM'S LENGTH ON-EDGE METHOD

The two-firefighter arm's length on-edge carry is shown in Figure 3.67. The evolution described, when used for extension ladders, assumes that the firefighters are positioned on the bed section (widest) side of the ladder when it is in the vertical position.

Figure 3.67 The two-firefighter arm's length on-edge carry.

Two-Firefighter Arm's Length On-Edge Carry

From Vertical Racking/Side Removal

Step 1: The two firefighters stand facing the ladder, one at each end. The ladder is grasped with both hands on rungs and lifted from the rack.

Step 2: The ladder is lowered as far as possible without bending over (Figure 3.68).

Figure 3.68 After being lifted from the rack, the ladder is lowered as far as possible without the firefighters having to bend over.

Step 3: While retaining the grasp of a rung with the hand nearest the butt end, each firefighter releases the hold with the other hand, reaches across the ladder, and obtains a new grip on the upper beam.

NOTE: On extension ladders the upper beam of the outermost fly section is grasped (Figure 3.69).

Step 4: The firefighters then release their hold on the rungs and let the ladder settle to full arm's length, simultaneously pivoting toward the butt end.

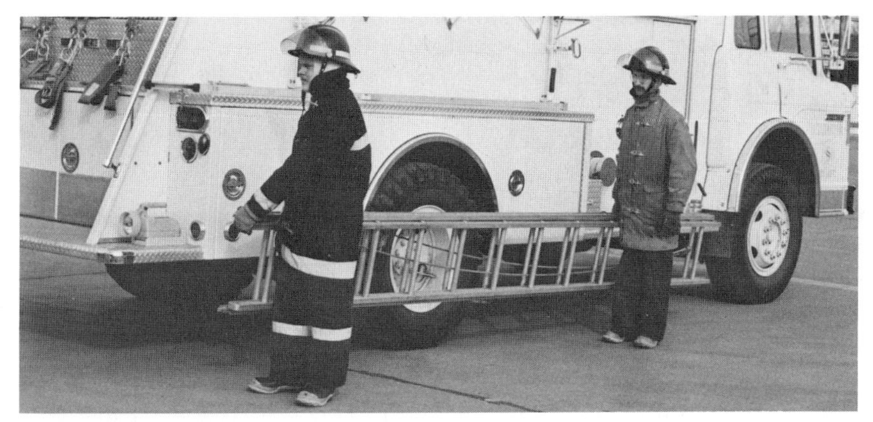

Figure 3.69 When extension ladders are carried in this manner, the upper beam of the outermost fly section is grasped.

Two-Firefighter Arm's Length On-Edge Carry

From Vertical Racking/Rear Removal

Step 1: Ladder is pulled most of the way out of the rack. The two firefighters position themselves on the wide side of the ladder (assuming an extension ladder), one at each end (Figure 3.70).

Step 2: The ladder is grasped with both hands on rungs and is lifted from the rack. The remaining steps are the same as steps 2, 3, and 4 of vertical racking/side removal.

Figure 3.70 When the ladder is nearly out of the rack, the two firefighters position themselves on the wide side of the extension ladder.

Two-Firefighter Arm's Length On-Edge Carry

From Flat Racking/Side Removal

Step 1: The firefighters position themselves facing the ladder, one at each end. Both grasp the beam and pull the ladder outward. As soon as the ladder is part way out, the hands are shifted to grip the rungs as shown in Figure 3.56 on page 133.

Step 2: The ladder is pulled clear of the rack and lowered. At the same time, the far beam is allowed to swing downward so that the ladder ends up being held vertical.

Step 3: Each firefighter then shifts the grip of the hand farthest from the butt to the upper beam. The grasp of the hands holding the rung is then released.

Step 4: Firefighters simultaneously lower the ladder to arm's length and pivot towards the butt.

Two-Firefighter Arm's Length On-Edge Carry

From Flat Racking/Rear Removal

Step 1: The ladder is pulled nearly out of the rack. Both firefighters position themselves on the same side, one at each end.

Step 2: Both firefighters grasp the ladder as shown in Figure 3.71, slide it clear of the rack, and then tilt it to vertical.

Step 3: The grasp of the hand farthest from the butt is shifted to the upper beam as the hand nearest the butt is released. The firefighters pivot toward the butt and lower the ladder to arm's length.

Figure 3.71 Each firefighter reaches over the beam with one hand and under it with the other to grip convenient rungs.

Two-Firefighter Arm's Length On-Edge Carry

From Flat on the Ground

Step 1: One beam is tilted up so that the ladder is resting on the other beam.

Step 2: The two firefighters position themselves on the same side of the ladder (on the bed section side of extension ladders), one near each end. They squat down slightly facing the butt end, and grasp the upper beam with the near hand (the beam of the outermost fly section on an extension ladder) (Figure 3.72).

Step 3: The firefighters then stand, lifting the ladder until it is at arm's length.

Figure 3.72 Both firefighters squat beside the ladder facing the butt and grasp the upper beam with the near hand.

3.14 Check the correct response.
Two firefighters utilize the __________ carry when carrying a roof ladder that is to be carried up another ladder to be used with the hooks over the peak of a sloped roof.

- ☐ A. Low-shoulder
- ☐ B. High-shoulder
- ☐ C. Hip or underarm
- ☐ **D.** Arm's length on-edge

TWO-FIREFIGHTER MULTIPLE-LADDER CARRY

The two-firefighter multiple-ladder carry is used where apparatus is unable to gain access to one or more sides of a property. Two or more ladders are removed and stacked together. The stack is carried to a convenient location where it is placed on the ground. Ladders are used individually as needed. If lightweight ladders are carried on the apparatus, the stack may consist of a 35-foot (11 m) extension ladder, a 24-foot (7 m) extension ladder, and a 14-foot (4 m) roof ladder (Figure 3.73), or it may simply be the ladder complement from a pumper.

The procedure for this carry is as follows:

Step 1: The desired ladders are removed and stacked one on top of the other.

Step 2: Both firefighters kneel, one facing the tip of the ladder and one facing the butt. Both grasp a rung of the bottom ladder with both hands.

Step 3: The firefighters then stand.

CAUTION: Since this is a heavy load, particular care must be taken to use the leg muscles and not to stress the back.

Step 4: The firefighter at the butt pivots so that the ladder is being grasped from behind, as shown in Figure 3.74.

Figure 3.73 The two-firefighter multiple ladder carry is used to efficiently move a variety of commonly used ground ladders to locations remote from apparatus.

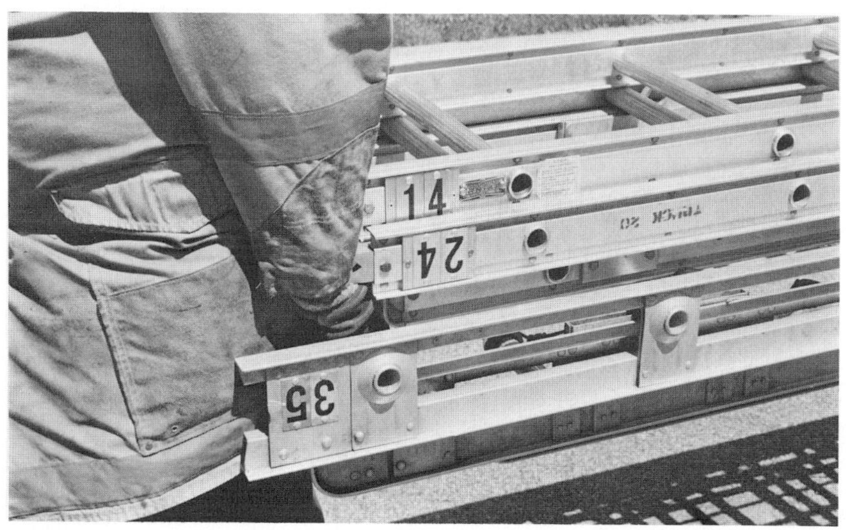

Figure 3.74 The firefighter at the forward end pivots so that the first rung of the bottom ladder is being grasped with the palms toward the rear.

Three-Firefighter Carries

3.14 A

There are four commonly used methods when three firefighters carry ladders; the flat-shoulder method, the flat arm's length method, the low-shoulder method, and the arm's length on-edge method.

FLAT-SHOULDER METHOD

The flat-shoulder carry has two firefighters on one side at each end and one firefighter on the other side in the middle (Figure 3.75).

Figure 3.75 The three-firefighter flat-shoulder carry.

Three-Firefighter Flat-Shoulder Carry

From Vertical Racking/Side Removal

Step 1: Two firefighters face the ladder, one at each end, and grasp convenient rungs with both hands. The third firefighter, who is not involved in removal of the ladder, takes a position beside the apparatus adjacent to one end of the ladder (Figure 3.76).

Figure 3.76 Two firefighters grasp the ladder by the rungs the same as for two-firefighter carries. The third firefighter stands beside the apparatus at one end of the ladder.

Step 2: The two firefighters grasping the ladder lift it off the rack and step back clear of the apparatus (Figure 3.77).

CAUTION: Care must be taken before stepping back to be sure that the way is clear and that there are no potholes or other hazards such as previously removed ladders.

Step 3: The third firefighter steps into the space between the apparatus and the ladder and faces the ladder at midpoint (Figure 3.78).

Figure 3.77 The ladder is lifted from the rack. The two firefighters step backwards.

Figure 3.78 The third firefighter walks between the ladder and the side of the apparatus. At midpoint the firefighter pivots to face the ladder and grasps the bottom beam with both hands, palms upward.

Step 4: The third firefighter, with assistance of the other two firefighters, lifts the lower beam, pivots toward the butt end, and places the beam on the shoulder. An alternate method is for the two firefighters to tilt the top beam down toward the third firefighter (Figure 3.79).

Step 5: With the ladder resting on the third firefighter's shoulder, the other two firefighters pivot toward the butt end and lift the near beam onto their shoulders (Figure 3.80).

Figure 3.79 The third firefighter lifts the lower beam and simultaneously pivots toward the butt (top photo). Steps of alternate method are shown in the middle and lower photos. The third firefighter proceeds to the ladder midpoint and facing the butt, then the two firefighters holding the ladder tilt the upper beam onto the third firefighter's shoulder.

Figure 3.80 The other two firefighters pivot toward the butt simultaneously lifting the near beam onto their shoulders.

Three-Firefighter Shoulder Carry

From Vertical Racking/Rear Removal

Step 1: One firefighter stands facing the butt and grasps the bottom rung. The other two firefighters stand adjacent to the rear of the apparatus facing each other on opposite sides of the ladder (Figure 3.81).

Step 2: The first firefighter pulls the ladder out of the rack with assistance from the other two firefighters (Figure 3.82).

Figure 3.81 One firefighter faces the butt and grasps the bottom rung. The other two firefighters are positioned one inboard and one outboard of the ladder.

Figure 3.82 All firefighters assist in pulling the ladder from the rack.

Step 3: When the ladder is almost clear of the rack, the first firefighter shifts to one side of the butt, faces the ladder, and grasps two rungs. The firefighter at the tip end, on the same side with the first firefighter, grasps two rungs. The third firefighter, on the side opposite the other two, shifts to the ladder's midpoint and faces the ladder (Figure 3.83).

Step 4: The third firefighter, with assistance of the other two firefighters, lifts the lower beam, pivots toward the butt end, and places the beam on the shoulder (Figure 3.84).

Step 5: With the ladder resting on the third firefighter's shoulder, the other two firefighters simultaneously pivot toward the butt end and place the near beam on their shoulders (Figure 3.85).

Figure 3.83 No. 1 shifts to the inboard side, faces the ladder, and grasps two rungs. No. 2 grasps two rungs. No. 3 shifts to midpoint, faces the ladder, and grasps the bottom beam.

Figure 3.84 No. 3 lifts the lower beam to the shoulder simultaneously pivoting towards the butt.

Figure 3.85 No. 1 and No. 2 pivot toward the butt and place the near beam on their shoulders.

Three-Firefighter Flat-Shoulder Carry

From Flat Racking/Side Removal

Step 1: Two firefighters stand facing the ladder, one at each end, and grasp the near beam. The third firefighter stands at the side of the apparatus near one end (Figure 3.86).

Step 2: The two firefighters slide the ladder part way out of the rack, and each shifts the grip of the hand nearest the tip to a rung. The ladder is then pulled clear of the rack and the near beam is brought onto their shoulders. The ladder is kept in the horizontal plane (Figure 3.87).

Step 3: The third firefighter proceeds into the space between the apparatus and the ladder, moves to the ladder's midpoint, pivots to face the butt, and places the near beam on the shoulder (Figure 3.88).

Figure 3.86 Two firefighters grasp the ladder by the beam. The third firefighter stands beside the apparatus at the end of the ladder.

Figure 3.87 The firefighters place the near beam on their shoulders and maintain the ladder in the horizontal plane by grasping a rung with the near hand.

Figure 3.88 The two firefighters step sideways away from the apparatus. The third firefighter proceeds to the midpoint of the far side of the ladder, pivots, and faces the butt, simultaneously placing the beam on the shoulder.

Three-Firefighter Flat-Shoulder Carry

From Flat Racking/Rear Removal

Step 1: One firefighter faces the butt and the other two firefighters face each other, one on each side at the butt (Figure 3.89). All three assist in pulling the ladder from the rack (Figure 3.90).

Figure 3.89 One firefighter stands facing the butt. The other two stand near the butt, one on each side facing the beam.

Figure 3.90 The three firefighters work together to pull the ladder outward.

Step 2: When the ladder is pulled nearly from the rack, the firefighter at the butt shifts to one side. This results in two firefighters on the same side, one at each end. The third firefighter is positioned at midpoint on the opposite side. All three face the ladder, place their hands palm up under the beams, and lift the ladder clear of the rack (Figure 3.91 on next page).

Step 3: The ladder is raised to shoulder height. Each firefighter then pivots toward the butt end and places the beam on the shoulder (Figure 3.92).

Figure 3.91 When the ladder is nearly out of the rack, the firefighters shift so that two are on one side at each end and one is at midpoint on the other side.

Figure 3.92 All firefighters pivot toward the butt simultaneously with raising the ladder and placing it on the shoulder.

Three-Firefighter Flat-Shoulder Carry

From Flat on the Ground

Step 1: Two firefighters kneel on one side of the ladder, one at either end, facing the tip. The third firefighter kneels on the opposite side at midpoint, also facing the tip end. In each case, the knee closer to the ladder is the one touching the ground (Figure 3.93).

Step 2: The ladder is lifted and the firefighters begin to stand. When the ladder is about chest high, the firefighters pivot toward the butt end and place the beam onto their shoulders.

Figure 3.93 The firefighters kneel beside the ladder facing the tip in the position shown.

FLAT ARM'S LENGTH METHOD

The three-firefighter arm's length carry is shown in Figure 3.94. Note that the positioning of the three firefighters is basically the same as the flat-shoulder carry.

Figure 3.94 The three-firefighter flat arm's length carry.

Three-Firefighter Flat Arm's Length Carry

From Vertical Racking/Side Removal

Steps 1, 2, 3: Same as steps 1, 2, and 3 for the three-firefighter flat-shoulder carry from vertical racking/side removal.

Step 4: The third firefighter grasps the lower beam (Figure 3.95) and brings it up so that the ladder approaches horizontal (Figure 3.96).

Step 5: Each firefighter shifts the grip of the hand nearest the tip from the beam to a convenient rung (Figure 3.97).

Step 6: The ladder is lowered the rest of the way to horizontal at arm's length. The grip of the hands on the beams is released as the firefighters pivot toward the butt.

Figure 3.95 The lower beam is grasped by the third firefighter.

Figure 3.96 The ladder is tilted toward horizontal.

Figure 3.97 The grip of the hand nearest the tip is shifted to grasp a rung.

Three-Firefighter Arm's Length Carry

From Flat Racking/Side Removal

Step 1: Two firefighters stand facing the ladder, one at each end, and grasp the near beam. The third firefighter stands at the side of the apparatus near one end (Figure 3.98).

Step 2: The two firefighters slide the ladder part way out of the rack. Each shifts the hands to grip two rungs.

Step 3: The ladder is removed from the rack and tilted to vertical. Then the two firefighters step back. This places them in a position similar to that shown in Figure 3.77 on page 146.

Figure 3.98 Two firefighters grasp the ladder by the beam. The third firefighter stands beside the apparatus at the end of the ladder.

Steps 4, 5, 6: Same as steps 4, 5, and 6 of the three-firefighter flat arm's length carry from vertical racking/side removal.

Three-Firefighter Arm's Length Carry

From Flat Racking/Rear Removal

Steps 1 and 2: same as steps 1 and 2 for three-firefighter flat-shoulder carry from flat racking/rear removal except that the three firefighters grasp the beams and lift the ladder clear of the rack and then let it settle to arm's length.

Three-Firefighter Flat Arm's Length Carry

From Flat on the Ground

Step 1: Two firefighters kneel beside the ladder facing the butt, one at each end. The third firefighter kneels on the opposite side at midpoint, also facing the butt. All firefighters grasp the beam with their near hand (Figure 3.99).

Step 2: All firefighters stand and simultaneously lift the ladder to arm's length.

Figure 3.99 The three firefighters kneel beside the ladder, as shown, facing the tip.

LOW-SHOULDER METHOD

The procedures for three firefighters to use the low-shoulder carry are the same as for the two-firefighter low-shoulder carry except that the third firefighter is positioned at midpoint (Figure 3.100). **NOTE:** In order for three firefighters to utilize this carry, they should be nearly the same height.

Figure 3.100 The three-firefighter low-shoulder carry.

ARM'S LENGTH ON-EDGE METHOD

The procedures for three firefighters to use the arm's length on-edge carry are the same as for the two-firefighter arm's length on-edge carry except that the third firefighter is positioned at midpoint between the other two firefighters (Figure 3.101).

Figure 3.101 The three-firefighter arm's length on-edge carry.

3.15 Check the correct response(s).
The commonly used methods by which ladders are carried by three firefighters are: ______________________________.

- ☐ A. Low-shoulder.
- ☐ B. High-shoulder.
- ☐ C. Flat-shoulder.
- ☐ D. Stacked.
- ☐ E. Flat arm's length.
- ☐ F. Arm's length on-edge.
- ☐ G. Overhead.

Four-Firefighter Carries

The same four methods used by three firefighters for carrying ladders are used by four firefighters except that there is a change in the positioning of the firefighters to accommodate the fourth firefighter.

FLAT-SHOULDER METHOD

When four firefighters use the flat-shoulder carry, two are positioned at each end of the ladder opposite each other (Figure 3.102).

Figure 3.102 The four-firefighter flat-shoulder carry.

3.16 True or False.

	True	False
The same four methods used by three firefighters for carrying ladders are used by four firefighters except that there is a change in positioning of the firefighters.	☐	☐

Procedure for Flat Overhead Racking on a Pumper/Rear Removal

An objective of this manual is to present examples of procedures for initiating the various carries from each of the basic racking arrangements. In the discussion of racking at the beginning of this chapter one example was of a pole ladder carried flat overhead on a pumper (Figure 3.103). The four-firefighter flat-

Figure 3.103 A pole ladder carried in a flat overhead rack on a pumper.

shoulder carry is particularly well suited for ladders carried in this location. The steps for the four-firefighter flat-shoulder carry from a pumper overhead rack are as follows:

Step 1: Two firefighters stand on the rear step, or as in Figure 3.104 on intermediate steps, reach up, and grasp a convenient rung with the near hand while maintaining a grasp on the apparatus with the other hand (Figure 3.105).

Step 2: The third and fourth firefighters stand facing the rear of the apparatus approximately 10 feet (3 m) from tailboard (Figure 3.106).

3.15 A, C, E and F

Figure 3.104 Two firefighters are positioned on the rear of the apparatus in such a manner that the ladder can be grasped for removal.

Figure 3.105 Each grasps a rung with the near hand.

Figure 3.106 The other two firefighters stand facing the rear of the pumper. They are positioned approximately 10 feet (3 m) from it.

Step 3: Firefighters one and two slide the ladder back. Firefighters three and four raise their inside arms in preparation for receiving the butt end of the ladder (Figure 3.107).

Step 4: When the ladder passes its balance point, the butt end will tilt downward and is grasped by the waiting firefighters (Figure 3.108).

Figure 3.107 The ladder is slid back. The firefighters on the ground prepare to grasp the butt end.

Figure 3.108 When the ladder is slid back beyond its midpoint the butt end will swing downward into the hands of the two firefighters on the ground.

Step 5: Firefighters one and two continue sliding the ladder back until just the tip is resting on the lip of the rack. Firefighters three and four place the ladder on their shoulders and step back as the ladder is being slid back (Figure 3.109).

Figure 3.109 The firefighters on the ground place the ladder on their shoulders and step backwards until the tip reaches the edge of the rack.

3.16 1. True

Step 6: Firefighters one and two release their grip, leaving the tip of the ladder resting on the lip of the rack. They then descend to the tailboard, turn and face the butt end, and reach up and grasp the ladder beam (Figure 3.110).

Step 7: Firefighters one and two lift the ladder from the rack, bring it down onto their shoulders, and step from the tailboard to the ground. Firefighters three and four pivot

Figure 3.110 The two firefighters at the tip stand on the tailboard and pivot 180 degrees to face the butt.

180 degrees to complete positioning for the carry (Figure 3.111).

Figure 3.111 The firefighters at the tip step to the ground and place the ladder on their shoulders, simultaneous with the firefighters at the butt stepping backward and then pivoting 180 degrees.

FLAT ARM'S LENGTH METHOD

When the flat arm's length carry is used by four firefighters they are positioned exactly the same as for the flat-shoulder carry except that the ladder is carried at arm's length (Figure 3.112).

Figure 3.112 The four-firefighter flat arm's length carry. Note that the firefighters at the butt end place one of their hands over the butt spur.

LOW-SHOULDER METHOD

This carry is not recommended for use with pole ladders. When used to carry an extension ladder, the four firefighters are spaced evenly along the length of the ladder (Figure 3.113). The evolution is performed in the same manner as for the two-firefighter low-shoulder carry.

Figure 3.113 The four-firefighter low-shoulder carry.

ARM'S LENGTH ON-EDGE METHOD

This carry is also not recommended for use with pole ladders. Four firefighters are spaced evenly along the length of the extension ladder, the same as for the low-shoulder method (Figure 3.114). The evolution is performed in the same manner as for the two-firefighter arm's length on-edge method.

Figure 3.114 The four-firefighter arm's length on-edge carry.

Five-Firefighter Carries

Five firefighters are normally required for a 40-foot (12 m) or longer pole ladder. Manning levels are often such that fewer than five firefighters respond with the apparatus. Because of this, some fire departments do not make an effort to use pole ladders and have stopped drilling with them. They are being carried by these departments only because NFPA 1901 requires it. However, consideration must be given to the fact that when aerial apparatus cannot gain access to a particular area, the pole ladder may be the only piece of equipment available to get the job done. If this happens, personnel will have to be recruited from other crews, making it important that all fire fighting personnel be familiar with handling pole ladders.

Due to the weight, bulk, and the presence of staypoles, the flat-shoulder and flat arm's length carries are used.

3.17 True or False.

	True	False
1. When fewer than five firefighters ride an apparatus, the pole ladder is not used.	☐	☐
2. The arm's length on-edge carry is used for carrying pole ladders.	☐	☐
3. The flat-shoulder and flat arm's length carries are used for carrying pole ladders.	☐	☐

FLAT-SHOULDER METHOD

The same procedures are used as for the three-firefighter flat-shoulder method except that the firefighters are positioned differently. Three are on one side, evenly spaced down the length of the ladder, and two are spaced evenly down the other side halfway between the three firefighters on the opposite side (Figure 3.115).

Figure 3.115 The five-firefighter flat-shoulder carry.

Procedure for Flat Overhead Racking on a Pumper/Rear Removal

The same basic procedure is used as for the four- firefighter flat overhead racking on a pumper/rear removal except that the firefighters on the ground are spaced differently. Initially, all three firefighters on the ground stand where the butt will come down — one at the end and one on each side (Figure 3.116). As the ladder is almost out of the rack, the firefighters on the sides begin to shift their positions (Figure 3.117), and are in place by the time the two firefighters up on the back of the apparatus are ready to

Figure 3.116 Position of firefighters when five firefighters remove a pole ladder from a flat overhead pumper rack.

Figure 3.117 When the ladder is nearly out of the rack, the firefighters on the ground begin to shift positions as indicated.

step down (Figure 3.118). The firefighter at the butt shifts to the outside as the evolution is completed (Figure 3.119).

Figure 3.118 The firefighters on the ground are nearly in position by the time those on the rear of the apparatus are ready to step down.

Figure 3.119 The firefighter at the butt shifts to the three-firefighter side to complete the position shift.

FLAT ARM'S LENGTH METHOD

The same procedures are used as for the three- firefighter flat arm's length carry except that three firefighters are positioned on one side — one at the tip, one at midpoint, and one at the butt. The other two firefighters are positioned on the opposite side — one halfway between the tip and midpoint, and one halfway between the midpoint and the butt (Figure 3.120).

3.17 1. False
2. False
3. True

Figure 3.120 The five-firefighter flat arm's length carry.

Six-Firefighter Carries

The ideal size crew for a pole ladder is six firefighters. With this number of individuals lifting and carrying, the risk of straining is nearly eliminated.

FLAT-SHOULDER METHOD

The procedures for six firefighters using the flat-shoulder carry are the same as described in detail in the section on the three-firefighter flat-shoulder carry except for the positioning of the firefighters. They are located at the butt, midpoint, and tip on each side (Figure 3.121).

Figure 3.121 The six-firefighter flat-shoulder carry.

Procedure for Flat Overhead Racking on a Pumper/Rear Removal

The procedure for six firefighters is the same as that for four firefighters except that there are four firefighters instead of two on the ground to receive the ladder as it is tilted down from the rack (Figure 3.122). Two firefighters assume a position at midpoint for the carry.

Figure 3.122 When there are six firefighters, those on the ground position themselves as shown.

FLAT ARM'S LENGTH METHOD

Procedures for the six-firefighter flat arm's length carry are the same as the six-firefighter flat-shoulder carry except that the ladder is lowered to arm's length rather than being placed on the shoulder (Figure 3.123).

Figure 3.123 The six-firefighter flat arm's length carry.

Procedures when Hydraulic Jacks are in Use

When aerial apparatus hydraulic jacks are being set, personnel should not attempt to remove ground ladders. The same rule applies when pumpers are equipped with aerial devices (articulated or telescoping master stream devices) and the jacks are being set. **NOTE:** Some adjustment may be required in positioning firefighters for removal of ground ladders when jacks are down (Figure 3.124).

Figure 3.124 When apparatus outriggers are in use, some adjustment of the firefighters' positions for ladder removal may be necessary.

Carrying Other Ladders

FOLDING LADDERS

One firefighter carries the folding ladder in the closed position, tip forward. The ladder is held horizontal while being grasped with both hands (Figure 3.125).

Figure 3.125 The folding ladder carry.

3.18 True or False.

	True	False
The one-firefighter low-shoulder carry is the recommended carry for folding ladders.	☐	☐

COMBINATION LADDERS

Combination ladders are usually carried by one firefighter. The arm's length on-edge method is used (Figure 3.126).

Figure 3.126 A carry for a combination ladder.

POMPIER LADDERS

Pompier ladders are carried by one firefighter using a shoulder carry (Figure 3.127).

Figure 3.127 A shoulder carry is used for the pompier ladder. The butt end is carried forward.

3.18 1. False

Special Carry for Narrow Passageways

When firefighters using the flat-shoulder carry and the flat arm's length carry encounter a narrow passageway an easy way to overcome this problem is for the firefighters to raise the ladder over their heads, assuming that there is space overhead. The procedure is as follows:

Step 1: The firefighter at the right tip position gives the signal to shift to the overhead carry.

Step 2: All firefighters lift the ladder up until their arms are extended straight overhead.

Step 3: The firefighter at the left butt steps back and under the ladder. The firefighter at the left tip steps forward and under the ladder.

Step 4: The firefighters on the right side shift under the ladder (Figure 3.128).

3.19 Fill in the blank.
The firefighter at the ____________ position gives the signal to shift to the overhead carry.

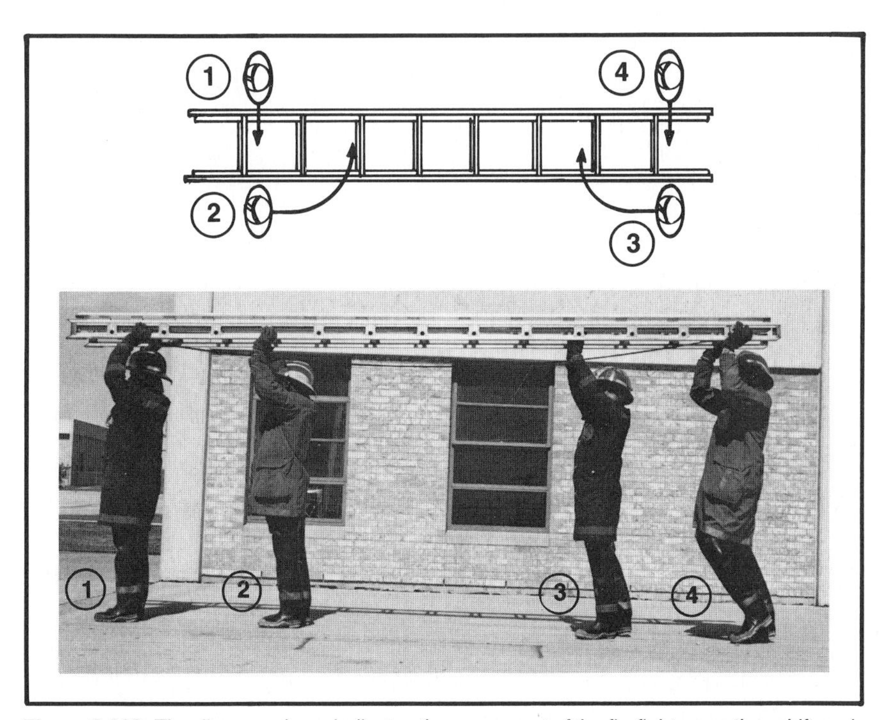

Figure 3.128 The diagram above indicates the movement of the firefighters as they shift positions after lifting the ladder overhead. The photograph shows the firefighters in position for the narrow passageway carry.

POSITIONING (PLACEMENT)

Proper positioning is important because it affects the safety and efficiency of ground ladder operations.

Responsibility for Positioning

Normally an officer will designate the general location where the ladder is to be positioned and/or the task to be performed. However, personnel carrying the ladder frequently decide on the exact spot where the butt is to be placed. The firefighter nearest the butt end is the logical person to make this decision because this end is placed on the ground to initiate raising the ladder. Where there are two firefighters at the butt, as in the four-and six-firefighter flat carries, the one on the right side is usually the one responsible for placement. However, this designation is an option as far as each individual department's policy is concerned.

3.20 Fill in the blank.
When four firefighters use the flat-shoulder carry and there is no officer, the firefighter at the ____________ position is responsible for placement.

Objectives

When placing ladders there are two objectives to be met: first, to place the ladder properly for its intended use; and second, to place the butt the proper distance from the building for safe and easy climbing.

FACTORS OF INTENDED USE AND HOW THEY AFFECT PLACEMENT

- If the ladder is to be used to provide a vantage point from which a firefighter can break out a window for ventilation, it should be placed alongside the window to the windward side. The tip should be about even with the top of the window (Figure 3.129).
- The same position is used when firefighters desire to climb into or out of narrow windows.
- If the ladder is to be used for rescue from a window, usually the ladder tip is placed even with or slightly below the sill. If the sill projects out from the wall, the tip of the ladder can be wedged up under the sill for additional stability (Figure 3.130). If the window opening is wide enough to permit the ladder tip to project into it and still allow room beside it to facilitate the rescue, the ladder should be placed so that two or three rungs extend above the sill (Figure 3.131).

Figure 3.129 A ladder used for window ventilation is placed on the windward side with the tip about even with the top of the window.

3.19 Right tip

Figure 3.130 The tip is placed at windowsill level when the ladder is used for rescue.

Figure 3.131 When there is a wide enough window opening, the ladder tip is projected into the opening to provide a hand grip as illustrated in the first photograph. This procedure is particularly suited to evacuation of conscious persons as depicted in the second photograph.

- The same position shown in Figure 3.131 is used for firefighters to climb into or out of wide window openings.
- When a ladder is to be used as a vantage point from which to direct a hose stream into a window opening and no entry

is to be made, it is raised directly in front of the window with the tip on the wall above the window opening (Figure 3.132).

Figure 3.132 To direct a stream of water into a window from a ladder, place the tip above the window opening.

NOTE: Care must be taken to keep flames from engulfing the tip of the ladder. If this cannot be avoided, the ladder is raised just to the sill as shown in Figure 3.130.

- When a ladder is to be used as a support for a smoke fan, it is raised directly in front of the window with the tip on the wall above the window opening (Figure 3.133).

3.20 Right butt

Figure 3.133 The ladder tip is placed against the wall above the window when the ladder is used to support a smoke ejector.

3.21 True or False.

	True	False
1. When a ladder is used as a vantage point for ventilation from a window, it is placed to the leeward side of the window.	☐	☐
2. Ladders being used for rescue from a narrow window are set with the tip at windowsill height.	☐	☐
3. Ladders from which hose streams are to be directed into window openings are set on the windward side of the window except when there are flames coming from the window.	☐	☐

OTHER FACTORS WHICH MAY AFFECT PLACEMENT

- Overhead obstructions such as wires, tree limbs, or signs
- Uneven terrain
- Obstructions on the ground, such as bushes, parked cars, or fountains
- Soft spots such as mud holes

LOCATIONS TO AVOID

- Main paths of travel that firefighters or evacuees will need to use
- Burning surfaces or openings that flames are coming out of
- Rounder surfaces where stability might be endangered
- On top of sidewalk elevator trapdoors
- Over sidewalk deadlights
- Unstable walls or surfaces

> **3.22** Fill in the blanks.
> There are four factors which may affect ladder placement; two are overhead and ground obstructions, and the other two are ____________________ and ____________________.

PROPER DISTANCE FROM THE BUILDING

When the ladder has been raised and lowered into place, the desired angle of inclination is approximately 75 degrees. This angle provides good stability and places stresses on the ladder properly. It also provides for easy climbing since it permits the climber to stand perpendicular at arm's length from the rungs (Figure 3.134).

Figure 3.134 When the ladder butt is placed correctly, the climber will stand perpendicular with the shoulders at arm's length from the rung.

If the butt of the ladder is placed too far away from the building, the load carrying capacity of the ladder is reduced and it has more of a tendency to slip. Placement at such an angle may be necessary, however, when there is an overhang on the building (Figure 3.135).

If the butt is placed too close to the building, its stability is reduced because climbing tends to cause the tip to pull away from the building.

A rule of thumb is to estimate the length of ladder needed to reach the particular support point and divide that number by four. The result will be the distance from the building where the butt should be placed. For example, if 20 feet (6 m) of ladder is needed to reach a window, the butt should be placed 5 feet (1.5 m) out from the building (20 divided by 4). This method relies on the ability of the firefighter to accurately estimate how much ladder length is going to be needed. Experience in using ground ladders will usually allow the firefighter to just "eyeball it" and be able to place a ladder correctly as a matter of routine.

> **3.23** Check the correct response.
> The desired angle of inclination for ground ladders is __________.
>
> ☐ A. 65 degrees. ☐ C. 25 degrees. ☐ E. 90 degrees.
> ☐ B. 80 degrees. ☐ D. 75 degrees.

3.21 1. False 2. True 3. False

3.24 True or False.	True	False
1. The rule for determining the proper distance for the butt from a building is to divide the designated length by four.	☐	☐
2. When the butt is placed too far from the building, the ladder is awkward to climb but the load carrying capacity is increased.	☐	☐

Figure 3.135 The facade pictured here made it necessary to place the ladder at an improper angle in which case loading must be minimized. *Courtesy of Hampshire Register, Hampshire, IL.*

Review

Answers on page 387

Fill in the blanks.

1. Proficiency in handling ground ladders will only be realized when ______________, ______________ is carried out.

2. All ladder racking schemes are variations of two basic options; these are:
   (1) ______________ or ______________ racking.
   (2) ______________ or ______________ removal.

True or False.

	True	False
3. NFPA 1901 requires one 16-foot (5 m) metal extension ladder be carried on tankers.	☐	☐
4. It is desirable to have the ladder extend one rung beyond the windowsill or roof edge when it is used to gain access.	☐	☐

Fill in the blanks.

5. When all factors are considered the ______________ extension ladder emerges as the most versatile of all extension ladders.

6. When lifting ladders, the lifting is done with the ______________ ______________.

7. There are three methods by which one firefighter may carry a ladder; the ______________ method, the ______________ method, and the ______________ method.

True or False.

	True	False
8. The high-shoulder carry is particularly well suited for use when carrying a roof ladder up another ladder.	☐	☐
9. The two-firefighter hip or underarm carry is not suitable for use with three-section extension ladders due to their bulk.	☐	☐

Check the correct response.

10. When two or more ladders are removed and stacked together and then the stack is carried to a convenient location by two firefighters, the procedure used is the ______________.
    - ☐ A. Two-firefighter ladder tote
    - ☐ B. Two-firefighter ladder stack
    - ☐ C. Two-firefighter multiple ladder carry

Sketch the answer.

Using circles to indicate firefighters, and a simple sketch of the ladder show the positioning of firefighters for the carries indicated.

11. Two-Firefighter Hip or Underarm Carry

12. Three-Firefighter Flat Carry

13. Three-Firefighter Low-Shoulder Carry

14. Four-Firefighter Low-Shoulder Carry

15. Four-Firefighter Flat-Shoulder Carry

16. Five-Firefighter Flat-Shoulder Carry

17. Six-Firefighter Flat Arm's Length Carry

3.22 Uneven terrain, Soft spots

3.23 D

3.24 1. False 2. False

True or False.

	True	False
18. Due to the tendency to come open while being carried, it is necessary for two firefighters to be used to carry combination ladders.	☐	☐
19. When firefighters using a flat-shoulder carry encounter a narrow passageway; if overhead space permits, the ladder is carried overhead.	☐	☐

Fill in the blank.

20. The firefighter located at the ________________ will make the decision as to where the butt is to be positioned for raising the ladder.

Check the correct response.

21. If 20 feet (6 m) of ladder is needed to reach a window, the butt should be placed ________________ from the building.

- ☐ A. Three feet (.9 m)
- ☐ B. Four feet (1.2 m)
- ☐ C. Five feet (1.5 m)

LADDERS

Chapter 4

Raising and Climbing/ Ladder Safety

NFPA STANDARD 1001
Fire Fighter I

3-9.3 The fire fighter, operating as an individual and as a member of a team, shall raise each type and size of ground ladder using several different raises for each ladder.

3-9.4 The fire fighter shall climb the full length of every type of ground and aerial ladder.

3-9.5 The fire fighter shall climb the full length of each type of ground and aerial ladder carrying fire fighting tools or equipment while ascending and descending.

3-9.6 The fire fighter shall demonstrate the techniques of working from ground or aerial ladder with tools and appliances, with and without a life belt.

Chapter 4

Raising and Climbing/ Ladder Safety

All the do's and don'ts, background knowledge, and supportive skills learned so far about ground ladders come together in this chapter when procedures and skills involved in raising and climbing are studied. A major goal of this chapter is to provide the knowledge necessary to develop smooth, proper, and safe ground ladder operations.

The raising and climbing phase is the most important and also the most dangerous because it occurs in a hazardous operating zone. This is a situation that occurs frequently, so particular attention needs to be given to the information provided.

In most cases more than one option is provided; the one best suited for the particular situation should be used. Procedures of individual fire departments may vary from those presented. This is not important as long as the variation is not an unsafe one and teamwork results.

This chapter concludes with a review of ladder safety. While firefighters may be inclined to skim over the information presented, it should be digested thoroughly. To quote a phrase developed elsewhere: "The life you save may be your own."

RAISING

Miscellaneous Procedures

RAISING WITHOUT LAYING THE LADDER ON THE GROUND

With the exception of pole ladders, it is not necessary to place a ladder flat on the ground prior to raising; only the butt end need be placed on the ground. The transition from carry to raise can and should be smooth and continuous.

LADDER OPERATIONS NEAR ELECTRICAL HAZARDS

A major concern when raising ladders is contact with live electric wires or equipment, either by the ladder or by the person who will have to climb it. The danger of metal ladders in this respect has been previously stressed. However, many firefighters do not realize that WET wood or fiber glass ladders present the same hazard. Care must be taken BEFORE BEGINNING A RAISE to be sure that this hazard is avoided.

FLY OUT ON EXTENSION LADDERS

The question of whether the fly on an extension ladder should be in (next to the building) or out (away from the building) has to be settled before starting the discussion of raises. This has been a matter of controversy in the fire service for many years.

IFSTA recommends that the FLY SECTION BE OUT (away from the building) except for wood truss ladders which have the rungs mounted in the top rail (Figure 4.1). This recommendation is based on information received from fire service ground ladder manufacturers, NFPA 1932's requirement that the fly be out, and a report of strength tests run on ground ladders by the National Bureau of Standards which states that extension ladders are stronger with the fly out.

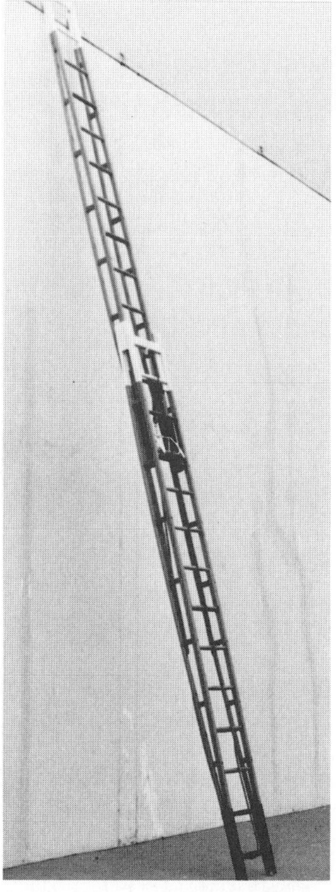

Figure 4.1 Extension ladders should be used with the FLY OUT as in the first illustration except for ladders which are manufactured with the rungs mounted in the top truss rail. These are designed to be used with the FLY IN as shown in the second illustration. *Second illustration Courtesy of ALACO Ladder Company.*

PIVOTING

Often when an extension ladder is raised to vertical, the firefighters will discover that the fly is in (next to the building). This occurs because of the method used to carry it and is to be expected. When this happens it is necessary to pivot the ladder so that the fly will be out prior to its being extended. Any ladder flat raised, parallel to a building, will also require pivoting to align it with the wall upon which it will rest. When pivoting a ladder the following procedures should be used.

One-Firefighter Pivot

NOTE: This procedure is NOT RECOMMENDED for use with the one-firefighter extension ladder raise. The procedure for the one-firefighter extension ladder raise details an alternative method of getting the fly section out. This method is therefore used only for single or roof ladders raised parallel to a building.

Step 1: Firefighter decides which beam to pivot on; the foot nearest that beam is placed between the beams against the butt spur.

Step 2: Firefighter reaches down as low as convenient and grasps the beam that the ladder is going to pivot on. The other hand is used to grasp the opposite beam up as high as convenient (Figure 4.2).

Figure 4.2 The beam that the ladder will be pivoted on is grasped as low as convenient. The opposite beam is grasped as high up as convenient.

Step 3: The ladder is then tilted up on the beam just enough for the butt spur of the other beam to clear the ground.

Step 4: The firefighter then simultaneously turns the body and shifts the other foot as the ladder is pivoted 90 degrees (Figure 4.3).

Step 5: The raised beam is brought back down, the firefighter shifts position so that the ladder is being faced in a ready-to-climb position (Figure 4.4).

Figure 4.3 The firefighter's body is rotated as the ladder is pivoted.

Figure 4.4 The firefighter continues to simultaneously pivot both the body and the ladder. When the pivot is complete the ladder is leveled and the firefighter is facing the ladder.

Two-Firefighter Pivot

Step 1: The two firefighters face each other through the ladder. They grasp the ladder with both hands and one firefighter places a foot against the beam that will be pivoted on (Figure 4.5). The ladder is then tilted onto one beam (Figure 4.6).

Figure 4.5 A foot is placed against the beam upon which the ladder will be pivoted. This will keep it from slipping as the ladder is pivoted.

Figure 4.6 The ladder is tilted onto the beam which is being heeled and the pivot is begun.

Figure 4.7 The firefighters adjust their positions as the ladder is pivoted 90 degrees.

Step 2: The ladder is then pivoted 90 degrees, with the firefighters simultaneously adjusting their positions (Figure 4.7).

Step 3: The process is repeated so that the ladder is turned a full 180 degrees and the fly is out.

SHIFTING IN A VERTICAL POSITION

Occasionally circumstances will require that ground ladders be moved while in a vertical position. Shifting a ladder in a vertical position should be limited to short distances, such as aligning ladders to a building or to an adjacent window.

4.1 True or False.

	True	False
1. Prior to raising, all ladders must be placed flat on the ground.	☐	☐
2. Contact with electrical wires presents the same hazard for all ladders.	☐	☐
3. Except for truss ladders with rungs mounted in the top rail, extension ladders should be placed with the fly in (next to the wall).	☐	☐
4. When one firefighter raises an extension ladder, the recommended procedure calls for pivoting the ladder to get the fly in the proper position.	☐	☐

Figure 4.8 Hand grip for vertically shifting a ladder.

One Firefighter

If the ladder is 20 feet (6 m) long or less, one firefighter can shift it in the vertical position.

Step 1: Firefighter faces the ladder, heels it, grasps the beams and brings it out to vertical.

Step 2: The grip on the ladder is shifted, one hand at a time, so that one hand is grasping as low a rung as convenient with the palm turned upward. The other hand grasps as high a rung as convenient with the palm turned downward (Figure 4.8).

Step 3: Firefighter turns slightly in the direction of travel, makes a visual check of the terrain and the area overhead, and proceeds with the ladder. The tip is watched as it is being moved.

CAUTION: This procedure should not be attempted close to live electrical wires as there will be some wobble, particularly at the top, when the ladder is shifted.

Two Firefighters

Extension ladders, because of their greater weight, may require two firefighters for the shifting maneuver.

Step 1: If the ladder isn't vertical, it is brought to vertical, and if extended, it is fully retracted.

CAUTION: The fly is ALWAYS FULLY RETRACTED before this evolution proceeds.

Step 2: The firefighters position themselves on opposite sides of the ladder. The hands are positioned as for the one-

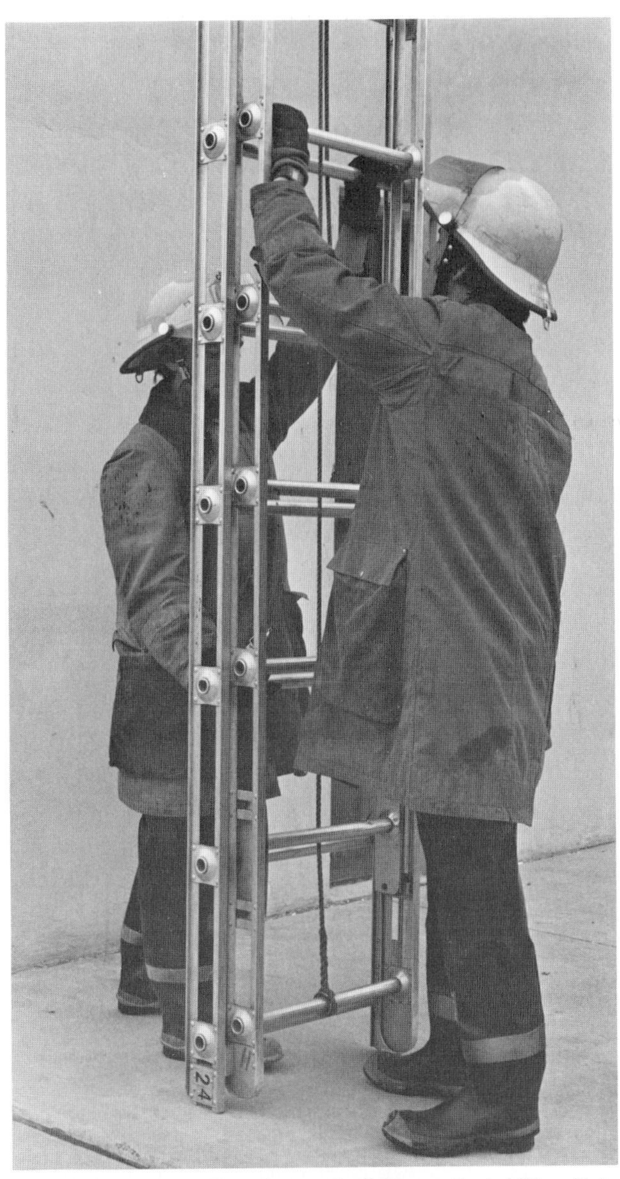

Figure 4.9 Hand positions for two-firefighter vertical shifting. Note that the extension ladder is fully retracted.

Figure 4.10 Ladder being shifted. Note that the firefighters watch the top of the ladder.

firefighter shift except that the side grasped low by one firefighter is grasped high by the other (Figure 4.9).

Step 3: The ladder is then lifted just clear of the ground. The firefighters, while watching the tip, shift it to the new position (Figure 4.10).

Shifting Pole Ladders

Pole ladders may also be shifted vertically. Four firefighters are required.

Step 1: The ladder is brought back to vertical and the fly(s) are fully retracted.

Step 2: The two heelmen grasp the ladder the same as two firefighters moving an extension ladder. The firefighters holding the poles are in the normal position for extending and retracting (Figure 4.11).

Step 3: The heelmen lift the ladder (Figure 4.12), the polemen steady it, and the ladder is carried to the new position (Figure 4.13). When there are six firefighters the additional two people help steady the ladder.

Figure 4.11 Positioning of firefighters for vertically shifting a pole ladder.

Figure 4.12 The ladder is lifted just enough for the butt spurs to clear the ground.

Figure 4.13 The heelmen lift and carry while the firefighters holding the staypoles steady it.

4.1 1. False
2. False
3. False
4. False

ROLLING A LADDER

NFPA 1932 states that "ground ladders shall not be rolled beam-over-beam to reach a new position." Manpower at the time of need may not permit vertical shifting of an extension ladder, which is the alternative. Individual fire department policy should dictate whether rolling is permitted or not. IFSTA feels that it is necessary to make one possible exception[1] to the NFPA standard. When one firefighter raises an extension ladder without assistance, the procedure, for reasons of safety, requires that the fly be in as it is extended. In order to get the fly out for climbing, a single roll is recommended rather than risking a 180 degree pivot with the fly extended. Therefore, the procedure for rolling a ladder is not described separately as it was in previous editions. It is, however, detailed as a part of the description of the *One-Firefighter Extension Ladder Raise.*

[1]The term "possible exception" is used because while the roll is used, it is not used "to reach a new position" as prohibited in the NFPA standard.

4.2 True or False.

	True	False
1. Shifting a ladder in a vertical position should be limited to short distances.	☐	☐
2. Extension ladders may be shifted vertically with the fly still extended.	☐	☐
3. The minimum number of firefighters needed to vertically shift a pole ladder is four.	☐	☐
4. NFPA 1932 does not permit rolling a ladder to reach a new position.	☐	☐

RAISE POSITION: RIGHT ANGLE OR PARALLEL?

The right angle raise position (Figure 4.14) should be given first preference because the ladder will ultimately have to be brought to this position to be placed against the building. The parallel raise position is used when circumstances, such as obstructions, prevent the use of the right angle raise (Figure 4.15). When this happens the ladder is brought to vertical and then pivoted to the right angle position (Figure 4.16).

Figure 4.14 The right angle raise position.

Figure 4.15 The parallel raise position.

Figure 4.16 After a parallel raise the ladder is pivoted to the right angle position.

One-Firefighter Raises

SHORT SINGLE AND ROOF LADDERS

Single and roof ladders of 14 feet (4 m) or less are light enough and have little enough bulk that one firefighter can usually place the butt at the point where it will be located for climbing without first heeling it against a building or other object, especially when the high-shoulder carry is used. The following steps are used:

Step 1: The butt is lowered to the ground at the proper distance from the building for climbing (Figure 4.17).

Step 2: Simultaneously the ladder is brought to vertical (Figure 4.18).

CAUTION: The area overhead should be visually checked for obstructions before initiating this action.

Step 3: The ladder is pivoted if necessary. The firefighter grasps both beams, heels the ladder, and lowers the tip into place (Figure 4.19).

Figure 4.17 One butt spur can be placed on the ground and the ladder tilted upward without butting against the building when the firefighter's stature permits.

Figure 4.18 The ladder is tilted to vertical.

Figure 4.19 After pivoting, the ladder tip is lowered into place and the butt is adjusted as necessary to obtain a proper climbing angle.

4.2 1. True
2. False
3. True
4. True

ALL OTHER LADDERS

The bulk and weight of other ladders makes it necessary to have some means of keeping the butt from slipping as the ladder is brought to vertical.

Long Single and Roof Ladders

Step 1: The butt end of the ladder is tilted to the ground with the butt spur against the wall of the building (Figure 4.20). The building, in this instance, will act in place of a heelman to prevent the ladder butt from slipping while the ladder is being brought to vertical.

Step 2: The ladder is turned so that it is flat. The firefighter pivots under the ladder facing the building. Both butt spurs are placed against the building (Figure 4.21).

Step 3: The firefighter then advances hand- over-hand down the rungs, simultaneously stepping forward until the ladder is vertical against the side of the building (Figures 4.22 and 4.23).

Figure 4.20 A butt spur is placed against the building.

Figure 4.21 The ladder is turned until it is flat with both butt spurs against the building.

Figure 4.22 The firefighter advances hand-over-hand down the rungs to bring the ladder to vertical. The building wall acts as a heelman.

Figure 4.23 The ladder is raised until it is vertical against the side of the building.

CAUTION: The terrain in front of the firefighter should be checked visually before stepping forward and the area overhead should be visually checked for obstructions before bringing the ladder to vertical.

Step 4: In order to shift the butt outward the firefighter grasps a rung up high so that the tip will be held against the building and another rung down low, palm up, so that the butt can be lifted off the ground (Figure 4.24).

Step 5: The butt is lifted and brought out until the proper angle of inclination is attained (Figure 4.25).

Figure 4.24 The firefighter grasps one rung down low and one rung up high and proceeds to shift the butt outward.

Figure 4.25 The butt is set for the proper climbing angle.

Extension Ladders

Step 1: The butt is tilted to the ground adjacent to the wall (Figure 4.26). Remember that the ladder will have to be rolled to get into its final position, so the butt should be placed such that the ladder will be one width further to the side than when it is in its final position.

Step 2: The ladder is tilted so that it will be flat with both butt spurs against the building (Figure 4.27).

Step 3: After having made a visual check for overhead obstructions, the firefighter walks the ladder to vertical while the building acts as the heelman (Figure 4.28).

Figure 4.26 The butt touches the ground adjacent to the wall.

Figure 4.27 The ladder is tilted so that it will be flat.

Figure 4.28 The ladder is walked up to vertical.

Step 4: With the ladder in a vertical position against the side of the building (Figure 4.29), the firefighter places one foot at the butt of one beam, and the instep, knee, and leg against the beam. These actions will steady the ladder while both hands operate the halyard. The halyard is held firmly with the thumbs up (Figure 4.30).

Step 5: While watching the tip for signs of pulling outward or for obstructions, the firefighter pulls the halyard downward hand-over-hand. The tip is allowed to pull slightly away from the building as the fly is raised (Figure 4.31).

CAUTION: The halyard must be pulled either straight downward or a very slight angle from vertical, or the tip will be pulled too far from vertical and the ladder will tend to fall on the firefighter. If at any time the firefighter feels that the ladder is about to fall, the body weight is shifted against the ladder and it is pushed back against the side of the building.

Step 6: At the desired height the pawls are engaged.

Figure 4.29 The ladder is raised until it is vertical against the side of the building.

Figure 4.30 The firefighter heels the ladder with one foot and grasps the halyard.

Figure 4.31 The halyard is pulled straight downward, hand-over-hand.

Step 7: The butt is brought out to approximately the proper angle of inclination (Figure 4.32). A roll is made to get the fly out. To roll the ladder the firefighter stands facing the beam upon which the roll will be made. Both beams are grasped. The near beam is heeled by placing the foot of the leg on the side of the direction of the roll against the outside of the butt spur (Figure 4.33).

Step 8: The beam away from the direction of the roll is brought up and over toward the firefighter. The heel and tip of the other beam remain in place (Figure 4.34).

Figure 4.32 The butt is brought out. Note that the fly is still in.

Figure 4.33 The hand grip is shifted to the beams. A foot is placed against the butt of the beam upon which the roll will be made.

Figure 4.34 The beam away from the direction of the roll is brought outward.

Step 9: Once the ladder is at 90 degrees to the building the firefighter shifts handholds and heels the same beam but with the other foot as in Figure 4.35.

Step 10: The beam being moved is brought the remaining way to complete the roll and bring the fly out (Figure 4.36). Some repositioning may be necessary after the roll has been completed.

Figure 4.35 When the roll is halfway completed, the firefighter shifts handholds as shown. The feet are shifted so that the ladder is heeled with the other foot.

Figure 4.36 The roll is completed so that the fly will be out.

4.3 True or False.

	True	False
1. The parallel raise is given first preference because it can be used in any circumstances.	☐	☐
2. The wall serves as the heelman in the one-firefighter extension ladder raise.	☐	☐
3. When one firefighter is raising an extension ladder, the ladder is rolled into position prior to moving the butt out from the building.	☐	☐
4. During the one-firefighter extension ladder raise, the firefighter has better control when extending the fly if the halyard is pulled out at about 45 degrees to the ladder.	☐	☐

Two-Firefighter Raises

The steps for raising ladders described in the following paragraphs presume that the raise is being made directly from the carry without first placing the ladder flat on the ground.

FLAT METHOD

Step 1: If the ladder is not already flat, it is turned to this position. The firefighter at the tip end places one beam on the shoulder. The firefighter at the butt end heels the ladder, crouches, and grasps a convenient rung or the beams with both hands. (Figure 4.37).

Step 2: The firefighter at the tip end swings under the ladder. With arms extended this firefighter advances hand-over-hand down the rungs, or the beams, bringing the ladder toward vertical (Figure 4.38).

Step 3: As the ladder comes toward vertical the firefighter heeling the ladder shifts hands several times to get a grip higher up the ladder (Figure 4.39).

Figure 4.37 Firefighters in position to begin the two-firefighter flat raise.

Figure 4.38 The firefighter at the tip advances down the ladder hand-over-hand.

Figure 4.39 The firefighter heeling the ladder grasps successively higher points on the ladder as it is raised to vertical.

Step 4: Both firefighters face each other through the ladder. If it is an extension ladder, they pivot the ladder to get the fly out (Figures 4.40 and 4.41).

NOTE: When raising single or roof ladders, skip step 5 and go directly to step 6.

Step 5: The firefighter at the outside position holds the ladder steady while the firefighter at the inside position extends the fly and locks the pawls (Figure 4.42).

Step 6: The firefighter at the outside then places one foot against each butt spur. Both firefighters lower the ladder into the building.

Figure 4.40 The ladder is pivoted 180 degrees.

Figure 4.41 The pivot gets the fly out.

Figure 4.42 The firefighter on the inside extends the fly while the other firefighter steadies the ladder.

BEAM METHOD

Step 1: If the ladder is not already on beam, it is turned to this position. The firefighter at the tip end has one beam resting on a shoulder (Figure 4.43).

Step 2: The firefighter at the butt end places the inside foot upon the lower beam at the butt spur, turns toward the ladder, and grasps the upper beam with the hands well apart. The other leg should be extended well back to act as a counterweight and balance the body (Figure 4.44).

Figure 4.43 The firefighter at the tip faces the butt with the lower beam resting on the shoulder.

4.3 1. False
2. True
3. False
4. False

Figure 4.44 One stance for heeling a ladder for a beam raise.

Alternate Method of Heeling for the Beam Raise

An alternate method of heeling the ladder for the beam raise is to stand parallel to the ladder at the butt. One foot is placed against the butt spur and the other is positioned forward toward the tip of the ladder (Figure 4.45).

Step 3: The firefighter at the tip raises the lower beam by extending the arms and advances hand-over-hand toward the butt end, raising the ladder toward vertical (Figure 4.46).

Step 4: Proceed as in Flat Method steps 4, 5, and 6.

Figure 4.45 Alternate stance for heeling a ladder for a beam raise.

Figure 4.46 The lower beam is elevated by advancing hand-over-hand toward the butt end.

4.4 True or False.

	True	False
1. The only difference between the regular beam method of raising a ladder and the alternate beam method is in the stance of the heelman.	☐	☐

Three-Firefighter Raise

As the length of the ladder increases, you will find that the weight also increases. If available, more personnel should be used to raise the heavier extension ladders.

When using three firefighters, it is the responsibility of the firefighter at the butt end of the ladder to select the location for the raise and to properly space the butt end from the building. Although raising the ladder is a team effort, for ease of operation the heelman should give the commands.

A flat raise is used when three firefighters are involved. Since there are a number of different methods for carrying the ladder, there will be some variation in step 1.

Step 1: The heelman, having selected the location for the raise, places the butt end on the ground. If the ladder is not already flat, the remaining two firefighters shift the ladder until it is flat (Figure 4.47).

Figure 4.47 The butt is placed on the ground with the firefighters positioned as shown.

Step 2: The firefighter at the butt end heels the ladder. The firefighter at midpoint shifts to the tip, assuming a position opposite the firefighter already there. The two firefighters at the tip then place the ladder on their shoulders. The firefighter at the butt crouches down, leans forward, and grasps a convenient rung with both hands (Figure 4.48).

Figure 4.48 The firefighter at the butt shifts to heel the ladder. The firefighter at the midpoint shifts to the tip to assist in raising the ladder.

Step 3: On command of the heelman the firefighters at the tip end lift the ladder overhead and swing under their respective beams. With their arms straight, they advance hand-over-hand down the beams toward the vertical position (Figure 4.49).

Step 4: If the fly is in, the ladder is then pivoted so that the fly will be out (Figure 4.50).

Figure 4.49 The firefighters at the tip lift the tip end overhead and walk down the beams to bring the ladder to vertical.

Figure 4.50 The ladder is pivoted to get the fly out.

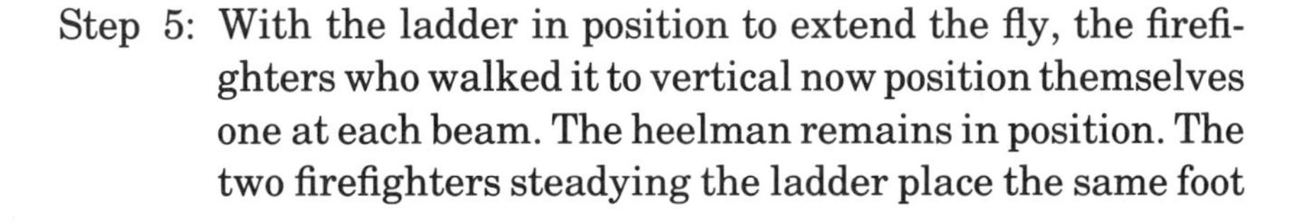

Step 5: With the ladder in position to extend the fly, the firefighters who walked it to vertical now position themselves one at each beam. The heelman remains in position. The two firefighters steadying the ladder place the same foot

(either right or left) on the side of the beam. They grasp the beam with their hands, watching fingers, as the former heelman extends the ladder to the desired position and locks the pawls. All three watch the tip as the fly is extended (Figure 4.51).

Step 6: The firefighter who operated the halyard grasps the ladder by each beam. The two firefighters who were steadying the ladder shift to the outside of the ladder and face it. Each places the nearest foot on the bottom rung and grasps the ladder. They both lower it into the building while watching the tip. The firefighter on the side next to the building assists with the lowering (Figure 4.52).

When the ladder is lowered, the reverse procedure is followed.

Figure 4.51 Position of the firefighters for extending the fly of an extension ladder.

Figure 4.52 The ladder is lowered into place.

4.5 Fill in the blank.
During a raise using three firefighters, the firefighter at the ________________ position is responsible for giving commands.

4.4 1. True

Four-Firefighter Raise

When manpower is available, four firefighters are desirable because they can handle the larger and heavier extension ladders; in this instance, the right hand heelman gives the commands. A flat raise is also used for the four-firefighter raise. As in the three-firefighter raise, there are a number of different methods for four firefighters to carry the ladder, so there will be some variation in step 1.

Step 1: The right side heelman selects the location for the raise and the butt end is lowered to the ground (Figures 4.53-4.54). If the ladder is not already flat, the remaining two firefighters shift the ladder until it is flat.

Figure 4.53 The firefighter at the right butt end position selects the location for the raise and gives the command for the butt to be lowered.

Figure 4.54 Ladder butt being lowered to the ground for the four-firefighter flat raise.

Step 2: The firefighters place themselves two at the butt heeling the ladder and two at the tip preparing to raise it.

Step 3: At the command of the firefighter at the right heel position, the firefighters at the tip lift the ladder overhead. With arms straight, they advance hand-over-hand down the beams toward the vertical position (Figure 4.55).

Figure 4.55 With arms straight the two firefighters at the tip walk hand-over-hand down the beams.

Step 4: Once the ladder is vertical it may be pivoted to get the fly out (Figure 4.56).

Step 5: Before extending the fly section, firefighters shift position so that there will be one firefighter on the outside of each beam facing the other. There will be one firefighter on the inside of the ladder and one firefighter on the outside, facing each other through the rungs.

Step 6: The fly is then extended by the firefighter at the halyard position as the other three firefighters steady the ladder (Figure 4.57).

Step 7: The firefighters on the outside of the ladder heel the ladder and grasp a convenient rung while the ladder is lowered to the building by all firefighters (Figure 4.58). The heelmen and the two firefighters at the beam watch the tip as the ladder is lowered in. To lower the ladder, reverse the raise procedure.

4.6 Fill in the blank.
During the four-firefighter extension ladder raise the firefighter at the ______________________ position is responsible for giving commands.

Figure 4.56 The ladder is pivoted to get the fly out.

Figure 4.57 Three firefighters steady the ladder while the firefighter on the inside raises the fly.

4.5 heel

Figure 4.58 The ladder is lowered into the building. Note that the firefighter on the inside does not look upward.

4.7 Check the correct response.
Both the three-and four-firefighter raises use the ____________ method raise.

- ☐ A. Beam
- ☐ B. Flat
- ☐ C. Parallel
- ☐ D. 90 degrees to the building
- ☐ E. It doesn't make any difference

Pole Ladder Raises

Pole ladders generally range from 40 feet (12 m) to 60 feet (18 m) in length. The weight of these ladders makes it hard for firefighters to raise them from flat to vertical and vice versa. The length tends to make them unstable during extension and retraction of the fly section(s), or when they are being lowered into or pulled away from a building. Staypoles are used to assist in overcoming these problems; they provide a means for two additional firefighters to apply lifting force when the ladder is being raised and to take part of the weight when it is being lowered. When the ladder is vertical and the fly is being extended or retracted, the staypoles are used to provide both sideways and in-and-out stability. The same use is made of them during the lowering in and pulling out of the ladder. Staypoles may be permanently attached to the bed section of the ladder or they may be removable. Currently the NFPA standard calls for them to be permanently attached.

When four firefighters raise a pole ladder, the operation must be performed at right angles to the building. Five-and six-firefighter raises may be performed at either right angles or parallel to the building. The right angle operation is preferred because when the pole ladder is raised parallel to a building a 90 degree pivot is necessary. This requires that the staypoles be shifted at the same time the ladder is pivoted. The procedure is detailed at the beginning of the section on five-firefighter raises.

A requirement to flip the ladder over prior to raising, to shift positions of the firefighters, and the need for passing the staypoles during raising makes it necessary that pole ladders be placed on the ground before beginning a raise.

PASSING UNATTACHED STAYPOLES

The text materials have been developed for pole ladders with permanently attached staypoles. Unattached staypoles are nested between the beams and upon the rungs of the top fly section of the pole ladder. The toggle end of the staypole is carried so that it is nested at the butt end of the ladder. Because of this, unattached staypoles are not passed overhead as permanently attached staypoles are. The following procedure is used instead of that described in the following sections.

Step 1: The firefighters who would normally pass attached staypoles proceed to the midpoint of the ladder and each grasps a staypole (Figure 4.59).

Figure 4.59 Unattached staypoles are grasped at midpoint.

Step 2: The staypoles are lifted as the firefighters turn and walk toward the tip end of the ladder carrying the staypoles. The ends with the spurs are directed into the hands of the respective firefighters who have moved out from the tip to receive them (Figure 4.60).

Step 3: The firefighters carry the poles forward until the toggles are opposite the toggle latches. Then they shift to that point and latch the toggles (Figure 4.61).

4.6 right heelman

4.7 B

Figure 4.60 The staypoles are carried toward the tip until the spur ends can be placed into the hands of the polemen.

Figure 4.61 The toggles are latched to attach the staypoles to the ladder.

4.8 Fill in the blanks.

1. When four firefighters raise a pole ladder, the raise must be performed at ______________________________ to the building.
2. The ______________________________ of pole ladders tends to make them unstable during extension and retraction of the fly section(s).
3. When five or six firefighters raise a pole ladder, the right angle raise is preferred because it avoids having to make a ______________________ that is required when a parallel raise is made.

4.9 Check the correct response.

Pole ladders have to be placed flat on the ground prior to raising because __________.

☐ A. The firefighters have to shift positions.

☐ B. The firefighters have to flip the ladder over.

☐ C. The firefighters have to pass the staypoles.

☐ D. All of the above.

☐ E. None of the above.

FOUR-FIREFIGHTER POLE LADDER RAISE

Step 1: The ladder is placed flat on the ground with the butt spurs almost against the building (Figure 4.62).

Step 2: The ladder is then turned over so that the fly will be down. Then the butt is placed against the building (Figure 4.63).

NOTE: The fly section is placed down so that the fly will be out when the ladder is raised.

Figure 4.62 The pole ladder is placed flat on the ground. The fly will be up at this point.

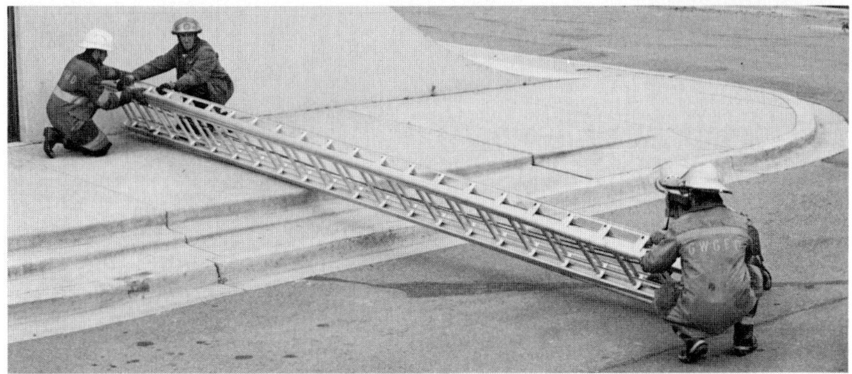

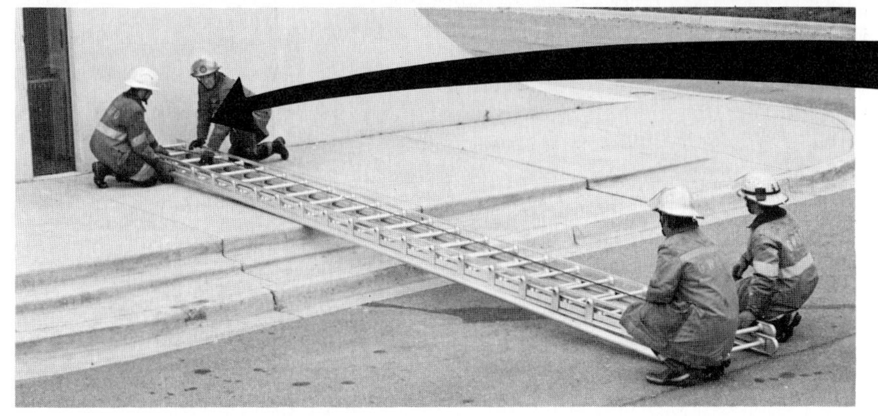

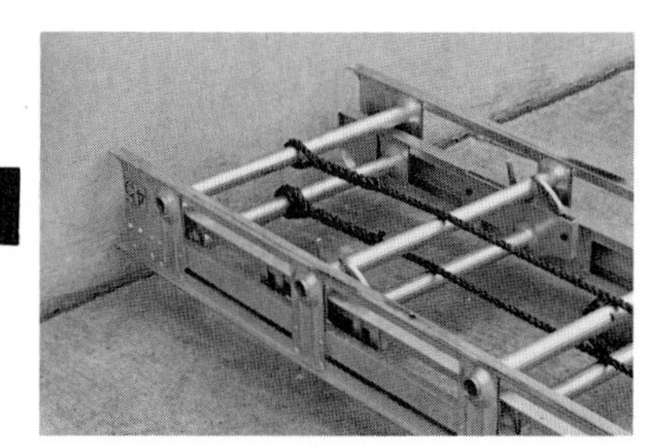

Figure 4.63 The ladder is turned over so that the fly will be out after it is raised. After it is turned the butt is placed against the building.

Step 3: The firefighters at the tip move outward in order to receive the staypoles. They stand about 5 feet (1.5 m) apart. The two firefighters at the butt unlatch the spur end of the staypoles and lift them to begin the passing evolution (Figure 4.64).

Figure 4.64 The firefighters who carried the tip move outward to receive the staypoles.

Step 4: The two firefighters with the staypoles walk toward the tip while grasping the staypoles hand-over-hand as they move the staypoles toward vertical. They check overhead for obstructions (Figure 4.65).

Step 5: The staypoles are brought to vertical and then lowered toward the two waiting firefighters who prepare to receive them (Figure 4.66).

Figure 4.65 The firefighters at the butt tilt the staypoles upward hand-over-hand.

Figure 4.66 After the staypoles pass vertical they are lowered to the other two firefighters.

Step 6: After the staypoles are passed, the two firefighters who passed them proceed to near the tip and kneel beside the beam facing the tip. Then each grasps a convenient rung with the near hand (Figure 4.67).

Step 7: The firefighters at the tip simultaneously lift the ladder tip and stand, using the leg muscles. They pivot under the ladder, grasp a convenient rung, and face the building (Figure 4.68). Then they walk toward the building, moving hand-over-hand down the ladder (Figure 4.69).

4.8 1. right angles
2. length
3. 90 degree pivot

4.9 D

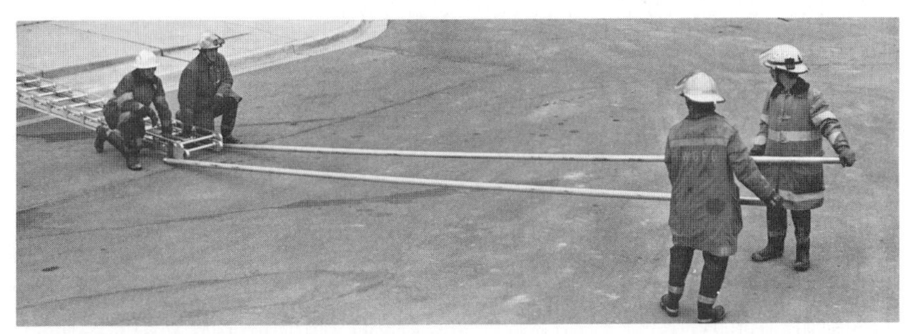

Figure 4.67 The firefighters who passed the staypoles kneel beside the tip facing the staypoles and grasp a rung with the near hand.

Figure 4.68 The tip is lifted simultaneous with the firefighters rising and pivoting toward the butt and stepping under the ladder.

Figure 4.69 The two firefighters at the tip then move hand-over-hand down the ladder to raise it to vertical. The building serves as a heelman.

As soon as the ladder is high enough, the firefighters on the staypoles push forward to assist with raising the ladder (Figure 4.70).

Step 8: All firefighters continue raising the ladder until it is vertical against the building (Figure 4.71).

Figure 4.70 When the ladder gets about one-fourth of the way up, the firefighters holding the staypoles push forward to assist with the raising. Note the hand grip used.

Figure 4.71 All firefighters continue raising the ladder until it is vertical against the building.

Step 9: The firefighters at the butt grasp a rung down low with their near hands, palm up. The other hands grasp the beams about head high (Figure 4.72).

Step 10: The butt is shifted outward to the point where the ladder will be positioned for climbing (Figure 4.73).

Figure 4.72 The firefighters at the butt position their hands as shown.

Figure 4.73 The butt is shifted outward.

Step 11: One of the firefighters on a staypole shifts toward the building until the staypoles now give four-way stability to the ladder. The two firefighters on the staypoles watch the tip of the ladder and control their respective forward and backward movement of the ladder (Figure 4.74 on next page).

Figure 4.74 One firefighter holding a staypole shifts to the side so that the ladder has four-way stability.

Figure 4.75 The ladder is heeled and then the tip is brought out until the ladder is vertical.

Step 12: One of the two firefighters at the butt shifts to a position behind the ladder and faces it. The other firefighter heels the ladder on the outside. The ladder is then brought to vertical (Figure 4.75).

Step 13: The firefighter on the inside then raises the fly while the other firefighter helps to steady the ladder. One of the firefighters on a staypole determines when the ladder is at proper height (Figure 4.76).

Step 14: The ladder is then gently lowered into position (Figure 4.77).

Step 15: The staypole firefighters check for proper positioning and alignment and adjustments are made as necessary.

Step 16: The staypole firefighters walk their staypoles toward the building and let them rest on the ground (Figure 4.78).

CAUTION: The staypoles must not be wedged. They are not designed to carry the stresses put on the ladder. When set in this position, they are used only to prevent side slippage.

To lower the ladder, reverse the raising procedure.

FIVE-AND SIX-FIREFIGHTER POLE LADDER RAISES

Step 1: The ladder is lowered to the ground at the approximate raise location. Note that the fly is on top.

Figure 4.76 The firefighter on the inside raises the fly.

Figure 4.77 The ladder is lowered into position.

Figure 4.78 Both staypoles are repositioned at about 45 degrees from the side of the ladder.

Step 2: The ladder is turned over so that the fly is down (Figure 4.79).

Step 3: The firefighters take their positions for passing and receiving the staypoles. Two stand off from the tip, ready to receive the poles. Two stand at midpoint of the beams to pass the poles. The fifth stands at the butt, ready to heel the ladder. When there are six firefighters the fifth and sixth both stand at the butt ready to heel the ladder (Figure 4.80).

Figure 4.79 The ladder is placed flat on the ground. Then it is turned over to get the fly down (so it will be out when the ladder is raised).

Figure 4.80 The firefighters are positioned as shown. The first illustration is for five firefighters. The second illustration is for a six firefighter crew. The two firefighters at the ladder midpoint grasp the staypoles and prepare to pass them.

Step 4: The firefighters at the beams pass the staypoles to the firefighters who moved out from the tip (Figure 4.81).

Step 5: The firefighters then take their positions for the raise. The firefighter at the butt crouches on the bottom rung and grasps convenient rungs with both hands (Figure

Figure 4.81 Passing staypoles.

4.82). The firefighters at the beams kneel (with inside knee on the ground) beside the ladder and grasp a rung just below the toggles (Figure 4.83). The firefighters with the staypoles stand at the outside of the poles, one hand holding the spur end with the spur extending between the fingers, and the other hand holding the staypole at a comfortable distance up from its spur end. The staypoles should be as nearly in line with the beams as possible (Figure 4.84). With six firefighters there are two firefighters heeling the ladder (Figure 4.85).

Figure 4.82 Heelman's position.

Figure 4.83 Positions of two firefighters at the tip.

Figure 4.84 Position of two firefighters with the staypoles.

Figure 4.85 When there is a six firefighter crew there are two heelmen positioned as shown.

Step 6: The firefighter at the butt gives the command to raise the ladder. The firefighters at the tip rise, bringing the ladder tip to shoulder level, and then pivot under the beams until they are facing the butt (Figure 4.86).

Step 7: The firefighters at the tip raise their arms upward and walk hand-over-hand down the beams, raising the ladder. The heelman leans back so that the body weight helps the raise (Figure 4.87).

Figure 4.86 The tip is lifted simultaneously with the firefighters there pivoting under the ladder to face toward the butt.

Figure 4.87 The two firefighters at the tip walk down the ladder hand-over-hand to raise it to vertical as in the first illustration. The heelman leans back so that body weight helps to raise it as in the second illustration.

Step 8: When the ladder is at an angle of about 45 degrees, the staypole firefighters assume most of the ladder's weight and push the ladder up. The firefighters walking the beams up continue to assist with the raise but act more to steady the ladder (Figure 4.88).

Step 9: When the ladder is vertical, one of the firefighters holding a staypole walks toward the building until the staypole is in line with the lateral plane of the ladder. The two staypoles now give four-way stability to the ladder. The two firefighters holding the staypoles watch the tip of the ladder and control their respective forward and backward movement of the ladder (Figure 4.89).

Figure 4.88 As the ladder rises the firefighters on the staypoles assume more and more of its weight.

Figure 4.89 After the ladder reaches vertical, one staypole is shifted to the side.

NOTE: If the ladder has been raised parallel to the building, it is pivoted as follows: (The numbers used to refer to firefighters are identified in the accompanying drawings.)

Step A: Firefighter 3 shifts to the side of the ladder away from the building. Firefighters 1 and 2 each place the near foot against the outside of the spur of the other beam (Figure 4.90).

Step B: Firefighters 1, 2, and 3 all work together to tilt the ladder slightly toward the building and to begin to pivot the ladder while continuing to steady it. Firefighter 5 shifts position as the ladder pivots keeping the staypole at 90 degrees to the building (Figure 4.91).

Step C: Firefighters 1, 2, and 3 realign themselves with the ladder and continue the pivot until the ladder is at right angles to the building. Firefighter 5 continues to shift as the ladder pivots, being careful to maintain the staypole at 90 degrees to the building (Figure 4.92).

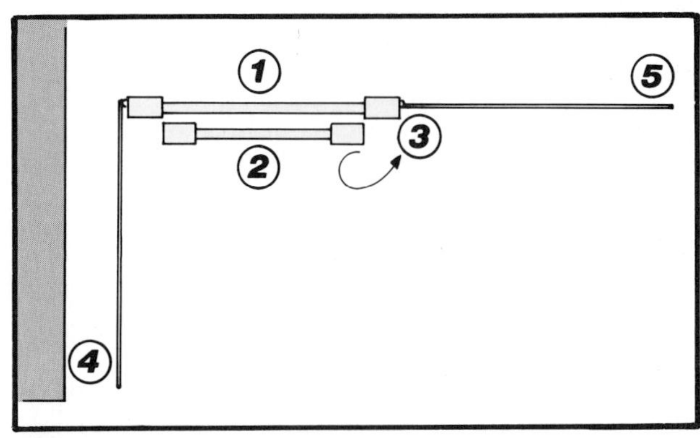

Figure 4.90 To begin the pivot No. 3 shifts to the side away from the building. No. 1 and No. 2 each place the near foot against the outside of the butt spurs.

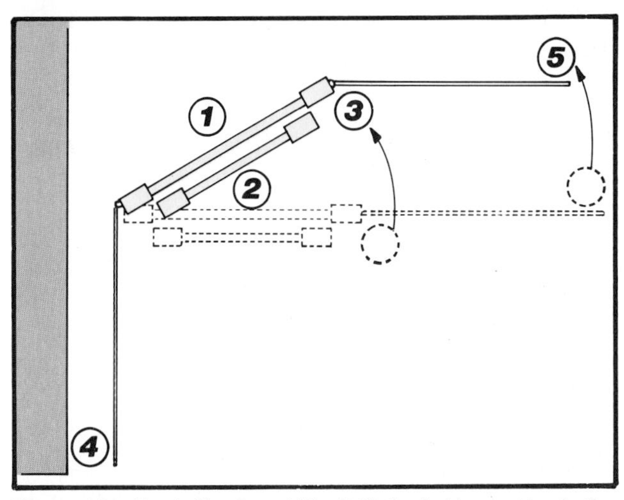

Figure 4.91 No. 1, No. 2, and No. 3 tilt the ladder and begin the pivot. No. 5 shifts as the ladder pivots.

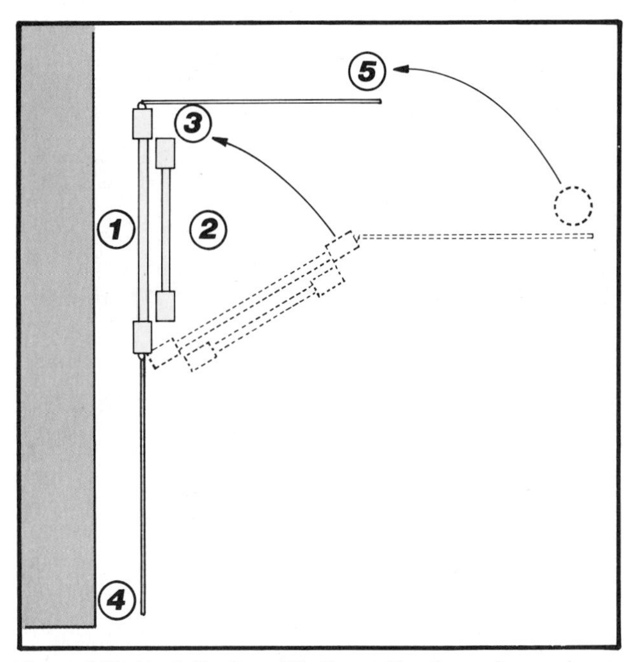

Figure 4.92 No. 1, No. 2, and No. 3 reposition themselves and continue the pivot bringing the ladder to right angles with the building.

When six firefighters perform the pivot, the additional firefighter places both feet on the bottom rung, grasps a convenient rung with the hands, and rides the ladder around as it is pivoted. This firefighter's action places extra weight on the spur being pivoted on to keep it from slipping.

Figure 4.93 Firefighters shift to the position shown while the fly is raised.

Step 10: One of the firefighters who walked down the beams to raise the ladder shifts to the side of the beam and steadies the ladder from this position. The remaining firefighter on the outside of the ladder also steadies it. The firefighter on the side of the ladder next to the building grasps the halyard (Figure 4.93).

Step 11: The ladder is extended by pulling down on the halyard as previously described. The firefighter holding the staypole out in front of the ladder determines when the ladder is at the proper height (Figure 4.94).

Step 12: The ladder is lowered into position (Figure 4.95).

Figure 4.94 The firefighter on the inside extends the fly while all others steady it.

Figure 4.95 The ladder is lowered against the building.

Step 13: The staypoles are set (Same as steps 15 and 16 of the four-firefighter pole ladder raise).

Beam Raise

Step 1: When the heelman reaches the location where the ladder is to be raised, the ladder is laid flat on the ground. A firefighter is positioned at each corner of the ladder and one firefighter is positioned out from the tip on the side of the ladder away from the building (Figure 4.96).

Step 2: The staypole on the side of the ladder away from the building is passed to the firefighter out from the tip (Figure 4.97).

Step 3: This staypole is placed on the ground. The firefighter at the butt on the side closest to the building grasps the remaining staypole and shifts it out from the beam a short distance (Figure 4.98).

Figure 4.96 The five firefighters assume the positions shown.

Figure 4.97 The staypole *farthest* from the building is passed to the firefighter out from the tip.

Figure 4.98 The heelman nearest the building grasps the staypole on that side and moves it aside a short distance. The other staypole is placed on the ground.

Step 4: The beam nearest the building is grasped by the other three firefighters and the ladder is tilted up so that it is resting on the other beam (Figure 4.99).

Step 5: The firefighter at the butt end heels the lower beam and grasps the upper beam. The firefighter out from the tip picks up that staypole and assumes a position to assist with the raise. The firefighter with the staypole next to the butt assumes a stance that will permit pulling on that staypole. The two firefighters at the tip then lift the tip and pivot under it so they are facing the butt one behind the other (Figure 4.100).

Figure 4.99 The beam nearest the building is grasped and the ladder is turned up so that it is resting on the other beam.

Figure 4.100 The firefighter at the butt end heels the ladder. The firefighters at the tip lift the ladder and pivot under the ladder to face the butt. Firefighters on the staypoles assist with raising.

Step 6: The firefighters holding the beam on their shoulders walk down the beam hand-over-hand, raising the ladder toward vertical. As soon as the ladder is at about 45 degrees, the firefighter with the staypole out from the tip takes most of the weight of the raise. The firefighter with the staypole next to the butt end pulls on that staypole (Figure 4.101).

Step 7: When the ladder reaches vertical (Figure 4.102), one of the firefighters on the staypoles shifts to face the outside of the ladder. From this point on the ladder is raised the same as the flat raise.

When there are six firefighters, the firefighter carrying the ladder at the midpoint of the side next to the building shifts the staypole out instead of the firefighter at the butt. Both of the firefighters at the butt end heel the ladder.

Figure 4.101 The firefighters on the staypoles assume more and more of the effort of raising as the ladder approaches vertical.

Figure 4.102 One firefighter on a staypole shifts to face the ladder when it reaches vertical.

4.10 Fill in the blank.

1. When four firefighters perform a pole ladder raise, the ____________________ serves as the heelman.
2. Before the flat pole ladder raises are begun the ladder is placed flat on the ground with the fly ____________________ so that when the ladder is raised the fly will be out.

4.11 True or False.

	True	False
1. During the raising of a pole ladder, the firefighters handling the poles watch the top for signs of movement away from vertical.	☐	☐
2. When pole ladders are in place against a building the staypoles are set to help take some of the weight load on the ladder.	☐	☐
3. The minimum number of firefighters required for a beam raise of a pole ladder is six.	☐	☐

4.12 Check the correct response.

During a beam raise of a pole ladder ______.

- ☐ A. Both staypoles are used to push the ladder upward.
- ☐ B. Both staypoles are used to pull the ladder upward.
- ☐ C. One staypole is used to push the ladder upward; one staypole is used to pull the ladder upward.
- ☐ D. One staypole is used to push the ladder upward; the other staypole is not used.
- ☐ E. Neither staypole is used.

FOLDING LADDER RAISE

Step 1: While folded, one beam projects further than the other. The foot pad of the projecting beam is placed on the floor or ground (Figure 4.103).

Step 2: The ladder is opened by pulling the beams apart until both beams rest firmly on the floor or ground and the rungs are level (Figure 4.104).

Figure 4.103 One foot pad is placed on the ground.

Figure 4.104 The ladder is opened by pulling the beams apart.

CAUTION: Care must be taken to keep from pinching the hands and fingers as this ladder is opened or closed.

Step 3: The brace is locked in place (Figure 4.105), and the tip is placed against the wall or the edge of the scuttle opening.

COMBINATION LADDER RAISES

Extension/A-Frame Combination Ladder

To use this ladder as an extension ladder:

Step 1: The firefighter, while grasping the ladder by both beams, positions the butt where the ladder is to be raised (Figure 4.106).

Step 2: One hand is shifted to grasp a rung of the fly section. The fly is extended by lifting it upward while grasping a rung (Figure 4.107).

Figure 4.106 The firefighter holds the ladder in a vertical position, butt on the ground.

Figure 4.107 One hand is shifted to grasp a rung of the fly section. The fly is extended by lifting upward while grasping the rung.

4.10 1. building wall
2. down

4.11 1. True
2. False
3. False
4. True

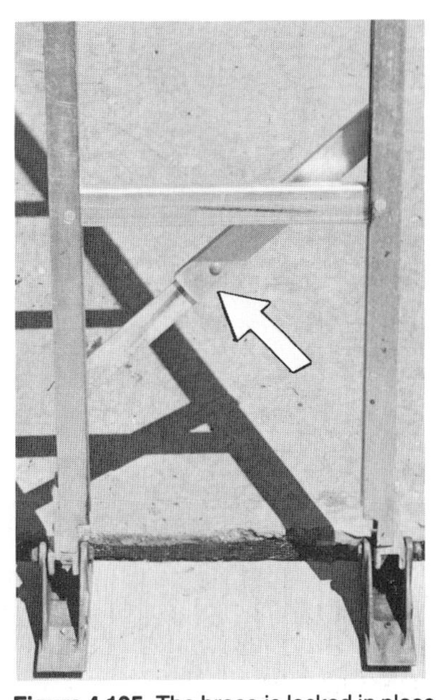

Figure 4.105 The brace is locked in place.

Step 3: The pawls are engaged and then the ladder tip is lowered into place.

To use this ladder as an A-frame:

Step 1: The fly is lowered until the steel rods engage slotted fittings at the top of the bed section (Figure 4.108).

Step 2: The lower part of the fly section is pulled outward away from its nested position inside the rails of the bed section (Figure 4.109).

Step 3: When the tip end of the fly section is the proper distance from the butt of the bed section, a locking device at the top of the A-frame prevents further spreading (Figure 4.110).

Figure 4.108 To use the ladder as an A-frame, the fly is lowered until the steel rods that project from the side of the beams engage the slots at the top of the bed section.

Figure 4.109 The two sections are spread apart to form the A-frame.

Figure 4.110 A locking device at the tip prevents further spreading.

Single/A-Frame Combination Ladder

4.12 C

A second type of combination ladder is a single ladder with each beam hinged at midpoint. It is usually racked and carried folded up.

To use this ladder as a single ladder:

Step 1: The top section is flipped up until it latches to form a single ladder (Figure 4.111).

Step 2: It is then raised like any other single ladder.

To use this ladder as an A-frame ladder:

Step 1: The butt is placed down. The tip half is shifted out to form an A-frame (Figure 4.112).

Step 2: The braces are locked into place (Figure 4.113).

Figure 4.111 The tip half of the ladder is swung up until it latches to form a single ladder.

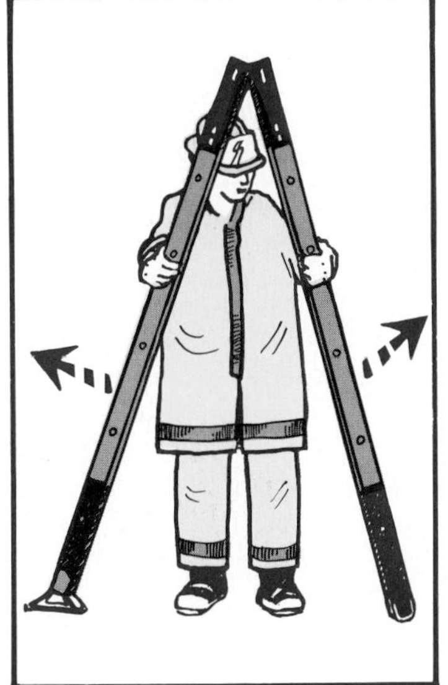

Figure 4.112 The two halves are spread to form an A-frame.

Figure 4.113 The sections are held in place by the braces shown in this illustration.

Telescoping Beam: Single/A-Frame Combination Ladder

A third type of combination ladder has two telescoping sections with the tips joined by a hinge. The rungs of the lower half of each section are mounted on the top of the beam rail so that the other part can slide inside the beam rail. It is carried folded up with the telescoping sections retracted.

To use this ladder as a single ladder:

Step 1: The ladder is placed on the ground; one side is rotated 180 degrees to form a single ladder. It is locked in place by a latching device that is incorporated into the hinge.

It is then raised the same as a single ladder (Figure 4.114). Each half has telescoping beams which allow length adjustment. The adjustment is secured by a pin through the beam. Figure 4.115 shows the telescoping sections fully extended.

Figure 4.114 In this view this ladder is used as a "single" ladder. Both halves are fully retracted so that it is at its minimum length. Note that the latching device which holds it in place as a "single" ladder is a part of the hinge.

Figure 4.115 In this view both sections are fully extended so that the "single" ladder is at maximum length. The arrows indicate the locking devices which hold the telescoping sections in position.

Step 2: It is raised the same as any single ladder.

To use this ladder as an A-frame ladder:

Step 1: The butts are placed on the ground and the two halves are spread to attain the A-frame configuration. A latch in the hinge prevents further spreading (Figure 4.116).

Step 2: The height of either or both sides can be adjusted, as in Figures 4.117 and 4.118. To do this, the pin on the beam is pulled outward, the section is pulled or pushed upward, and the pin is reinserted when the desired height is attained.

Figure 4.116 Here the ladder is being used as an A-frame ladder with the telescoping sections fully retracted. The latching device in the hinge keeps the two sections from spreading apart.

Figure 4.117 One side of the A-frame is extended more than the other to compensate for different elevations as would be the case with stairwell operations.

Figure 4.118 Here both sections are fully telescoped to form a taller A-frame ladder.

Extending A-Frame Combination Ladder

This is not a true combination ladder because it is not normally used as an extension ladder; the tip is not placed against the wall. The extension is freestanding while supported by the A-frame.

To use this ladder:

Step 1: The ladder butts are placed on the ground (Figure 4.119).

Step 2: The firefighter shifts to face the ladder, spreads the two A-frame sections and sets the brace latch (Figure 4.120).

Figure 4.119 The ladder is rested on the ground, hinge end up.

Figure 4.120 After the two outside sections are spread apart, the locking device is set to fix their position.

Step 3: The firefighter then shifts to the side, reaches in between the A-frame, grasps the fly section, and pushes upward to extend it (Figure 4.121).

Step 4: When the desired height is reached, the pawls are engaged (Figure 4.122).

NOTE: These pawls are different than normal. The ladder is shown fully extended in Figure 4.123.

Figure 4.121 The fly is extended by grasping one of its rungs and pushing straight upward.

Figure 4.122 The pawls are engaged at the desired height. This pawl is different from those normally used. It grasps the rung from underneath — the rung is cradled into a curvature of the hook. A counterweight on the other end of the hook keeps it in position. The finger is on top. It kicks the pawl back when the ladder is lowered.

Figure 4.123 The extending A-frame combination ladder fully extended.

4.13 Fill in the blank.
Folding ladders are opened by pulling the ______________.

4.14 True or False.

	True	False
1. All combination ladders can be used as an A-frame ladder.	☐	☐
2. None of the combination ladders are adjustable for use on stairways.	☐	☐
3. The extending A-frame combination ladder is used with the tip placed against the wall.	☐	☐
4. There is no heelman for pompier ladder raises.	☐	☐

POMPIER LADDER RAISE

The pompier ladder is mainly used for training, but a few fire departments still use them in fireground operations. The steps for raising described in the following paragraphs are for drill tower use.

Step 1: The butt is placed against the building.

Step 2: The firefighter raises the ladder to vertical by walking toward the building while gripping the beam hand-over-hand (Figure 4.124).

NOTE: The gooseneck hook is turned away from the building during this procedure.

Step 3: When the ladder reaches vertical, the gooseneck hook is turned into the window opening (Figure 4.125) and set over the windowsill (Figure 4.126).

Step 4: The firefighter tests the security of the hook by pulling down on the ladder before beginning the climb (Figure 4.127).

Figure 4.124 The pompier ladder butt is placed against the building and it is raised to vertical by walking hand-over-hand down the beam.

Figure 4.125 The hook is turned into the window opening.

Figure 4.126 The hook is set over the windowsill.

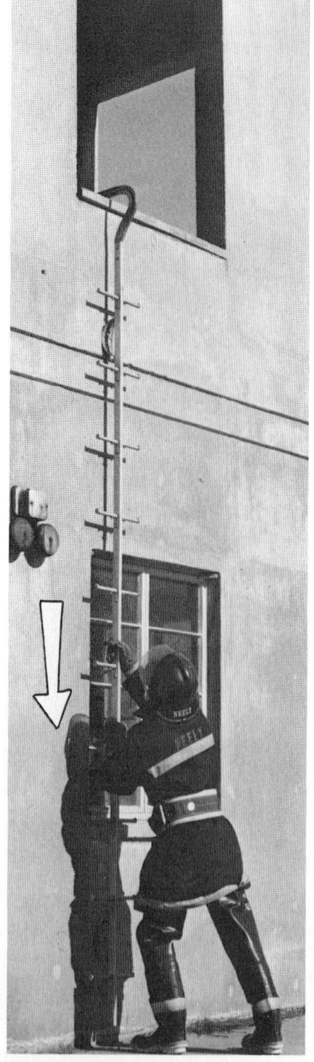

Figure 4.127 After the hook is set, the firefighter pulls downward to be sure that it is securely in place.

4.15 Fill in the blank.
During the pompier ladder raise, the gooseneck hook faces ________________ the building.

4.13 Beams apart

4.14 1. True
2. False
3. False
4. True

SPECIAL RAISES

Dome (Auditorium) Raise

The dome raise is a method of using a ground ladder to reach places where there is no means of supporting the tip of the ladder and where the height requires a ladder longer than an A-frame combination ladder or single ladder supported by firefighters will reach. This situation is found in some public buildings, auditoriums, arenas, gymnasiums, skating rinks, etc,. where high ceilings are required. Light fixtures suspended from high ceilings can also be reached with this raise.

Four guy lines are rigged from the tip and are held by firefighters to provide the necessary support for the ladder. Two 125-foot (38 m) lifelines are used and a minimum of six firefighters are required: eight are preferred.

The procedure for six firefighters to raise a ladder in this manner is described as follows:

Step 1: The ladder is placed flat on the floor, fly up. The top fly section is pulled out slightly to facilitate attaching the guy ropes (Figure 4.128).

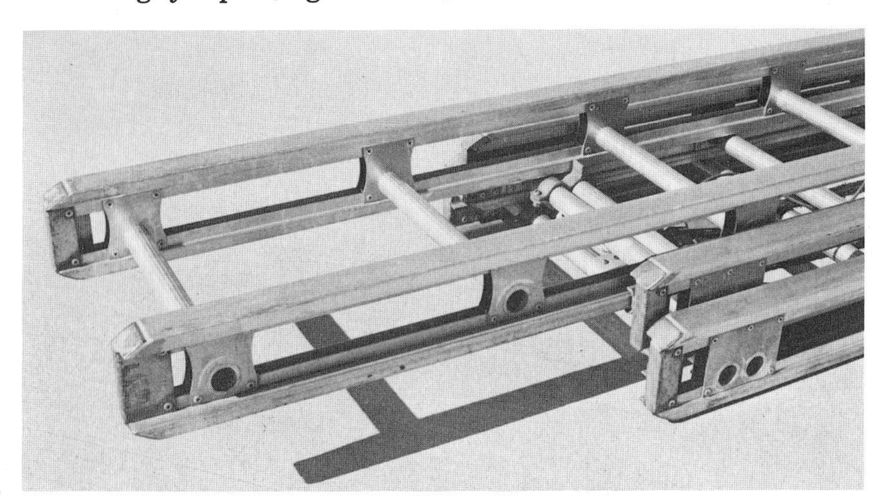

Figure 4.128 The top fly is pulled out so that two rungs are clear of the rest of the ladder.

Step 2: The ropes are strung out, one from the tip of each beam, with the rope midpoint next to the ladder and the free ends away from it.

Step 3: The rope loop thus formed is passed over the top of the beam, down between the two rungs, then under the beam to bring it back to the outside of the ladder. The loop end is then passed over the rope to bring it toward the tip

(Figure 4.129). The loop is placed over the tip end of the beam (Figure 4.130). The completed rope arrangement is shown in Figure 4.131.

Figure 4.129 The loop of rope is passed around the beam between the first and second rungs.

Figure 4.130 The loop is then passed over the tip end of the beam.

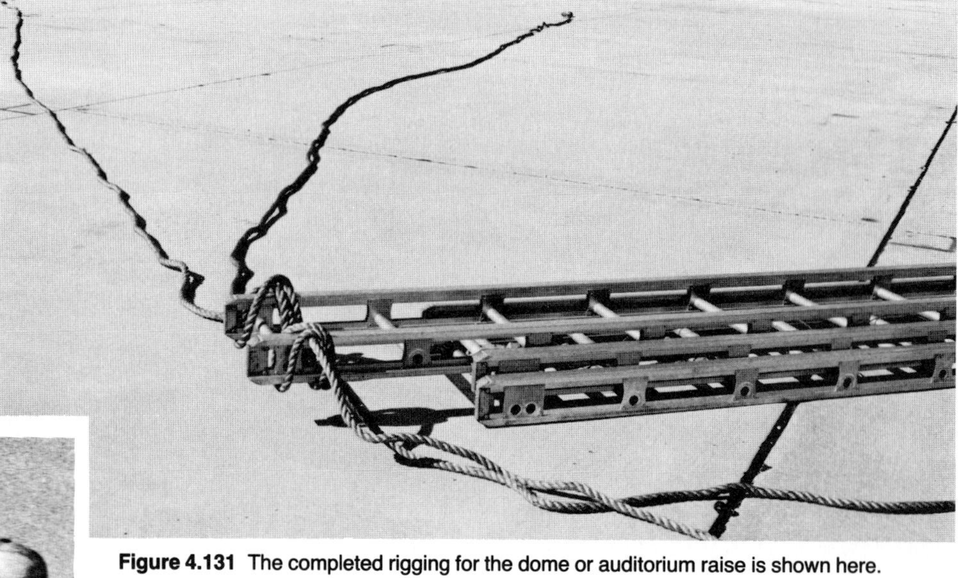

Figure 4.131 The completed rigging for the dome or auditorium raise is shown here.

Figure 4.132 The guy ropes are anchored by passing them around the rump as shown above.

Step 4: Four firefighters are positioned, one near the end of each guy line. Each passes the rope behind the body at the rump so that body weight can be used to support the weight of the ladder on the guy lines (Figure 4.132).

Step 5: The ladder is raised to vertical by normal means. The firefighters on the guy lines adjust their positions as the ladder is brought to vertical.

NOTE: If a pole ladder is being used, the guy lines must be outside the poles before the ladder is raised to vertical.

4.15 away from

Step 6: The fly is extended while the firefighters on the guy lines feed out sufficient rope and shift position as necessary to maintain loose tension on the top of the ladder (Figure 4.133).

Step 7: When the ladder is extended to the desired height and the pawls have been engaged, the firefighters on the guy lines take up the slack on the ropes and stand with their feet braced so that any sudden stresses put on the rope they are holding will not cause them to move suddenly and lose tension on the rope (Figure 4.134).

Figure 4.133 It will be necessary for firefighters to slack off on the guy lines as the fly is extended.

Figure 4.134 The completed dome or auditorium raise. Firefighters on the guy lines take up the slack on the ropes and stand with their feet braced.

Hotel or Factory Raise

This raise is designed to allow several persons to evacuate from different levels of a building simultaneously; it serves as an emergency fire escape. The pole ladder is usually used because it will reach more floors at once and the staypoles are used to provide extra stability.

It is necessary that windows be in line with each other on successive floors. The ladder is raised at a steep angle close to the side of the building and in line with the windows. The steps are detailed below.

Step 1: The ladder is placed on the ground at right angles to the building with the butt 2 feet (.6 m) to 3 feet (.9 m) from the side of the building and in line with the windows.

Step 2: While the ladder is flat on the ground, the fly is pulled out one rung's distance and one guy rope is attached to each beam (Figures 4.135 and 4.136).

Figure 4.135 The top fly is pulled out so that two rungs are clear of the rest of the ladder.

Figure 4.136 A guy line is tied to each beam tip as in the three illustrations above.

Step 3: The ladder is raised following normal procedures. The four-firefighter pole ladder raise is illustrated. Six firefighters are required because two are needed to operate the guy lines. They assist in raising the ladder by pulling on the ropes as the ladder comes to vertical (Figure 4.137).

Step 4: When the ladder is vertical, the firefighters holding the staypoles both remain at 90 degrees to the building, while the guy ropes provide lateral stability (Figure 4.138).

Step 5: After the ladder has been lowered into position, both the firefighters holding the staypoles remain in position to

Figure 4.137 A four-firefighter pole ladder raise is used. The firefighters holding the ropes help raise it once the ladder gets about halfway up.

Figure 4.138 When the ladder reaches vertical, both firefighters holding the staypoles REMAIN in a position at 90 degrees to the building. The firefighters on the guy lines provide lateral stability.

keep the ladder against the building — THE STAYPOLES ARE NOT SET — the firefighters holding the staypoles remain in this position during the whole operation. The firefighters on the guy ropes also assist in keeping the tip of the ladder against the side of the building (Figure 4.139).

Figure 4.139 The firefighters on the staypoles remain in this position after the tip is lowered in and throughout the rescue operation. The firefighters on the guy lines assist in keeping the ladder stable against the wall.

4.16 True or False.

	True	False
1. The dome raise is used in areas with high ceilings when the tip cannot be supported.	☐	☐
2. A minimum of five firefighters are required for the dome raise.	☐	☐
3. When a hotel or factory raise is used the window alignment is not important.	☐	☐
4. The hotel or factory raise permits multiple person evacuation; more than one evacuee is on the ladder at one time.	☐	☐
5. During the hotel or factory raise, the staypoles are set just as for other pole ladder raises.	☐	☐

RAISING LADDERS UNDER OBSTRUCTIONS

If hanging signs, overhead wires, or tree limbs prevent a normal ladder raise, it is still possible to raise a ladder under these obstructions. The length of the ladder and the number of available firefighters will determine which of the following raises will be used.

Two-Firefighter Single or Roof Ladder Obstructed Raise

Step 1: The ladder is placed on the ground at 90 degrees to the building with the tip toward the wall but approximately 3 feet (.9 m) from it (Figure 4.140).

Step 2: One firefighter faces the building and kneels beside the ladder at the tip. The other firefighter heels the ladder (Figure 4.141).

Figure 4.140 The ladder is placed flat on the ground at 90 degrees to the building with the tip forward and about three feet (.9 m) from the wall.

Figure 4.141 The two firefighters are positioned as shown.

Step 3: The firefighter at the tip lifts it and pivots underneath, grasping a beam with each hand. When the pivot is complete, the arms are fully extended (Figure 4.142).

Step 4: The heelman shifts to one side of the ladder, crouches enough to grasp the second rung from the butt (Figure 4.143).

Figure 4.142 The tip is lifted. The firefighter pivots underneath and grasps a beam with each hand. The arms are fully extended.

Figure 4.143 The heelman shifts position to crouch beside the ladder and grasp the second rung.

Step 5: The firefighter at the tip remains stationary while passing the beams upward with the hands as the other firefighter walks forward, pulling the butt end along the ground. A slight downward pressure is maintained to keep the butt from kicking up (Figure 4.144).

Step 6: The firefighter at the butt end stops pushing the ladder upward when it reaches the proper angle of inclination (Figure 4.145).

4.16 1. True
2. False
3. False
4. True
5. False

Figure 4.144 The firefighter at the tip remains stationary while passing the beams upward with the hands. Momentum is provided by the other firefighter who walks forward pulling the butt end along the ground.

Figure 4.145 The ladder is moved forward and up until the correct angle of inclination is attained.

Step 7: The ladder is then set into the building (Figure 4.146).

Figure 4.146 When the ladder is about at the correct location, the tip is set against the building and adjustments are made as necessary.

Three-Firefighter Single or Roof Ladder Obstructed Raise

When three firefighters perform this raise, the same procedures are followed as for the two-firefighter single or roof ladder obstructed raise except that there are two firefighters at the tip raising it (Figure 4.147).

Figure 4.147 Where there are three firefighters, two are located at the tip as shown here.

Alternate Three-Firefighter Single or Roof Ladder Obstructed Raise

Step 1: The ladder is placed flat on the ground, tip toward the building. One firefighter stands at the butt, facing it. The other two firefighters stand, one beside each beam, facing the butt at a point four rungs from it.

Step 2: The two firefighters standing beside the beams kneel and grasp the same rung (Figure 4.148). They stand and simultaneously lift the ladder, pivot under the beams, and place them on the shoulder. Their free hands are placed on the beam (Figure 4.149).

Figure 4.148 When the alternate method is used, two firefighters stand four rungs up from the butt facing it.

Figure 4.149 The ladder is lifted, the firefighters pivot under, and place the beams on their shoulders.

Step 3: The third firefighter reaches up and grasps the butt end, pulling it downward (Figure 4.150).

Step 4: As the firefighter at the heel continues to pull downward, the firefighters on the beams allow their hands and shoulders to act as a fulcrum.

Step 5: When the butt is waist high, all step forward (Figure 4.151).

Figure 4.150 The third firefighter reaches up and grasps the butt end and pulls it downward.

Figure 4.151 When the butt is waist high, all step forward. NOTE: Firefighter kneeling due to obstruction.

Step 6: The heelman continues to push downward on the butt. This brings the tip upward as the ladder approaches the building (Figure 4.152).

Step 7: When the butt reaches the proper distance from the building, all halt and the heelman places the butt on the ground. The other two firefighters place the tip against the building (Figure 4.153).

Figure 4.152 The butt is pulled downward bringing the tip upward as the building is approached.

Figure 4.153 The butt is placed on the ground and the other two firefighters set the tip against the building.

Four-Firefighter Obstructed Raise of an Extension Ladder

The same procedure is followed as for the two-firefighter single and roof ladder obstructed raise except that there are two firefighters at the tip and two adjacent to the beam at the butt end (Figure 4.154), and the ladder is brought to vertical, pivoted, and the fly extended before being lowered into the building.

Figure 4.154 When there are four firefighters, the evolution is the same except that there are two firefighters positioned beside the butt end.

4.17 Fill in the blanks.

1. All of the variations of the obstructed raises require that the ladder be placed on the ground ________________ to the building with the ________________ toward the building.
2. During the two-firefighter single or roof ladder obstructed raise, the firefighter that walks forward with the ladder maintains a ________________ pressure to prevent the ________________ from kicking up.

SECURING THE LADDER

Heeling (Also called Footing or Butting)

NFPA 1932 requires, and good common sense dictates, that ground ladders be heeled or otherwise secured whenever firefighters are climbing or working from them. There are several methods of properly heeling a ladder. One method, as illustrated in Figure 4.155, places the firefighter beneath the ladder and leaves the climbing side unencumbered. The steps for this method are as follows:

Step 1: The firefighter, wearing full protective clothing, stands on the side of the ladder towards the building and grasps the ladder beams about eye level.

CAUTION: Do not grasp rungs.

Step 2: The body is maintained in an erect posture with the feet spread wide apart to give the firefighter a solid footing. A small amount of tension is maintained on the beams by pulling gently on them.

CAUTION: The pressure applied to heel a ladder by this method should not stress the ladder. If the firefighter pulls hard or hangs from the underside of the ladder, the weight applied is considered a part of the weight loading of the ladder. For this reason, only enough pressure should be applied to keep the ladder from slipping.

CAUTION: It is important that the firefighter heeling the ladder by this method not look upward, as doing so makes the firefighter very susceptible to eye and other face injuries.

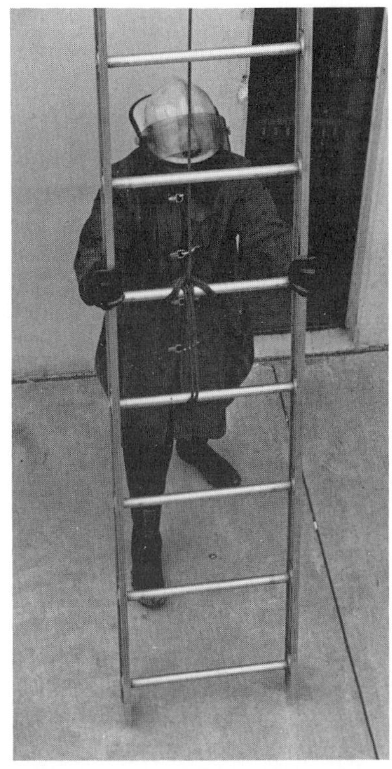

Figure 4.155 Heeling from the backside of the ladder. It is important that the firefighter not look upward during the performance of this task.

Another method of heeling a ladder is illustrated in Figure 4.156. This method does not place stress on the ladder but the

Figure 4.156 Heeling from in front of the ladder.

firefighter heeling the ladder may encumber travel on and off the ladder. The steps for this method are

Step 1: The firefighter stands in line with and facing the ladder and places the toes against the butt spur.

Step 2: The ladder is then grasped by either the beams or the rungs, whichever is the most convenient.

4.18 Fill in the blanks.
When a firefighter heels a ladder from underneath, the ______________ are grasped and a ______________ is maintained by gently pulling on them.

4.19 True or False.

	True	False
1. A firefighter heeling a ladder from underneath should watch the tip for any sign that it is pulling away from the building.	☐	☐
2. Heeling a ladder using the technique of facing the ladder with the toes against the butt spurs does not weight load the ladder.	☐	☐

Tying the Ladder In

Whenever possible, ladders should be tied in at both the top and bottom. This will prevent the tip from pulling away from the side of the building and the butt from slipping. A rope hose tool can be used as in Figures 4.157 and 4.158, a safety strap as in Fig-

Figure 4.157 The two illustrations show the tip of a ladder secured to a cornice using a rope hose tool.

4.17 1. at 90 degrees, tip
2. slight, butt

Figure 4.158 Here a rope hose tool has been used to secure the tip of the ladder to a pipe projecting through the roof.

ure 4.159, or a short length of rope as in Figure 4.160. Tying in also frees personnel who would otherwise be holding the ladder in place.

Figure 4.159 In this instance, a safety strap has been used to secure the tip to a pike pole handle that was placed across a window opening.

Figure 4.160 The butt of this ladder has been secured by tying a short length of rope from a rung to a garden hose fitting on the outside of a building; however, any available object of sufficient strength will do.

Tying the Halyard

Before climbing the ladder, any excess halyard rope should be tied to a rung as a safety measure to prevent the fly from slipping and to keep anyone from tripping over it. The method for making the tie will depend on whether the halyard is open ended or forms a continuous loop.

To make the tie with an open ended halyard:

Step 1: Take up the excess rope by wrapping it around a pair of convenient rungs and pulling it taut (Figure 4.161).

Step 2: Pass free end over top of the two rungs and behind the taut part of the rope. Grasp with the other hand (Figure 4.162). Bring the end back through between rungs and grasp the free end with the other hand (Figure 4.163). Again pass the free end over the top of the two rungs while keeping tension on the line with the other hand (Figure 4.164). Bring the free end through the newly formed loop, then pull it towards you to snug up the clove hitch (Figure 4.165).

Figure 4.161 Excess halyard rope is wrapped around two rungs.

Figure 4.162 Pass free end over the top rung and behind the taut part of the rope. Grasp end with other hand.

Figure 4.163 Bring end back through, grasp free end with other hand.

Figure 4.164 Again pass free end over the top rung. Keep tension on line with other hand.

4.18 beams, slight tension

4.19 1. False
2. True

Figure 4.165 Bring free end through the newly formed loop, then pull it towards you.

To make a closed loop tie:

Step 1: Grasp the slack rope below the two rungs that you want to use. Wrap the slack rope around the two rungs and pull it to you until the halyard is tight (Figure 4.166).

Figure 4.166 The loose rope is grasped below the two rungs where it will be tied.

Step 2: Double up the slack to obtain a loop end (Figure 4.167). Pass the newly formed loop end over the top of the two rungs and go behind the taut part of the halyard (Figure 4.168).

Step 3: Grasp the loop end and pull it toward you (Figure 4.169); hold this line taut with one hand while passing the loop end over the same rung as before (Figure 4.170).

Figure 4.167 The slack is doubled up to obtain a loop.

Figure 4.168 The loop is passed over the top of the two rungs and behind the taut part of the halyard.

Figure 4.169 The loop end is pulled toward the person doing the tying.

Figure 4.170 The line is held taut while the loop end is passed over the upper rung again.

Step 4: Grasp the loop end and pull it through the new large loop (Figure 4.171). Continue pulling until tight to form the completed clove hitch (Figure 4.172).

Figure 4.171 The loop end is pulled through the new large loop.

Figure 4.172 The rope is pulled tight to form the completed clove hitch.

4.20 True or False.

	True	False
1. It is sufficient to tie in a ladder either at the top or bottom.	☐	☐
2. An extra benefit of tying in a ladder is that it frees personnel who would otherwise be holding the ladder in place.	☐	☐
3. Continuous halyards are not tied off.	☐	☐

CLIMBING

Ladder Climbing Skills

CHECKING FOR PROPER ANGLE OF INCLINATION

A key factor for proper climbing is having the ladder at the proper angle of inclination. A quick and easy way to check for this just before climbing is detailed as follows:

Step 1: The firefighter stands erect facing the ladder with the toes touching the butt spur.

Step 2: The arms are held straight out from the body as if to grip a rung.

- If the palms fall naturally on the rung nearest shoulder height, then the angle of inclination is correct (Figure 4.173).
- If the hands will not reach the rung, then the butt needs to be moved in toward the building (Figure 4.174).
- If the hands extend beyond the rung, the butt needs to be brought out away from the building (Figure 4.175).

Figure 4.173 The palms fall naturally upon the rung when the angle of inclination is correct.

Figure 4.174 When the hands will not reach the rung while standing erect, as shown here, the butt is too far out. It should be moved in.

Figure 4.175 When the hands project beyond the rung, the butt is too close to the building. It should be moved out.

4.20 1. False
2. True
3. False

CLIMBING TECHNIQUES

The climb may be started when the ladder is properly secured. The movements made to climb should be smooth and rhythmical. Climbing in this manner will minimize ladder bounce and sway. This is accomplished by using the bending action of the knee to ease the climber's weight onto each rung and by not making any sudden or jerky movements.

The body is kept erect by keeping the arms straight during the climb (Figure 4.176). The climber should avoid reaching much above shoulder level to grasp a rung because this action will bring the body inward towards the ladder. This erect stance is needed for balance and to permit unencumbered knee movement during the climb (Figure 4.177).

Figure 4.176 Climbing is easiest when the body is erect and the arms are straight.

Figure 4.177 When the proper stance is used, knee movement is unencumbered.

The hands should grasp the rungs with the palms down and the thumb on the underside of the rung (Figure 4.178). Some find it natural to grasp every rung with alternate hands; others prefer

Figure 4.178 The proper method of grasping a rung is shown.

to grasp every other rung while climbing. Each individual should try both methods and select the one that feels most natural. Upward progress should not be the result of pull of the arm muscles; it should be caused by action of the leg muscles.

CAUTION: When reaching the top of the bed section of an extension ladder, the firefighter should pause before climbing onto the fly section, making a visual check of the pawls to be sure that they have engaged properly (Figure 4.179). The same procedure is followed before climbing onto additional fly sections, if there are any.

Figure 4.179 The pawls are checked before climbing onto the fly section to be sure that they are properly engaged.

Practice climbing should be done slowly to develop form rather than speed. Speed will be developed as the proper technique is mastered. Too much speed results in lack of body control, and quick movements cause the ladder to bounce and sway.

Many times during fire fighting a firefighter is required to carry equipment up and down a ladder. This procedure interrupts the natural climb, either by the added weight on the shoulder or by the use of one hand to hold the item. If the item is to be carried in one hand, it is desirable to slide the free hand along the underside of the beam while making the climb. This method permits constant contact with the ladder as shown in Figure 4.180. When time permits, a rope can be used to hoist tools and equipment.

Figure 4.180 These two illustrations show the proper procedure for grasping the ladder when climbing with tools or appliances. Note that the hand grasps the beam instead of the rungs.

4.21 Fill in the blanks.
Climbing movements should be ____________________ and ____________________. This is accomplished by using the ____________________ to ease the climber's weight onto each rung.

4.22 Check the correct response.
The body is kept ____________________ when climbing.

- ☐ A. Bent forward towards the ladder
- ☐ B. Erect
- ☐ C. Bent outward beyond vertical
- ☐ D. In whatever posture is comfortable

4.23 True or False.

	True	False
1. If the halyard is tied, it is not necessary to check the pawls before climbing onto the fly section.	☐	☐
2. When equipment is carried in one hand while climbing a ladder, the other hand is used to maintain a firm grip on a convenient rung.	☐	☐
3. When checking for proper climbing angle; if the palms fall beyond the rung, the ladder butt has been set too close to the building and it needs to be moved outward.	☐	☐

WORKING FROM LADDERS

Firefighters are routinely required to perform tasks while standing on ladders. These tasks require that the hands be free, so it is necessary to use some means to prevent the firefighter from falling off the ladder. There are two recommended methods:

- Safety Belt Method
- Leg Locking Method

Safety Belt Method

There are a number of types of safety belts; those which provide some distance between the hook and the belt are preferred because they allow the wearer to move more freely and stand erect (Figure 4.181). These belts are commonly called ladder belts because of the extension feature. Specially made truckman's belts are also used and rope hose tools can be used (Figure 4.182). The steps for using a safety belt are as follows:

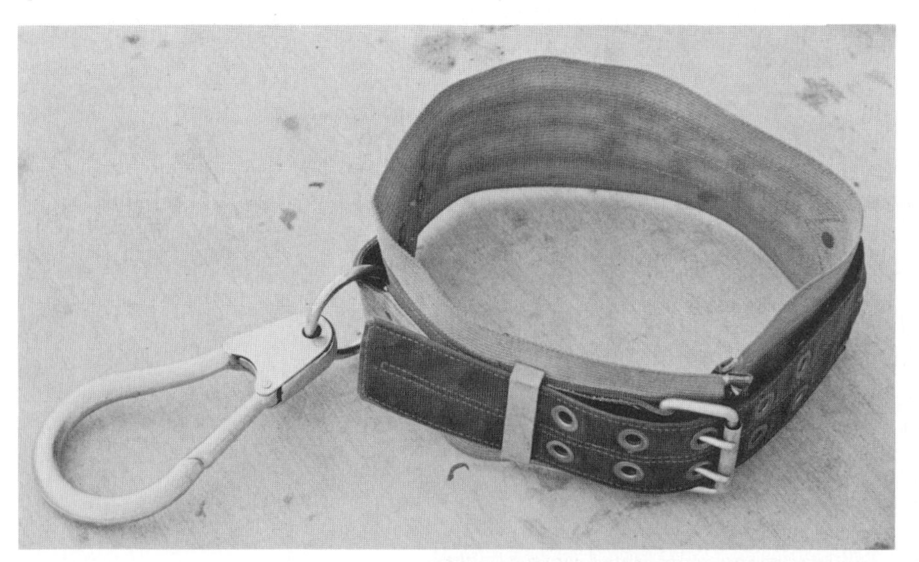

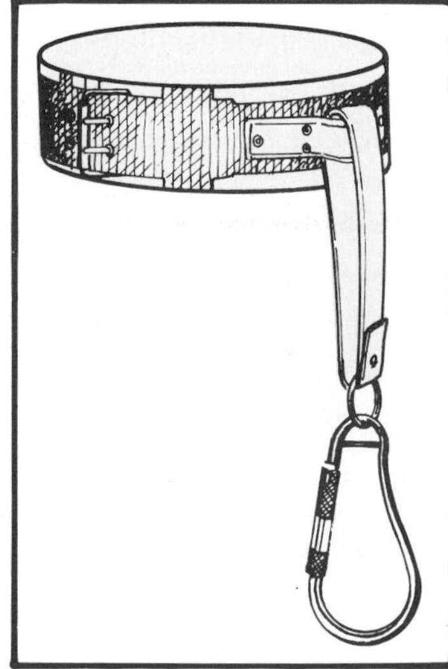

Figure 4.181 Two models of the same type safety belt are illustrated. The first has the hook attached directly to the belt. The second has an extension between the belt and the hook.

4.21 smooth, rhythmical, bending action of the knee

4.22 B

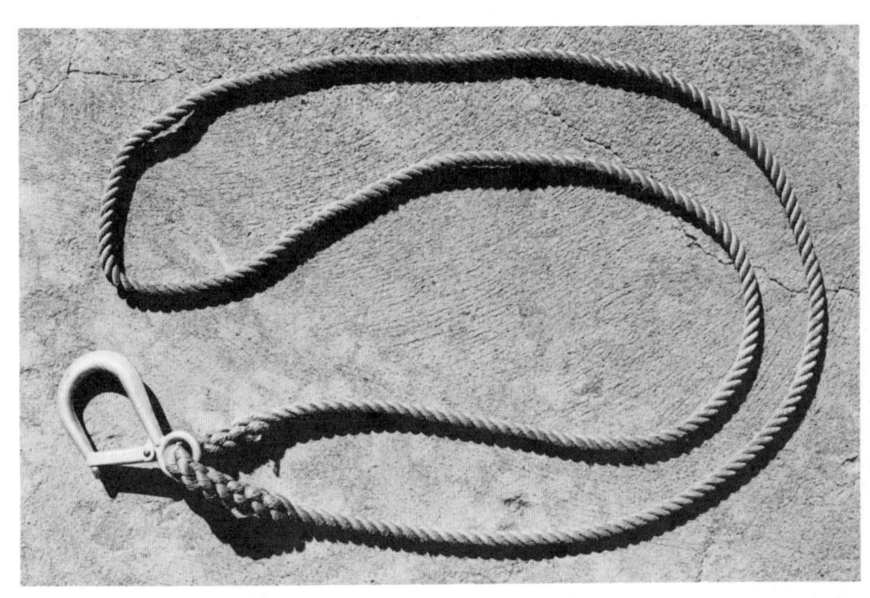

Figure 4.182 A rope hose tool.

Step 1: The safety belt, strap, or rope hose tool is placed snugly around the waist prior to climbing.

Step 2: When the desired height is reached, the hook is placed over a rung or the strap is placed around the ladder and clipped to a ring depending on the type used (Figure 4.183).

Safety belt.

Rope hose tool.

Truckman's belt.

Diaper Sling with hook.

Figure 4.183 Several methods of securing the firefighter on a ladder are shown in this series of illustrations.

Leg Lock Method

When a safety belt, strap, or rope hose tool is not being worn and it is necessary to work from a ground ladder, the proper procedure is for the firefighter to climb to the point where the work will be performed and leg lock in using the leg opposite from which the work will be performed. To apply a leg lock:

Step 1: The firefighter stands with both feet on the same rung and grasps a rung at a point below shoulder height (Figure 4.184).

Step 2: The foot of the leg that will be locked in is lifted up and inserted between the second and third rungs higher up the ladder. The other leg is kept straight. There is a tendency for the firefighter to stand too close to the ladder, making it difficult to get the leg up into the space between the body and the ladder. This awkwardness can be avoided by moving the hip back while keeping the arms and legs straight (Figure 4.185).

Figure 4.184 The firefighter stands with both feet on the same rung and grasps a rung that is below shoulder height.

Figure 4.185 The hip is moved back while keeping the arms and legs straight. Then one foot is lifted and inserted between the second and third rungs higher up the ladder.

Step 3: After the foot is inserted between the rungs, the body is brought in close to the ladder. This allows most of the leg to pass between the two rungs and hang between the ladder and the building (Figure 4.186).

Step 4: The foot is then brought back through the space between the two rungs immediately below where the back of the leg is resting (Figure 4.187).

The completed leg lock is shown in Figure 4.188. Some firefighters prefer to complete the leg lock by placing the toe over the beam (Figure 4.189). An extremely long or short legged person may need to alter this procedure to make the leg lock comfortable.

4.23 1. False
2. False
3. True

Figure 4.186 The leg is brought between the two rungs by moving the body close to the ladder.

Figure 4.187 The foot is brought back through between the rungs as shown.

Figure 4.188 The completed leg lock is shown.

Figure 4.189 Some firefighters additionally place the toe over the outside of the beam.

Figure 4.190 A leg lock facing away from the ladder.

If it is necessary to face away from the ladder, the same procedure is followed. Then, the body is turned and the leg that is locked in is shifted to the opposite beam while the other foot is shifted to the next lower rung, where it is placed with the toe pointing outward (Figure 4.190).

4.24 Fill in the blanks.

1. The two recommended methods of preventing a firefighter from falling off a ladder while performing tasks from it are the ______________________________ method and the ____________________ method.
2. When a leg lock is used, the leg lock is made with the leg ____ ____________________ the work will be performed.

Climbing to Place a Roof Ladder

There are a number of ways to get a roof ladder in place on a sloped roof. Two methods are described for one firefighter to accomplish this task and three methods are described for two firefighters to do it.

ONE-FIREFIGHTER METHOD

One-Firefighter: Low-Shoulder Carry

As the name implies, the low-shoulder method of carrying is utilized. It is particularly well suited for instances where there is a heelman for the ladder to be climbed because it allows the firefighter carrying the ladder to go directly from the carry to climbing the other ladder; the roof ladder does not need to be placed on the ground or shifted from another carry method.

Step 1: If the hooks are not already open, they are opened when the firefighter gets to the other ladder (Figure 4.191).

Figure 4.191 The hooks are opened prior to climbing with the roof ladder.

NOTE: If assistance is not available to open the hooks, it will be necessary to set the ladder down, using the reverse procedure for picking it up from flat on the ground. After setting the ladder down, the firefighter proceeds to the tip, squats and lifts it, and opens the hooks (Figure 4.192). The tip is then placed back down. The ladder is then picked up to resume the low-shoulder carry.

Step 2: As the firefighter carrying the roof ladder mounts the other ladder, the body is turned slightly outward to bring the roof ladder out away from the side of the building (Figure 4.193).

Step 3: The hand not holding the roof ladder is used to grasp the opposite beam of the ladder being climbed. As the firefighter climbs, the hand is slid up the beam (Figure 4.194).

CAUTION: As the firefighter climbs, it is necessary to check for electrical wires, eave overhangs, or other obstructions.

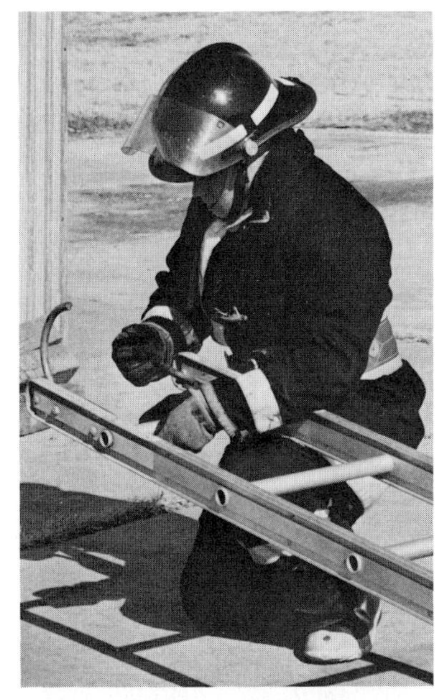

Figure 4.192 The tip is lifted. The firefighter grasps a rung with one hand while opening the hooks with the other hand.

Figure 4.193 The body is turned slightly to bring the roof ladder tip end out away from the side of the building.

Figure 4.194 The free hand grasps the beam while climbing.

Figure 4.195 The ladder is tilted to vertical when the tip reaches the eaves.

Step 4: As the firefighter approaches the tip of the ladder being climbed, the roof ladder is swung toward vertical to clear eaves, etc. (Figure 4.195).

Step 5: Upon reaching the top of the ladder, the firefighter either leg locks in, or uses a safety belt as in Figure 4.196.

Figure 4.196 The firefighter is secured at the top of the ladder prior to placing the ladder upon the sloping roof.

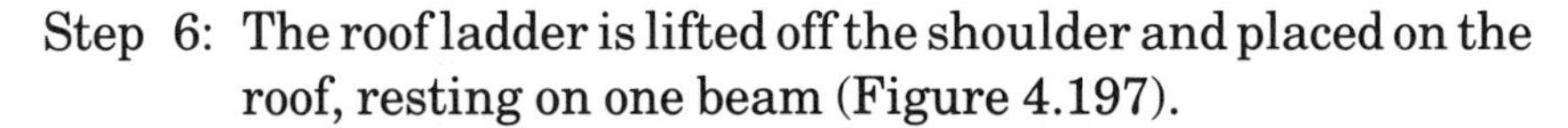

Step 6: The roof ladder is lifted off the shoulder and placed on the roof, resting on one beam (Figure 4.197).

Figure 4.197 The roof ladder is placed on the roof resting on one beam.

Step 7: The ladder is slid up the roof on the beam until the balance point is reached. Then it is turned flat, hooks down, and slid the remaining distance to the roof peak on the hooks (Figure 4.198).

Step 8: The hooks will drop over the peak and then the firefighter should pull back on the roof ladder to snug it in (Figure 4.199).

4.24 1. safety belt, leg lock
2. opposite from which

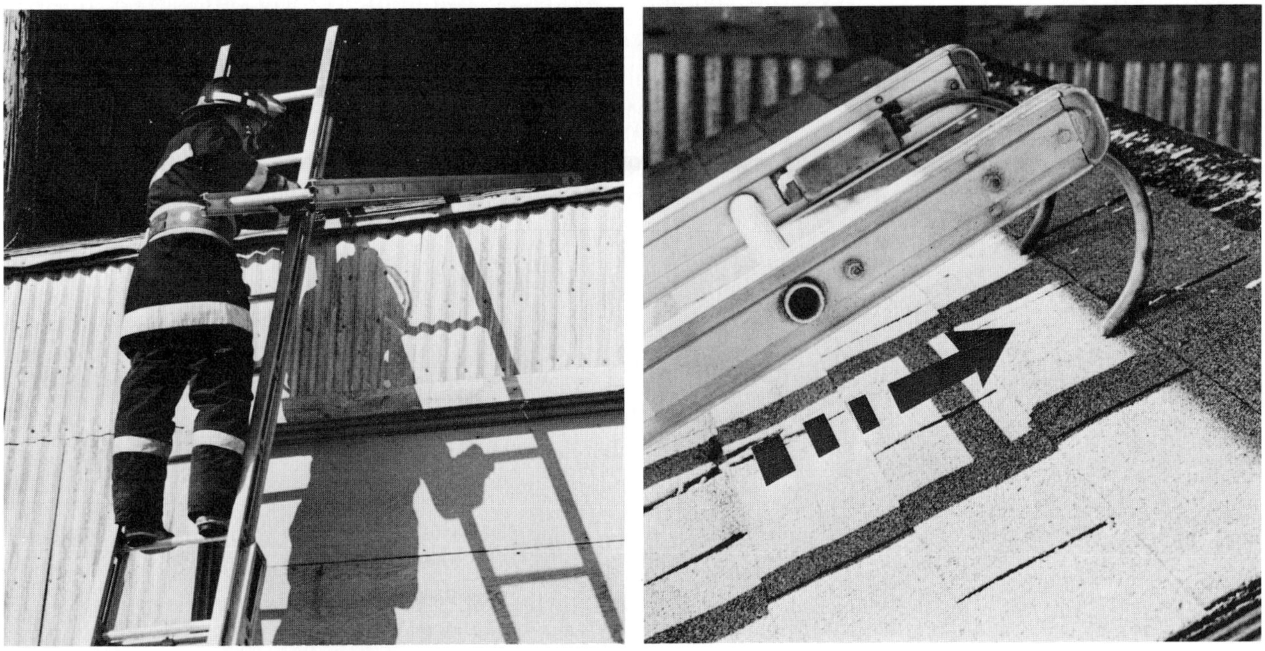

Figure 4.198 The ladder is slid up the roof on the beam until the balance point is reached. Then it is turned flat, hooks down, so that it can be slid the rest of the way up the roof on the hooks.

Figure 4.199 After the hooks drop over the roof peak, the firefighter pulls downward to snug it in place.

One-Firefighter: Modified Shoulder Carry

When this method of getting a roof ladder to the roof is used, any carry may be used to get the roof ladder from the apparatus to the point where the other ladder has been raised. The procedures from this point are

Step 1: The ladder is placed on the ground and the hooks are opened.

Step 2: The ladder is then picked up and placed vertical against one beam of the other ladder (Figure 4.200).

Step 3: The firefighter climbs the other ladder to a point one or two rungs above the midpoint of the roof ladder. The firefighter pauses and slips the near arm between two rungs. The higher of the two rungs is placed on the shoulder (Figure 4.201).

Step 4: The firefighter then proceeds as in steps 5 through 8 in the preceding section detailing the low-shoulder carry.

Figure 4.200 The roof ladder is stood against one beam of the other ladder.

Figure 4.201 The firefighter climbs to where the arm can be inserted between the two roof ladder rungs just above its midpoint and it is brought onto the shoulder.

TWO-FIREFIGHTER METHODS

It is much easier to climb another ladder and place the roof ladder using two firefighters. There are two methods, both named for the way the ladder is carried from the apparatus: Hooks First Method and Butt First Method.

Figure 4.203 The two firefighters ascend the ladder while grasping the beam with their free hand.

Two-Firefighter: Hooks First Method

Step 1: The ladder is carried using the low-shoulder method, hooks (tip) first. As the firefighters reach the other ladder, the firefighter at the tip opens the hooks in such a manner that the hooks face outward (Figure 4.202).

Figure 4.202 The firefighter at the tip opens the roof hooks outward.

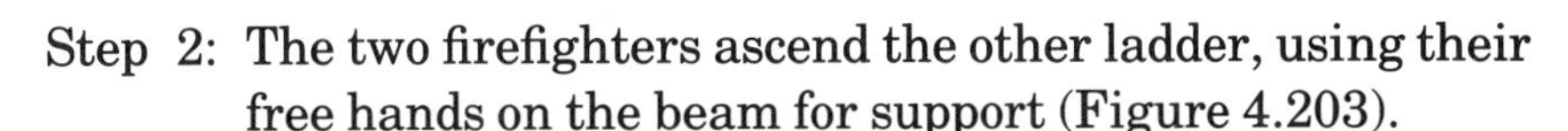

Step 2: The two firefighters ascend the other ladder, using their free hands on the beam for support (Figure 4.203).

Step 3: The firefighter at the roof edge leg locks in or uses a safety belt.

Step 4: The ladder is removed from the shoulder and both firefighters push the roof ladder on its beam up onto the roof (Figure 4.204).

Step 5: Proceed as in steps 7 and 8 of the one-firefighter low-shoulder method.

Figure 4.204 The ladder is pushed up onto the roof while resting on one beam.

Two-Firefighter: Butt First Method

When a roof ladder has been carried to the scene butt first, there is no need to waste valuable time turning it around. Open the hooks and proceed as follows:

NOTE: This procedure is intended for use with one and one-and-one-half story buildings where the eaves are less then 14 feet (4 m) off the ground. For multiple story buildings the modified butt first method may be used.

Step 1: The butt end of the roof ladder is lowered to the ground adjacent to the ladder which has been raised to the eaves.

The firefighter at the tip maintains the carry position (Figure 4.205).

Step 2: The firefighter who carried the butt end assumes a beam raise heelman position. The firefighter at the tip shifts out of the carry position and raises the ladder to vertical. The roof ladder is laid on beam edge alongside the other ladder. The hooks point away from the other ladder. It is steadied by one of the firefighters (Figure 4.206).

Figure 4.205 The butt end is lowered to the ground adjacent to the other ladder.

Figure 4.206 The roof ladder is raised to vertical and laid against one beam of the other ladder.

Step 3: One firefighter heels the other ladder while holding the roof ladder, and the other firefighter climbs to a point near the tip and leg locks in (Figure 4.207).

Step 4: Both firefighters then grasp a convenient rung of the roof ladder and push it upward. The firefighter at the tip then grasps the roof ladder and slides it up the roof on one beam (Figure 4.208).

Step 5: Proceed as in steps 7 and 8 of the one-firefighter low-shoulder method.

Figure 4.207 One firefighter heels the other ladder. The second firefighter climbs it to a point near the tip and leg locks or uses a safety belt, etc.

Figure 4.208 Both grasp the roof ladder and help shove it up the roof on one beam.

Two-Firefighter: Modified Butt First Method

Step 1: The butt of the roof ladder is lowered to the ground approximately one foot (30 mm) from the side of the other ladder, and approximately one foot (30 mm) closer to the building than the butt of the other ladder (Figure 4.209).

Figure 4.209 The butt is lowered to the ground.

Step 2: The firefighter who carried the butt end of the roof ladder becomes the heelman for raising it. The roof ladder is raised to vertical. If a flat raise has been used, the ladder is then pivoted. The roof ladder is laid against the outside of one beam of the other ladder. The hooks point away from the other ladder (Figures 4.210 and 4.211).

NOTE: It is important that the roof ladder be placed so that it extends several rungs' length above where it rests on the beam of the other ladder.

Figure 4.210 The roof ladder is raised to vertical as shown. Note that the hooks are open and point outward.

Step 3: When there are only two firefighters, the firefighter who heeled the roof ladder as it was raised shifts position and heels the other ladder.

NOTE: It is desirable to secure the butt of the other ladder with a rope or to use a third firefighter as the heelman, in which case this step may be omitted.

Step 4: The uncommitted firefighter climbs the other ladder until parallel to a point just above the midpoint of the roof ladder. The near arm is extended between the rungs of the roof ladder and the upper rung is placed on the shoulder (Figure 4.212).

Step 5: The firefighter then continues climbing to the eaves. The heelman helps by feeding the roof ladder up as far as possible (Figure 4.213).

Figure 4.211 The roof ladder is then laid against the beam of the other ladder.

Figure 4.212 One firefighter climbs to just above the midpoint of the roof ladder. A shoulder carry is then executed.

Figure 4.213 The heelman helps lift the ladder upward.

NOTE: If there is a third firefighter heeling the other ladder or if it has been tied in, the second firefighter may climb the other ladder to assist.

Step 6: The firefighter climbing the other ladder leg locks in or uses a safety belt at the top of the other ladder and proceeds as in steps 7 and 8 of the one-firefighter low-shoulder method.

4.25 Check the correct response.
When roof ladders are to be carried up another ladder for placement on a sloped roof, the hooks are normally opened ______.
- ☐ A. Before climbing the other ladder.
- ☐ B. At the time the ladder is removed from the apparatus.
- ☐ C. When the ladder is placed on the roof.
- ☐ D. It doesn't make any difference.

4.26 Fill in the blank.
When a roof ladder is laid over flat to be pushed up a sloping roof the hooks are ______________________.

Climbing a Pompier Ladder

The pompier ladder climbing procedures described are for drill tower use. The main difference between drill towers and fireground use is that the windows in drill towers have been removed. In real fire situations, windows may be wholly or partly intact. It may be necessary to break through the screen and to break out glass sufficiently to let the gooseneck hook through. When climbing, it would be necessary to hook in at the top of the pompier ladder and then clear the window opening with some type of forcible entry tool.

The section on raising ladders dealt with getting the pompier ladder in place initially. The following steps detail correct climbing procedures.

Step 1: The pompier ladder is climbed with the hands grasping the beam and the feet on the rungs close to the beam (Figure 4.214).

Step 2: The firefighter steps into the window opening by straddling the windowsill while grasping the gooseneck hook (Figure 4.215).

Step 3: While the firefighter is astride the windowsill, the gooseneck is removed and turned outward (Figure 4.216).

Figure 4.214 The pompier ladder is climbed by grasping the beam hand-over-hand and placing the feet on the rungs as close to the beam as possible.

Figure 4.215 It is necessary to enter each floor. To do so, the firefighter grasps the gooseneck hook and straddles the windowsill.

Figure 4.216 The gooseneck hook is removed. The ladder is lifted a short distance and then it is turned so that the hook is out. This allows the ladder to be elevated to the floor above.

Step 4: The beam is grasped and the ladder is raised vertically hand-over-hand until the gooseneck hook is opposite the next window up (Figure 4.217).

Step 5: The gooseneck hook is then turned into the window opening and is secured over the windowsill (Figure 4.218).

Step 6: The firefighter then stands on the windowsill, grasps the beam, and steps out onto the pompier ladder for the climb to the next floor (Figure 4.219).

Figure 4.217 While astride the windowsill, the beam is extended upward by passing it hand-over-hand.

Figure 4.218 The gooseneck hook is then again turned into the window.

Figure 4.219 The firefighter then remounts the ladder for the climb to the floor above.

4.27 True or False.

	True	False
1. The feet should be spread as far apart as possible when climbing the pompier ladder.	☐	☐
2. When the pompier ladder is being raised from one windowsill to the next above, the gooseneck hook is turned away from the building.	☐	☐

4.25 A

4.26 down

LADDER SAFETY

Throughout this and the preceding chapters various safety precautions have been stressed. In order to review and re-emphasize certain points, these and some additional precautions are listed below.

- There is no substitute for the use of good common sense during all operations involving ladders.
- Full protective gear, including gloves, should always be worn.
- The proper number of firefighters should be used for carries and raises.
- Leg muscles, NOT BACK MUSCLES, are used for lifting and lowering ladders.
- Care must be taken whenever ladders are used around live electrical equipment. Metal ladders are good conductors and all other ladders are conductors when wet.
- A heelman located between a ladder and a building should not look up during this task.
- Before climbing a ladder, a check for the proper angle of inclination should be made. The ladder should be adjusted if it is incorrect.
- Anytime it is suspected that a metal ladder has been subjected to excessive heat, it should be subjected to a hardness test.
- Before climbing, a check should be made that the ladder is secure from unwanted movement.
- When climbing extension ladders, a check should be made that the halyard rope is tied off. Before climbing onto a fly section, check that the pawls are securely latched.
- Do not move a ladder that another firefighter has used to gain entry. It will probably be used as a line of retreat.
- When raising ladders and extending fly sections, frequent visual checks should be made of the tip, both for obstructions and indication of ladder instability.
- Climbing should be smooth and rhythmic.
- Ladders should be inspected periodically and after each use for signs of damage.
- When working from a ladder, always tie in with a leg lock or use a safety belt, strap, or rope hose tool.

- When it is necessary to perform a rescue using a ladder, all other loads and activities should be removed from the ladder, which must be securely anchored at both tip and butt.
- Firefighters should know the maximum weight load for the length ladder they are using. See Table 2.3.

Fire service ladders are subject to considerable abuse because of the situations under which they are used. It is not always practical or feasible to be over cautious concerning ladder damage during an emergency situation. Proper training and practice in handling ladders will familiarize firefighters with the techniques of safe ladder practices under "normal" and emergency situations.

4.28 True or False.

	True	False
1. It is better not to wear gloves while climbing a ladder.	☐	☐
2. Leg muscles are used for lifting and lowering ladders.	☐	☐
3. Metal ladders suspected of having been subjected to excessive heat should be rewaxed before any further use.	☐	☐
4. Ladders used to gain entry may be moved to another location once firefighters have entered or gotten off onto the roof.	☐	☐
5. When it is necessary to perform a rescue using a ladder to which a hoseline has been secured, the hoseline should be removed from the ladder.	☐	☐

Review

Answers on page 387

Check the correct responses.

1. Of the ladders listed, check those which must be placed flat on the ground prior to raising.

☐ A. Single
☐ B. Roof
☐ C. Folding
☐ D. Extension
☐ E. Pole
☐ F. Combination
☐ G. Pompier

2. The fly on extension ladders is placed ________________.

☐ A. Out (Away from the building).
☐ B. In (Next to the building).

3. In reference to question 2, the fly is placed in this position because ________________.

☐ A. It is more convenient to raise it in this manner.
☐ B. It is more convenient to climb with the fly in this position.
☐ C. The ladder is stronger with the fly in this position.
☐ D. The ladder will extend further with the fly in this position.
☐ E. It has always been done this way (tradition).

4. When one firefighter raises an extension ladder it is ________________ to get the fly into the proper position.

☐ A. Rolled
☐ B. Pivoted
☐ C. Shifted
☐ D. Flipped over on the ground prior to raising

5. When five firefighters raise the pole ladder it is ________________ to get the fly into the proper position.

☐ A. Rolled
☐ B. Pivoted
☐ C. Shifted
☐ D. Flipped over on the ground prior to raising

6. When two or three firefighters raise an extension ladder it is ________________ to get the fly into the proper position.

☐ A. Rolled
☐ B. Pivoted
☐ C. Shifted
☐ D. Flipped over on the ground prior to raising

4.27 1. False
2. True

Short Answer Essay.

7. The one-firefighter extension ladder raise is being used. When extending the fly section; if the firefighter sees that the ladder is tilting away from the building past vertical and is about to fall, the firefighter should ______________________________

____

____

____

True or False.

	True	False
8. The utility rope is used for the guy lines when the dome raise is made.	☐	☐
9. When the hotel or factory raise is made, the butt of the ladder is NOT placed one-fourth the extended distance from the building.	☐	☐
10. When the hotel or factory raise is made, the ladder is placed in front of the windows.	☐	☐

Check the correct response.

11. When two firefighters are shifting a ladder vertically; if one firefighter grasps the ladder with the right hand up high and the left hand down low, the other firefighter grasps the ladder ______.

☐ A. With the left hand high, right hand low.
☐ B. With both hands high.
☐ C. With the right hand high, left hand low.
☐ D. With both hands low.

12. The correct angle of inclination when the dome raise is used is ______.

☐ A. 75 degrees.
☐ B. 60 degrees.
☐ C. 90 degrees.
☐ D. 45 degrees.
☐ E. 110 degrees.

13. The stance of a firefighter climbing a ladder should be ______.

☐ A. Vertical (erect).
☐ B. Parallel to the ladder surface.
☐ C. Is of no significance.

Fill in the blanks.

14. Climbing should be smooth and rhythmical to avoid ________ and ______________________________ of the ladder.

15. Upward progress when climbing should be the result of ____.

4.28
1. False
2. True
3. False
4. False
5. True

Short Answer Essay.

16. Whenever a firefighter is to work while standing on a ladder, what safety precautions are taken? ______________________________.

17. When a ladder is heeled from behind, why should the heelman not look upward? ____.

Check the correct response.

18. The main advantage of tying a ladder in compared to using a firefighter to heel a ladder is ______________.
- ☐ A. The ladder is stronger when it is tied in.
- ☐ B. There is no advantage.
- ☐ C. The firefighter needed to heel the ladder is freed to do other tasks.
- ☐ D. It is quicker.

19. If a ladder is set at too steep an angle ___________.
- ☐ A. It will not carry the prescribed load.
- ☐ B. There will be a tendency for the tip to pull away from the building.
- ☐ C. It will be easier to climb.
- ☐ D. A longer reach is needed.

True or False.

	True	False
20. When climbing while carrying an object, the ladder rung is grasped with the other hand.	☐	☐
21. For safety, the ladder tip and butt should both be tied off.	☐	☐
22. Climbing should be done at as fast a pace as possible.	☐	☐
23. After pole ladders are in place, the staypoles are placed in such a manner that they will help carry the weight load placed on the ladder during climbing.	☐	☐
24. The primary reason that the halyard rope on an extension ladder is tied off is safety; it is a back up in case the pawls have not engaged properly.	☐	☐

LADDERS

Chapter 5

Special Uses

NFPA Standard 1001

FIRE FIGHTER I

3-9.6 The fire fighter shall climb the full length of each type of ground and aerial ladder and bring an "injured person" down the ladder.

Chapter 5
Special Uses

> While the members of the Ladder Committee, the IFSTA Validation Conference, the manufacturers and/or their representatives, and Fire Protection Publications staff of/and Oklahoma State University may have varying degrees of knowledge that the ladders may be or may have been used in methods described in Chapter 5, they do not suggest, encourage, endorse, or sanction such actions or uses and shall not be held responsible for failure of equipment or any injury or death as a direct or indirect result of a failure proximately caused by using ladders for any purpose not considered good ground ladder practices by the original manufacturer for the product.

The text so far has dealt with the "routine" uses of ground ladders. While some of the more difficult and specialized raises have been covered, there have been only a few references to rescue operations and other uses. Chapter 5 will cover these two categories in depth.

It is important that firefighters be proficient in carrying, raising, and climbing skills before attempting many of the following operations. It is also important that firefighters understand that some of the uses discussed in this chapter are not recommended by the manufacturer. When a ladder is used for one of these special applications, considerable reliance is placed on the fact that the ladder has been designed and constructed with a substantial safety factor. Therefore, it MUST be in first-class condition.

While it is necessary to train in these operations, extreme care must be taken not to damage ladders while training with them. Whenever ground ladders have been subjected to severe or unusual temperatures, loads, stresses, or impacts, the ladders have to be removed from service and tested.

LADDER RESCUE OPERATIONS

When a rescue situation occurs, the rescue operation gets first priority in the allocation of apparatus, manpower, equipment, and communications channels, even to the point that other fireground operations stop. Fire fighting operations that continue are in support of rescue operations.

Assisting Conscious and Physically Able Persons Down Ground Ladders

Evacuation of persons from a building by ground ladders is difficult and slow. If possible, it is better to evacuate down stairways, through connecting buildings, over connecting roofs, or by using elevating platform or aerial ladder apparatus. Persons being evacuated are generally unaccustomed to climbing down a ladder and must be accompanied by firefighters to prevent injury while descending. If a person panics and becomes violent during this operation, the recommended procedure is for the firefighter to pin the person to the ladder until that person calms down.

The firefighter's actions during evacuation of conscious, physically able persons are as follows:

Step 1: If the situation and manpower permit, one or more firefighters ascend the ladder and enter the building. Another firefighter climbs to the top of the ladder. All assist persons onto the ladder (Figure 5.1).

Figure 5.1 The person being evacuated is assisted onto the ladder by firefighters. It is particularly important that a firefighter be at the top of the ladder during this phase of the operation.

Step 2: As soon as the person is on the ladder, the firefighter on the ladder places the arms around that person under the armpits. The firefighter's hands grasp the ladder rung in front of the person's face. One knee is placed between the person's legs to provide support in case a rung is missed or unconsciousness occurs (Figure 5.2).

NOTE: The firefighter should explain what is occurring and reassure the person.

Step 3: The person and the firefighter descend the ladder together (Figure 5.3).

Figure 5.2 The firefighter stands behind the evacuee, places the hands under the person's armpits, and grasps a rung. One knee is placed between the person's legs to provide support if it becomes necessary.

Figure 5.3 The evacuee and the firefighter descend the ladder together in a coordinated manner. The firefighter directs the moves and reassures the evacuee.

Bringing Unconscious Persons Down Ground Ladders

One or more firefighters are assumed to have entered the building, located an unconscious person, and brought the individual to the window.

Step 1: A firefighter climbs to near the top of the ladder. One foot is placed on a rung in such a way that the knee is bent 90 degrees and the upper leg is in a horizontal plane.

Step 2: The firefighters inside pass the victim out the window in such a manner that the individual will be facing the ladder while being taken down.

NOTE: This position is preferred in case the person regains consciousness. They can then assist in their own rescue. Also, if they should panic, it is difficult for them to grapple with the firefighter.

The victim is lowered until the crotch is resting on the horizontal part of the firefighter's leg with the victim's legs astride it. The feet are placed over the outside of the ladder beams.

Step 3: The firefighter reaches under the person's armpits and grasps a rung in front of the person's face (Figure 5.4).

Figure 5.4 This grasp allows the firefighter to keep the victim from falling off of the ladder and provides a means of controlling the descent.

Step 4: The firefighter then holds the victim to the ladder with the arms under the armpits while the knee is positioned one rung lower. The victim is then let down onto the knee again and the firefighter readjusts the grasp one rung lower.

Step 5: This procedure is repeated one rung at a time down the ladder (Figure 5.5).

Figure 5.5 As the firefighter descends, a new knee rest is formed at each rung level and the victim is eased down on it, then the hand grip is readjusted to the next lower rung. Note that the victim's feet have been placed outside the beams.

Step 6: When the ground is reached, the victim is grasped under the armpits and pulled clear of the danger area (Figure 5.6). If there is another firefighter, the person can be carried to a safe location (Figure 5.7).

Figure 5.6 The victim is dragged clear of the danger area.

Figure 5.7 When there are two firefighters available, the victim is carried clear of the danger area.

ALTERNATE METHOD

An alternate method of getting an unconscious person down a ground ladder uses the same procedure except that the victim is turned around to face the rescuer. This position lessens the chance of the victim's limbs becoming caught between the rungs but leaves the firefighter vulnerable to being choked or assaulted if the person regains consciousness and becomes violent (Figures 5.8 and 5.9).

Figure 5.8 The victim is placed on the ladder facing the firefighter with their arms on the shoulders, crotch on the firefighter's knee, and feet outside the beams.

Figure 5.9 The firefighter proceeds down one rung level at a time maintaining the stance shown.

5.1 Check the correct responses.

1. If a victim becomes violent while being assisted down a ladder, the firefighter should first ______________.

- ☐ A. Call for assistance.
- ☐ B. Let the person descend the ladder alone.
- ☐ C. Physically pin the person to the ladder until they calm down.

2. During a ladder rescue operation, all other operations should be ______________.

- ☐ A. Closely monitored.
- ☐ B. Either stopped or be in support of the rescue operation.
- ☐ C. Proceed as normal.

3. The arms of the firefighter bringing an unconscious person down a ladder are positioned ___________.

- ☐ A. Under the person's armpits, grasping the rung in front of their face.
- ☐ B. One arm around the person's waist, the other grasping the ladder beam.
- ☐ C. At the person's waist, grasping the beams.

Removing an Unconscious Firefighter Leg Locked on a Ground Ladder

Occasionally a firefighter who has leg locked in on a ladder is knocked out or becomes unconscious (Figure 5.10). After being given any necessary first aid, the firefighter must be removed from the leg lock in order to be brought down the ladder. The steps are as follows:

Step 1: Another firefighter climbs the ladder and pins the unconscious firefighter against the ladder by placing the hands under the armpits, grasping a rung, and then leaning against the victim (Figure 5.11).

Figure 5.10 Unconscious firefighter leg locked on a ladder.

Figure 5.11 A firefighter climbs to the victim and pins that person to the ladder as a second ladder is raised.

Step 2: Another ladder is raised and placed next to the one where the unconscious firefighter is located. It is placed on the same side as the unconscious firefighter's leg lock (Figure 5.12).

NOTE: If available, a third ladder is raised on the other side. If the unconscious firefighter is a large person, the third ladder is necessary.

Step 3: Still another firefighter ascends the second ladder to a level where the unconscious firefighter's leg locked foot is located. The foot is freed and pushed back through the space between the rungs so that it is hanging free. The boot is removed (Figure 5.13).

Step 4: The firefighter on the second ladder then ascends the ladder to a point slightly higher than the unconscious firefighter.

Step 5: The two firefighters then slide the unconscious firefighter up the ladder until the leg and foot can be pushed back through from between the rungs (Figure 5.14).

5.1 1. C
2. B
3. A

Figure 5.12 The second ladder is placed beside the first one on the side where the leg lock is.

Figure 5.13 A firefighter ascends the second ladder stopping where the unconcious firefighter's leg lock is. The leg lock foot is freed and pushed back through. The boot is then removed to make further movement easier.

Figure 5.14 The two rescuers slide the victim up the ladder until the leg can be pulled back through.

Step 6: The unconscious firefighter is then lowered down onto the rescuing firefighter's knee, the feet are positioned free of the ladder, and the procedure for bringing an unconscious person down a ladder is followed (Figure 5.15).

Figure 5.15 The victim is then lowered onto the first rescuer's knee. Then the standard procedure for bringing down an unconscious person is used. The turnout coat will make it more difficult to slide the victim down.

ALTERNATE METHOD

When speed is a factor in freeing the unconscious firefighter or when the firefighter is a large person and the rescuers have been unable to get the leg free, the alternative is to cut the rung that was used in the leg lock.

The normal position is taken to pin the unconscious firefighter to the ladder and the second ladder is raised next to the one holding the unconscious firefighter just as in the previous procedure. The firefighter on the second ladder clears the unconscious firefighter's foot but no attempt is made to elevate the unconscious firefighter to clear the leg. The alternate procedure of cutting the rung is used. The cutting is done by the firefighter on the second ladder while the unconscious firefighter is being supported as before (Figure 5.16).

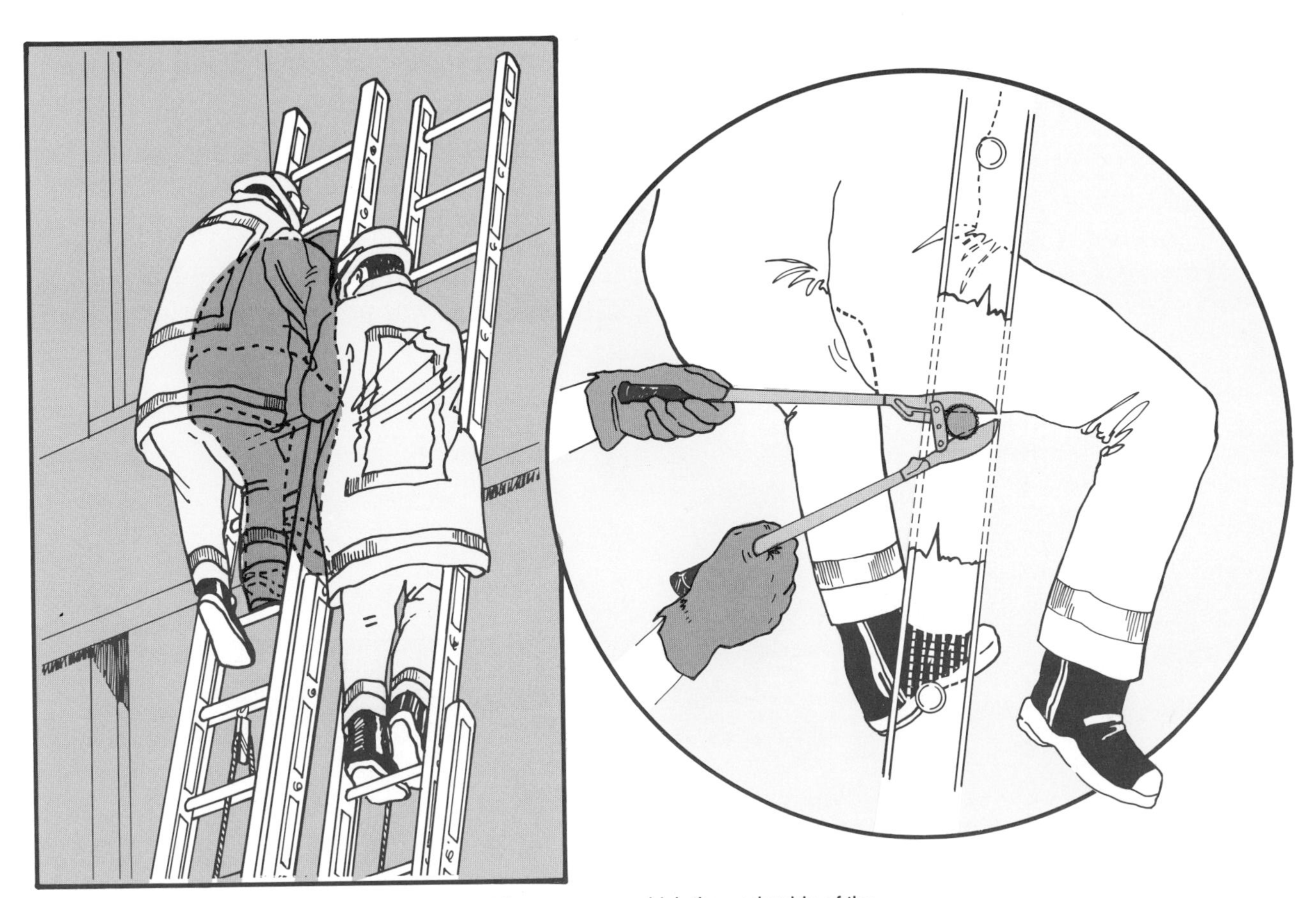

Figure 5.16 It will sometimes be necessary to cut the rung upon which the underside of the victim's leg is resting.

When the rung is removed, the leg can be pulled through and the previous procedure used to get the unconscious firefighter down the ladder.

5.2 Fill in the blanks.

1. When rescuing an unconscious firefighter who is leg locked in on a ground ladder, a second ladder is placed ____________ ________________________________.
2. When speed in freeing an unconscious firefighter leg locked in on a ground ladder is essential, the best procedure to free the leg that is locked in is to ________________________ ______________________________.

Lowering Extension Ladders to Below-Grade Locations

Sometimes a ladder is required for access to a below-grade location. An extension ladder may be required, in which case the ladder is lowered after having been extended. The steps are

Step 1: The ladder is laid flat on the ground. The fly is extended by having firefighters anchor the bed section while others grasp a rung of the fly section and pull it outward (Figure 5.17).

Step 2: When the desired length is attained, the pawls are latched.

Step 3: A rope is tied to the tip of the bed section. The ladder is tilted up on one beam. The rope is strung along the ground where the ladder was previously resting (Figure 5.18).

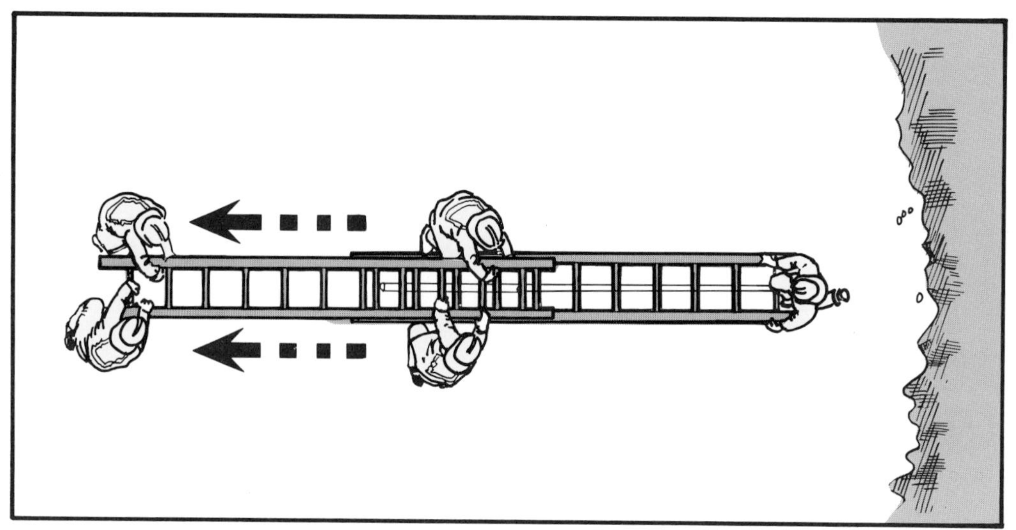

Figure 5.17 The bed section is anchored while the fly section is pulled outward by grasping a rung and walking out with it.

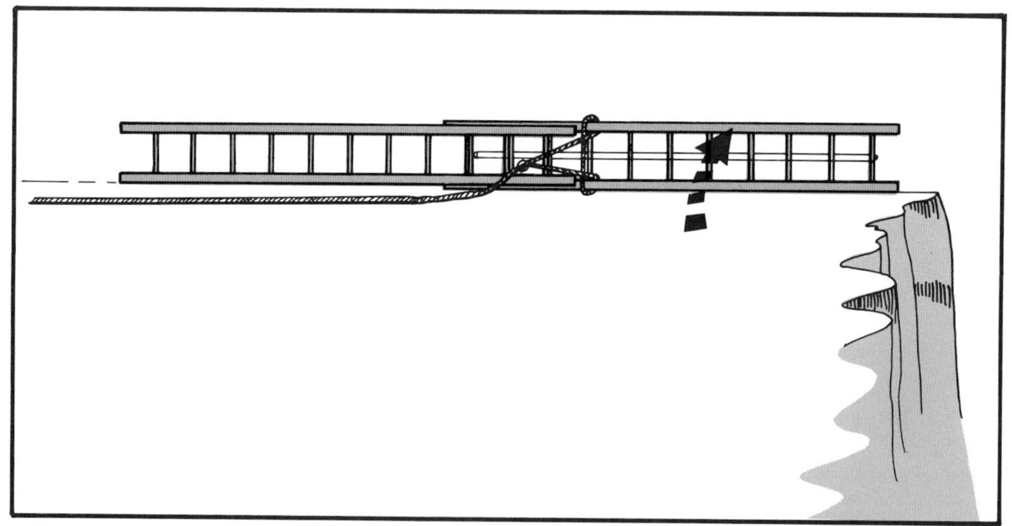

Figure 5.18 A rope is tied to the bed section. The ladder is tilted onto one beam so that the rope can be strung along the ground to and beyond the tip.

Step 4: The ladder is laid back flat. The rope should now be under the ladder. The remaining rope is strung out from the tip (Figure 5.19).

Step 5: At least two firefighters grasp the free end of the rope and take up the slack. The ladder is picked up and carried, butt first, to the edge of the precipice. The firefighters holding the rope follow (Figure 5.20).

Step 6: The butt is placed on the ground at the edge of the precipice and the ladder is shoved outward until the balance point is reached. The rope is slacked off just enough to allow the ladder to move forward. Then the ladder is allowed to swing downward, at which time the firefighters

5.2 1. On the side where the leg lock is
2. Cut the rung

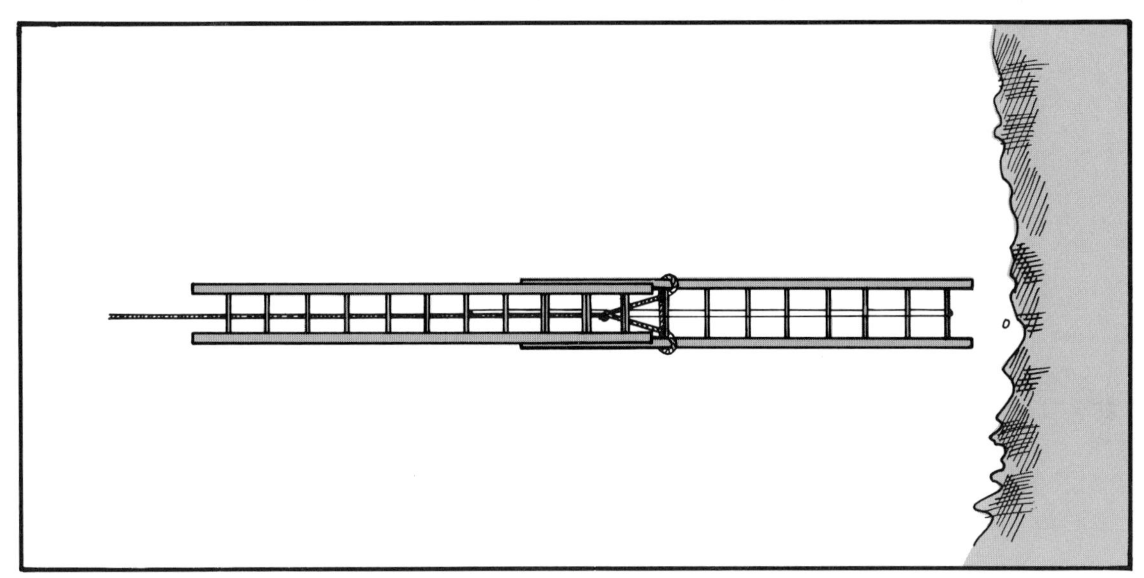

Figure 5.19 The ladder is laid back atop the rope.

Figure 5.20 Firefighters grasp the rope. Others lift the ladder and carry it to the edge of the precipice.

on the rope take the weight of the ladder. The firefighters who were carrying the ladder now steady it against sideways movement (Figure 5.21).

Step 7: Just before the ladder butt reaches the ground the firefighters steadying the ladder pull the top toward them to cause the butt to kick outward enough to obtain a better angle for climbing (Figure 5.22).

NOTE: It may be necessary to reset the butt when the first firefighter reaches the bottom.

5.3 Fill in the blank.

1. When lowering an extension ladder to below-grade locations, the rope used to lower it is tied to the ____.

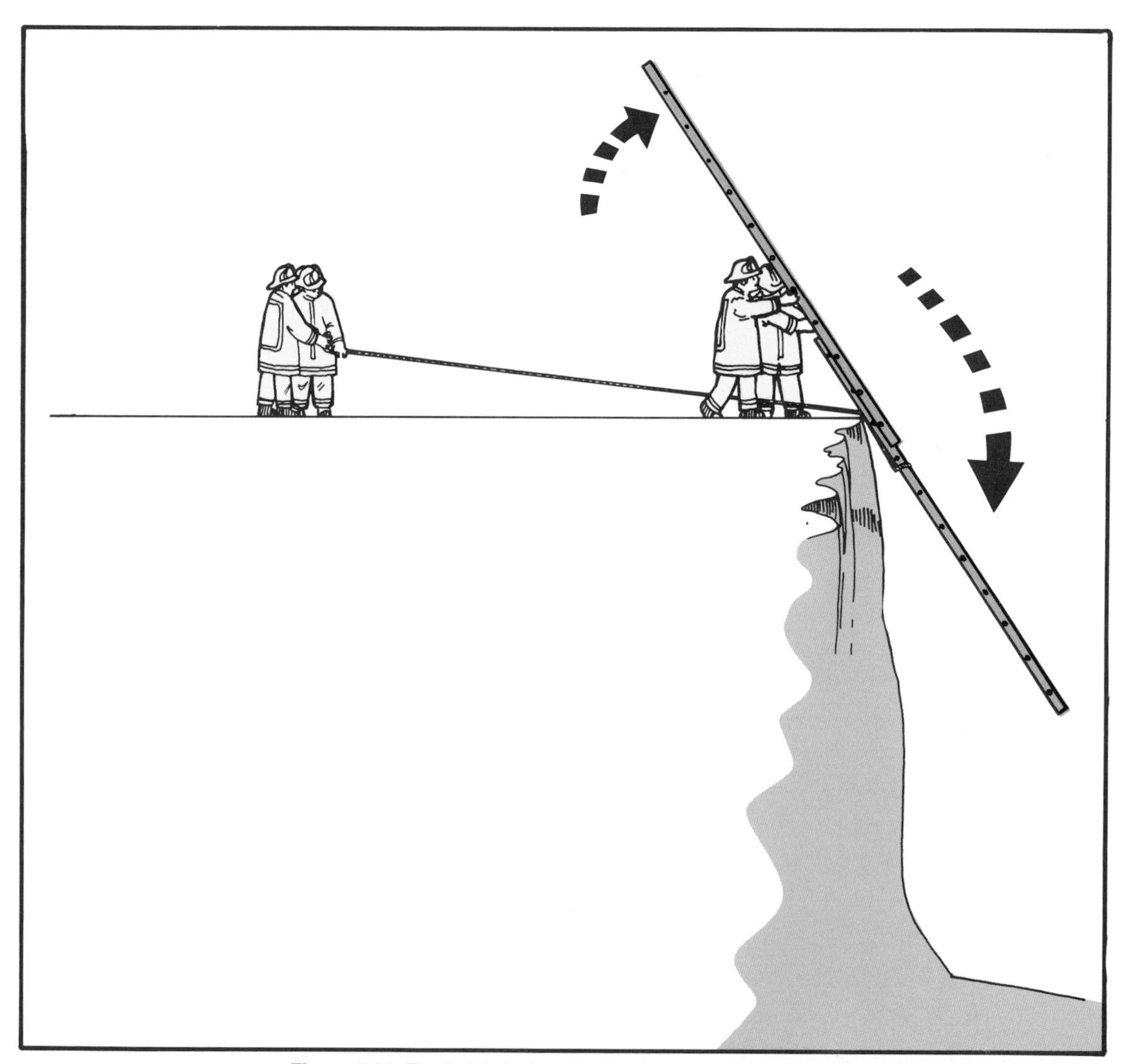

Figure 5.21 The butt is lowered to the ground and then the ladder is shoved outward. When the balance point is reached, it will swing downward.

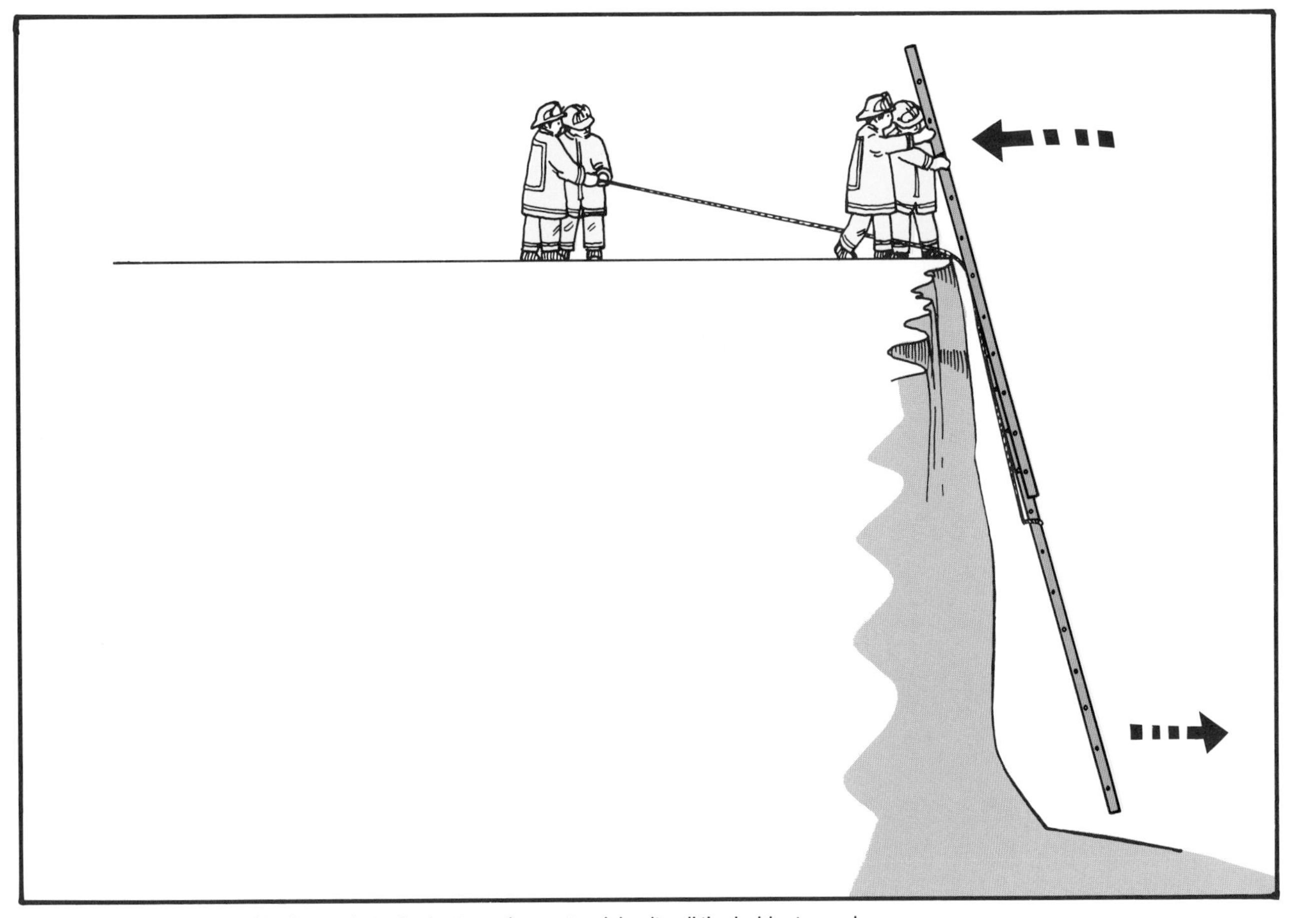

Figure 5.22 When the ladder is nearly to the bottom, those steadying it pull the ladder toward them to cause the butt to kick outward slightly.

Bridging with Ladders for Rescue

Occasions may arise where persons must be rescued from locations where ground ladders or aerial apparatus cannot be used or will not reach, but where access can be gained by using ground ladders to horizontally bridge from an adjacent point. There are two different types of locations where bridging may be practical.

- Two buildings separated by a narrow alley or space. The uninvolved building has windows or a roof opposite the windows or roof of the involved building.
- A narrow drainage canal, ditch, or similar excavation where it is necessary to span the open space for rescue.

CAUTION: Bridging is not a manufacturer's recommended use for a ground ladder: LOADING MUST BE KEPT TO A MINIMUM.

Extension ladders are only used if no single or roof ladder is available because pawls may not hold properly.

After the ladder has been placed across the open space, firefighters should crawl across one at a time, then use extreme care to get victims back across to safety. It will probably be necessary to accompany each person across, giving constant reassurance and encouragement. They should be advised to look straight ahead and not down.

METHODS OF BRIDGING

There are four methods of placing a ladder across an open space: the Three-Firefighter Shoulder Carry Method, the Three-Firefighter Flat Arm's Length Method, the Hoisting Method, and the Ladder Drawbridge Method. The following paragraphs detail each method.

Three-Firefighter Shoulder Carry Method

This procedure is used to bridge from one roof across an alley to a window of a taller building, to reach another roof, or for spanning ditches or other open spaces.

Step 1: The ladder is placed flat on a surface. Two firefighters position themselves six rungs from the butt end, one on each side beside the beam, and face the butt. They kneel and grasp the rung, palms forward. The third firefighter, who is positioned at the butt facing the tip, kneels and grasps the bottom rung (Figure 5.23).

Step 2: All firefighters then pick up the ladder. The firefighters at the beams pivot under, placing the ladder on their shoulders (Figure 5.24).

Step 3: The heelman then pulls downward to lift the tip of the ladder to a horizontal position (Figure 5.25).

Step 4: The firefighters then step forward to extend the ladder over the open space (Figure 5.26). Note that if the ladder was originally placed parallel to the wall it will be necessary for the firefighters to turn after picking it up.

Figure 5.23 The ladder is laid down flat with three firefighters positioned as shown. Note that the gap between buildings would be greater than that pictured here.

5.3 bed section

Figure 5.24 When the ladder is lifted, the two firefighters pivot under and place the beams on their shoulders. The other firefighter grasps the bottom rung.

Figure 5.25 The butt is pulled downward until the ladder is level.

Figure 5.26 All step forward to the edge. The ladder is extended across and lowered to rest on the two surfaces and a firefighter proceeds across while another steadies the ladder.

Three-Firefighter Flat Arm's Length Method

This method is used for the same purposes as the Three-Firefighter Shoulder Carry Method.

Step 1: The ladder is placed flat on the surface. One firefighter kneels at the butt. The other two firefighters kneel facing the ladder, one on each side near the tip (Figure 5.27).

Step 2: The three firefighters stand, lifting the ladder to arm's length as they do so, and carry the ladder to the space to be spanned (Figure 5.28).

Step 3: The ladder is projected across the open space by being fed through the hands of the two firefighters at the tip. The heelman moves forward and at the same time bears down to counterbalance the weight of the extended part of the ladder (Figure 5.29).

Figure 5.27 Two firefighters kneel facing the ladder at its tip. A third firefighter kneels at the butt facing it and grasps the beams.

Figure 5.28 The ladder is lifted to arm's length and carried to the edge.

Figure 5.29 The ladder is pushed across by the firefighter at the butt while the other two firefighters support it by letting it slide across the upturned palms.

Hoisting Method

This method is used to span from a window across to another window or to the roof of a lower building.

Step 1: A ladder of sufficient length to bridge the gap is selected and a rope is passed down to the ground from the floor or roof above the proposed bridging site. A hose roller should be used at this location, both to protect the rope and to lessen the workload of the firefighters who will hoist the ladder. The ladder is secured with a clove hitch on each beam at about the middle of the ladder (Figure 5.30). The ladder is turned over to bring the ropes outside the ladder beams, and a bowline is tied about six feet (2 m) from the clove hitch. A half hitch is placed around the end of the ladder at the butt as a guide (Figure 5.31).

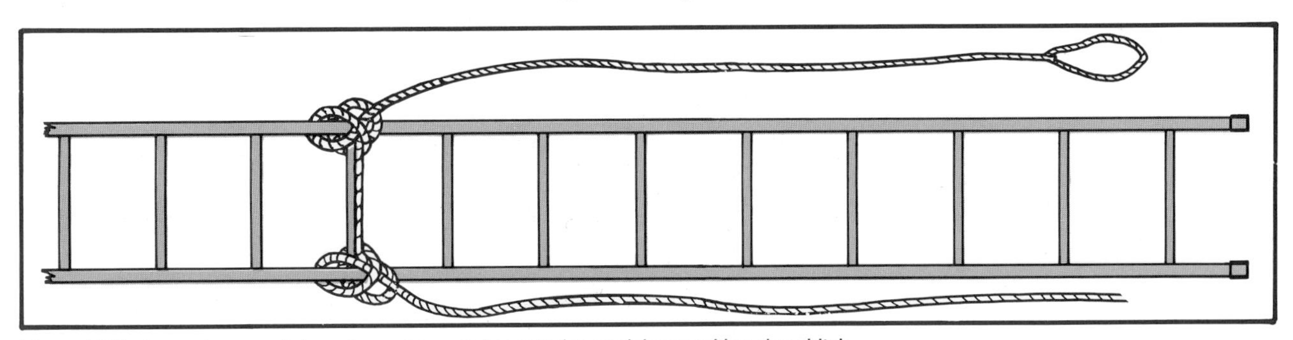

Figure 5.30 A rope is passed down from above and secured to each beam with a clove hitch.

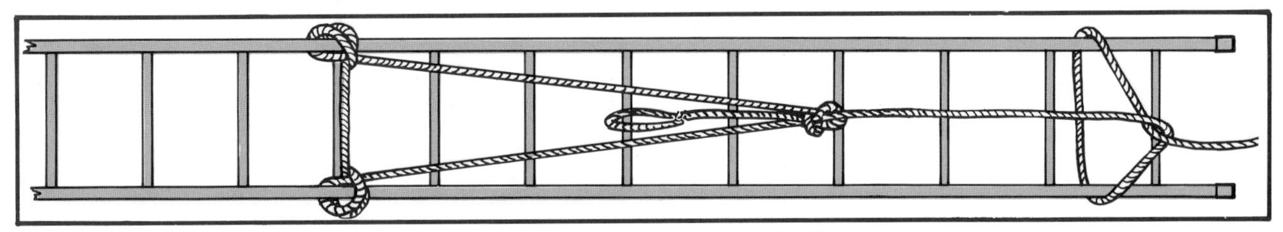

Figure 5.31 The ladder is turned over. A bowline is tied about six feet (2 m) from the clove hitch. Then a half hitch is placed around the end of the ladder.

Step 2: The ladder is hoisted by the firefighters above the bridging level until the butt can be grasped by firefighters below them (Figure 5.32). The firefighters at bridging level remove the half hitch guide while maintaining control of the ladder.

Step 3: The firefighters above the bridging level hoist the ladder into the horizontal position as the firefighters below control the butt. The tip is elevated slightly higher than the windowsill upon which it will rest (Figure 5.33).

Step 4: The firefighters at the butt push the ladder ahead as the firefighters above slack off on the hoist rope enough that the tip will come to rest on the windowsill. Once the ladder is in position, the firefighters above take up tension on the rope to give added support to the ladder. The ladder butt should either be tied in or a firefighter should be assigned to ensure stability (Figure 5.34).

To Lower: To lower the ladder at the completion of operations, the firefighter on the bridging level pulls the butt back into the window, takes slack on the line from above, reties the half hitch guide around the butt, eases the ladder out the window, and gives the signal to lower the ladder to the ground.

Figure 5.32 The ladder is hoisted until it can be grasped by those at the bridging level.

Figure 5.33 Those above hoist the ladder to horizontal with the tip elevated slightly.

Figure 5.34 The ladder is pushed across into the opposite opening, then the tip is lowered into place.

Ladder Drawbridge Method

This procedure is used to span streams, culverts, trenches, or pits. When time permits its rigging, it can also be used for ice rescue. A long ladder and a short single (or roof) ladder are lashed together. The short ladder must be narrower than the long ladder. Two 125 foot (38 m) ropes are required. The procedure is as follows:

Step 1: The two ladders are placed on the ground butt to butt. The butt spurs of the shorter ladder are placed against the bottom rung of the longer ladder. A rope hose tool is used to secure the joint (Figure 5.35).

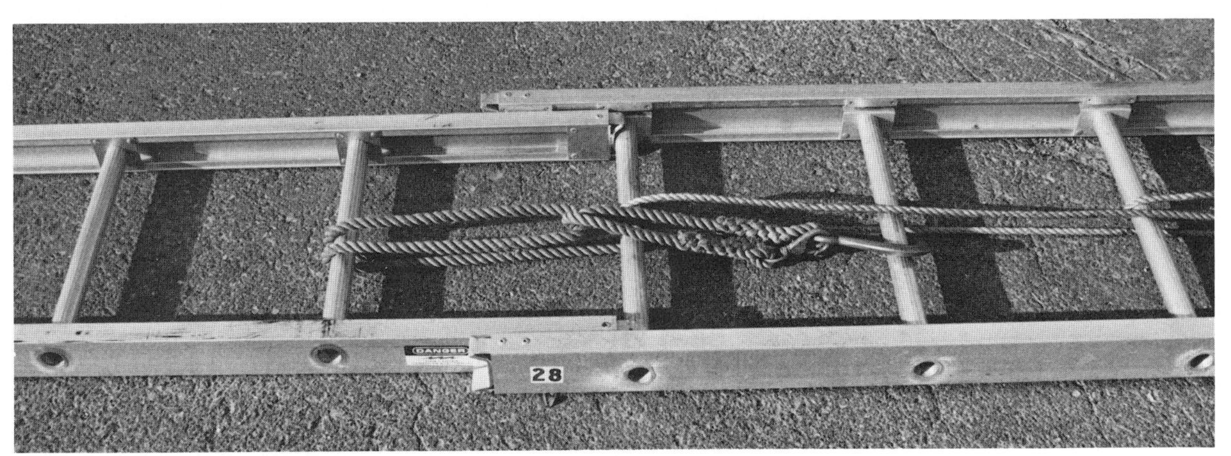

Figure 5.35 The butts of two ladders of unequal length and width are lashed together, in this instance with a rope hose tool. The tip of the longer ladder is placed at 90 degrees to the bridging site.

Figure 5.36 Ropes are tied to the tip of the longer ladder in the manner shown.

Step 2: Ropes are attached to each beam of the longer ladder. They are tied at the top rung (Figure 5.36).
To tie this hitch

- Make a loop in the rope about three feet (1 m) from one end.
- Pass the loop over the beam between the first and second rungs, then under the beam and across the rope.
- Pass the loops around the top end of the beam and slide it down to the rung.
- Cinch up tight.

Step 3: One firefighter grasps both ropes and heels the tip of the shorter ladder. Another firefighter heels the longer ladder while two others raise it to vertical. Both heelmen assist with raising by pulling on the ropes (Figure 5.37).

Figure 5.37 The longer ladder is raised to vertical. At first its heelman assists by pulling on the ropes. Then as the ladder nears vertical, the ropes are released and the firefighter at the tip of the short ladder assists. No pressure is put on the ropes when the longer ladder reaches vertical.

Step 4: The two ropes are tied to the beam ends of the shorter ladder using clove hitches (Figure 5.38).

Step 5: The ladders are now rigged and ready for lowering after being dragged or carried into place at the edge of the trench, culvert, or pit (Figure 5.39).

Figure 5.38 The ropes are tied to the tip of the shorter ladder using a clove hitch.

Figure 5.39 The assembly is moved to the edge of the bridging site.

Step 6: The two firefighters who raised the long ladder to vertical now heel it while lowering is taking place (Figure 5.40).

Figure 5.40 The long ladder is then lowered over the culvert, trench, ditch, or other space.

5.4 True or False.

	True	False
1. Bridging is a manufacturer's approved procedure.	☐	☐
2. During bridging weight loading must be kept to a minimum.	☐	☐
3. When evacuating a person by bridging, the firefighter assists that person onto the ladder but does not accompany the person across the ladder.	☐	☐

Using a Ladder as a Fulcrum (Hinge) for Lowering Injured Persons

This procedure can be used when a victim needs to be lowered in a horizontal or near horizontal position. They are secured in a Stokes stretcher or similar arrangement. At least three firefighters are needed and four are preferred, two or three at the top of the ladder, and one on the ground. The procedure is

Step 1: A ladder is positioned at the correct climbing angle. A Stokes stretcher or similar device is taken up and the victim is lashed into it.

Step 2: At least two firefighters are required to balance the Stokes stretcher on the windowsill or roof edge while another lashes the foot of the Stokes stretcher securely to the top rung of the ladder (Figure 5.41).

Step 3: A lifeline is attached to the head of the Stokes stretcher (Figure 5.42).

Figure 5.41 The Stokes stretcher is lashed to the top rung of an in-place ladder.

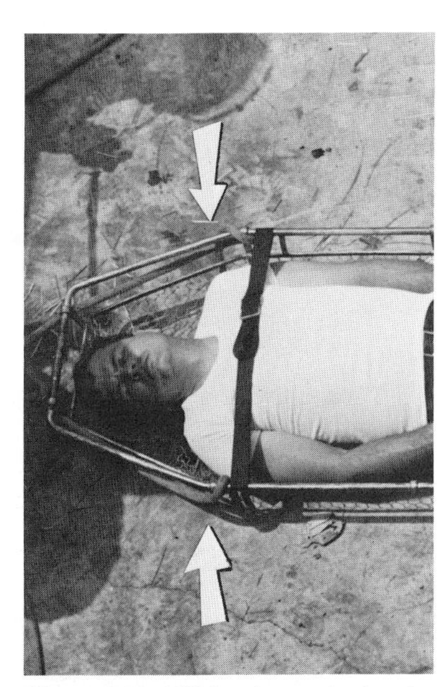

Figure 5.42 A lifeline is attached to the other end of the Stokes stretcher.

Step 4: One firefighter takes the standing part of the rope and assumes an anchoring stance (Figure 5.43).

Step 5: The ladder heelman moves the butt of the ladder inward until the butt spurs are against the wall (Figure 5.44).

Figure 5.43 Firefighters anchor the lifeline.

Figure 5.44 The ladder butt is lifted and moved inward until it is against the building and the ladder is vertical.

Step 6: The Stokes stretcher is then slid out the window or off the roof as the anchoring firefighter slowly feeds out rope (Figure 5.45). In the meantime, the firefighter on the

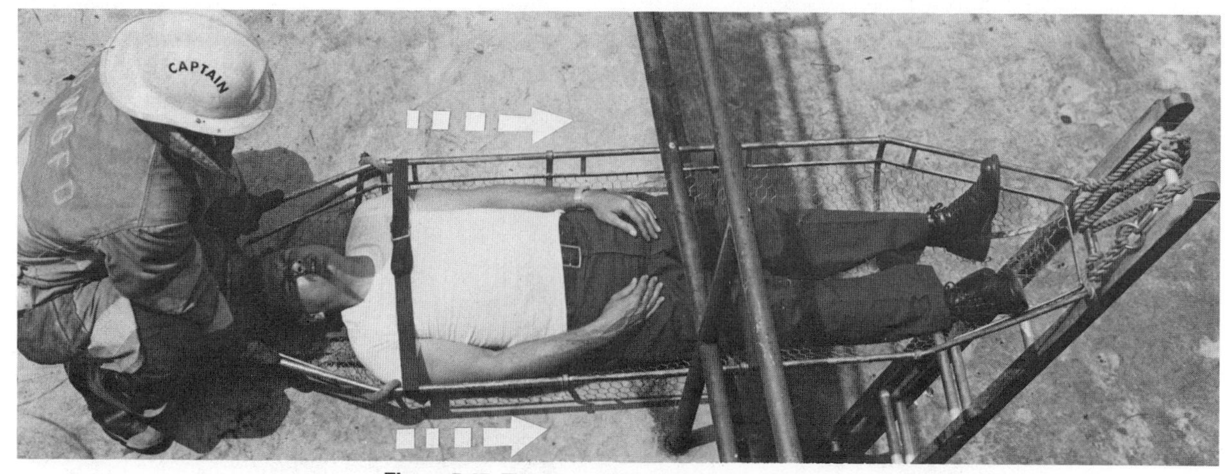

Figure 5.45 The Stokes stretcher is slid outward.

ground begins to lower the ladder by walking hand-over-hand down the rungs. The butt is heeled by the building (Figure 5.46).

Step 7: When the Stokes stretcher clears the windowsill or roof edge, the firefighters there assist on the anchor rope (Figure 5.47), while the firefighter on the ground continues lowering the ladder (Figure 5.48). The firefighters

5.4 1. False
2. True
3. False

Figure 5.46 The firefighter on the ground controls the downward movement of the ladder by gripping the rungs hand-over-hand.

Figure 5.47 When the Stokes stretcher clears the edge, the firefighters on the lifeline control the elevation of their end.

Figure 5.48 The firefighter on the ground continues to allow the ladder to tilt outward as the firefighters on the lifeline slack off just enough to let it come down smoothly.

on the anchor rope keep just enough tension on the rope that the Stokes stretcher will remain level (Figure 5.49).

Step 8: When the firefighter on the ground reaches the tip of the ladder (Figure 5.50), the top rung is grasped and the tip is lowered to the ground (Figure 5.51).

Step 9: When the ladder gets to the ground, the Stokes stretcher is untied and the victim removed to a medical facility. The ladder is raised back onto the building so that the firefighters there can descend. The butt is reset for the proper climbing angle.

Figure 5.49 The firefighters on the lifeline maintain just enough tension to keep the Stokes stretcher level as it descends.

Figure 5.50 The firefighter on the ground will reach the tip while the Stokes stretcher is still at about shoulder height from the ground.

Figure 5.51 The tip is grasped with both hands and the ladder is lowered until it is flat on the ground with the Stokes stretcher atop it. The firefighter on the ground has most of the weight load during this part of the operation.

WHEN ADDITIONAL FIREFIGHTERS ARE AVAILABLE

It is desirable to have guy lines attached to the top of the ladder so that side sway can be avoided. When additional firefighters are available this can be done. Two additional firefighters take positions on the guy lines at about 45 degrees to the ladder.

Using a Ladder as a Slide for Removing Injured Persons

An injured person on a stretcher or in a Stokes stretcher can be slid down a ladder without much trouble. The victim is secured to the stretcher using normal procedures.

Step 1: Handles of common tools such as short pike poles and shovels, or other short tools such as straight bars, halligan tools etc., are run through the bottom D-rings of army-style stretchers; or they are lashed across the bottom of a Stokes stretcher or backboard from handhold to handhold (Figure 5.52).

Figure 5.52 A Stokes stretcher with two tools tied to its bottom. The tools will act as skids for sliding down the ladder beams.

Figure 5.53 A bowline is used to secure a lifeline to the head of the Stokes stretcher in the manner shown.

Step 2: A lifeline is secured to the head of the stretcher with a bowline (Figure 5.53).

Step 3: If the ladder tip is not already positioned just below the windowsill, it is repositioned so that it is.

Step 4: One firefighter climbs onto the ladder to guide the stretcher down; or if manpower is short, a tag line is attached so the heelman can guide the stretcher down.

Step 5: One of the firefighters in the building anchors the rope about 20 feet (6 m) from the window by taking the rope around the body slightly below the buttocks or passing it around some fixed object.

Step 6: The two firefighters then gently slide the stretcher out of the window and onto the ladder as the anchorman keeps tension (Figure 5.54).

Step 7: When the stretcher is on the ladder, the two firefighters help on the anchor rope. The firefighter on the ladder or the firefighter holding the tag line guides the stretcher as it descends (Figure 5.55).

Figure 5.54 The Stokes stretcher containing the victim is slid out of the window while the firefighter anchoring the lifeline maintains tension on it.

Figure 5.55 All firefighters in the building anchor the lifeline while the firefighter holding the tag line guides the Stokes stretcher as it descends.

Using a Ladder Leaning Against a Building to Support Lowering of an Injured Person (Ladder Sling)

This arrangement may be used to lower someone upon whom the rescue knot (also called a rescue basket) has been tied.

Step 1: A ladder is raised directly in line with and above the window opening where the victim is located.

Step 2: One end of a lifeline has been used to tie the rescue knot on the victim. The standing end of the rope is then run over and down across four or five rungs. Then it is threaded back through between rungs so that it is on the building side of the ladder. The firefighter at the foot of the ladder grasps the rope and comes under the bottom rung.

Step 3: This firefighter then places one foot on the bottom rung and draws the rope tight.

Step 4: Another firefighter then lifts the victim clear of the windowsill with the assistance of the firefighter pulling on the rope.

Step 5: The firefighter on the rope gently lowers the victim. A tag line attached to the victim is used by another firefighter to steady the victim during lowering (Figure 5.56).

Figure 5.56 A lifeline is threaded through a ladder as shown. The end is tied to the victim using a rescue knot. A tag line is also used. The victim can then be eased out of the window and slowly lowered to the ground.

5.5 Check the correct response.

1. To lower a person in a Stokes stretcher in a horizontal position, the ________________ method is used.
   - ☐ A. Slide
   - ☐ B. Sling
   - ☐ C. Fulcrum

2. To lower a person lashed to a stretcher in a vertical position, the ____________________ is used.
   - ☐ A. Slide
   - ☐ B. Sling
   - ☐ C. Fulcrum

3. Lowering a victim with the ladder leaning against the building utilizes the ____________________ method.
   - ☐ A. Slide
   - ☐ B. Sling
   - ☐ C. Fulcrum

4. During the ____________________ method of lowering a victim, the building wall acts as the heelman.
   - ☐ A. Slide
   - ☐ B. Sling
   - ☐ C. Fulcrum

Using a Ladder as a Stretcher

One section of the short extension ladder, a short single ladder, or a roof ladder can readily be adapted for use as a stretcher. A salvage cover, blankets, turnout coat, or similar materials are laid upon the ladder rungs to provide support for the victim (Figure 5.57).

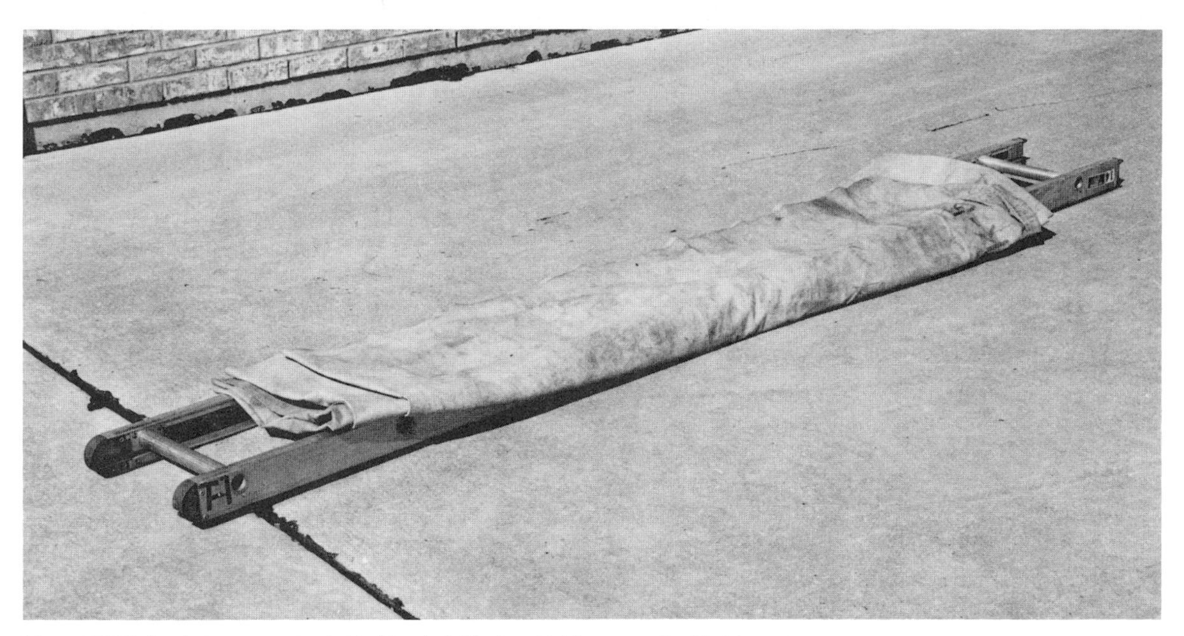

Figure 5.57 A salvage cover or similar item is folded and laid upon a short ladder.

The victim is secured to the ladder following the procedure shown in **Fire Service Rescue Practices:** *Rescue Situations Involving Elevation Differences* or as shown in Figure 5.58.

There are a number of variations in the way the ladder is used:

- As a substitute litter for two firefighters to carry an injured person (Figure 5.59)
- As a sledge for one firefighter to drag an injured person (Figure 5.60)
- As support for lowering a victim horizontally (Figure 5.61)
- As support for lowering a victim vertically (Figure 5.62)

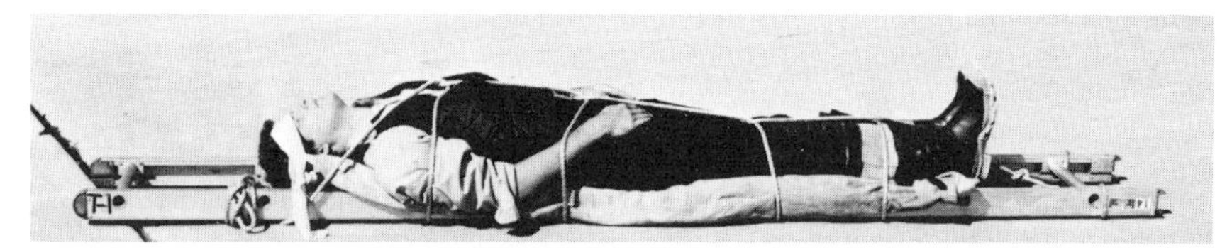

Figure 5.58 The victim is lashed to the ladder.

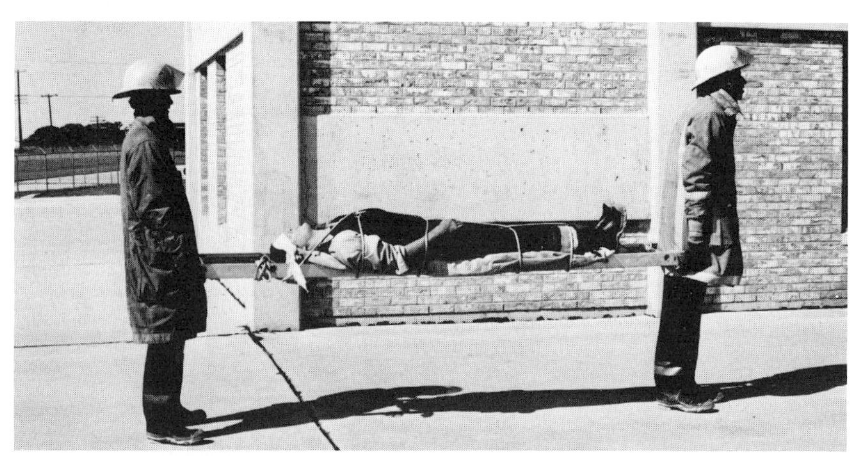

Figure 5.59 Two firefighters carry the ladder as if it were a litter.

Figure 5.60 The ladder sledge can be dragged by one firefighter.

5.5 1. C
2. A
3. B
4. C

Figure 5.61 Here the ladder is being used as a support for lowering a victim lying horizontally.

Figure 5.62 The ladder is used to support a victim being lowered vertically.

Using a Ladder for Ice Rescue

The ability of people to survive for extended periods of time in cold water is very limited. The loss of body heat in water is about 25 times greater than the loss of body heat in the normal atmosphere, so fast action is necessary to prevent death. The immediate goal is to keep a victim from sinking and drowning, then to retrieve the victim.

CAUTION: It is important that firefighters not complicate the problem by taking foolish risks and also falling through the ice.

When the ice is intact between the shore and the spot where the person fell into the water, a ladder is particularly well suited to the job because it provides a means of reaching the victim from a relatively safe location, the victim can get hold of it, and then be retrieved with it. The pressure on the surface of the ice caused by drawing the victim from the water is distributed over a large area by the ladder being used. This may prevent further ice breakthrough. If the ladder will not reach from the shore to the victim, a rope is also used. The steps needed to implement this procedure are as follows:

Step 1: A rope is tied to the shore end of the ladder (Figure 5.63).

Step 2: The ladder is slid out across the ice to the victim (Figure 5.64). It may be necessary for the firefighters handling the ladder to lie on the ice to get additional reach.

Figure 5.63 A lifeline is secured to the shore end of the ladder.

Figure 5.64 The ladder is pushed across the ice towards the victim.

CAUTION: If firefighters handling the ladder have to advance onto unstable ice, lifelines are attached to them using the rescue knot.

Step 3: The firefighters with the ladder steady and guide it (Figure 5.65), while another firefighter holds the rope (Figure 5.66).

Step 4: After the victim grasps the ladder, the firefighter holding the rope pulls the ladder and the victim toward shore. The firefighters handling the ladder assist.

Figure 5.65 Firefighters steady and guide the ladder.

Figure 5.66 A firefighter on shore mans the anchor rope.

5.6 True or False.

	True	False
1. In cold water the loss of body heat is 5 times greater than the loss of body heat in normal atmosphere.	☐	☐
2. The first objective of an ice rescue is to keep the victim from drowning.	☐	☐
3. When a ladder is used for ice rescue, it not only provides a grab hold for the victim, but also distributes some of the pressure of the victim being pulled from the water over a large area of the ice, preventing further breakthrough.	☐	☐

Figure 5.67 An auto tire mounted on a rim can be tied to the end of a ladder to provide buoyancy for either ice or water rescue. *Courtesy of Springfield, Mo. Fire Department.*

Using a Ladder Float Drag for Ice and Water Rescue

A modification of the procedure for sliding a ladder onto ice to reach a person who has fallen through the ice is applicable both for thin ice or open water rescue: A mounted regular spare tire or other buoyant object is lashed to one end of a ladder (Figure 5.67). The ladder, with the spare tire attached, is slid tire end first out to the victim. The operation is otherwise the same as for the ice rescue detailed in the preceding paragraphs.

Prying with a Ground Ladder

It may be necessary to remove persons from beneath an overturned vehicle, fallen wall, collapsed roof or floor, etc. In some instances, a ground ladder can be used as a lever to accomplish this (Figure 5.68).

CAUTION: This may damage the ladder. All further use of the ladder should be discontinued until it can be tested.

Figure 5.68 A ladder can be used to pry a heavy object off a victim.

Using Ground Ladders for Shoring and Other Uses During Trench Rescue

SHORING

Twenty-four foot (7 m) or longer extension ladders in the retracted position may be used as a part of the shoring, especially when sidewall collapse has occurred.

NOTE: Shoring is a special topic of its own. For complete information on how to shore, consult texts developed specifically on the subject. The intent here is only to provide knowledge of how the ladder is used during these operations.

One Sidewall Collapse

When one side of the ditch has fallen in the rescuers have to create an artificial wall to shore against. The procedure is

5.6 1. False
2. True
3. True

Step 1: A four foot (1.3 m) to five foot (1.5 m) section of two foot (.6 m) by eight inch (250 mm) plank is secured to the midpoint of the ladder by driving nails through the plank adjacent to rungs (Figure 5.69) and then turning the ladder up on beam and bending the nails over the rungs (Figure 5.70).

Figure 5.69 A fully retracted extension ladder is used. A four foot (1.3 m) to five foot (1.5 m) length of two foot (.6 m) by eight inch (250 mm) plank is secured to the midpoint of the ladder using nails driven beside rungs.

Figure 5.70 After the nails are driven through the plank, they are toed over the rungs to hold them in place.

Step 2: Ropes are attached to each end of the ladder and it is lowered into the trench (Figure 5.71).

Figure 5.71 After ropes are attached to each end, the ladder is lowered into the trench on the cave-in side.

Step 3: Firefighters maintain the ladder in this position by holding onto ropes until a timber or shoring plank is placed between the plank that is attached to the ladder and shoring that is placed on the intact side of the ditch (Figure 5.72).

Step 4: Shoring panels are placed against the ladder on either side of the plank and braced. Then the plank is removed and a shoring panel put in its place (Figure 5.73).

Figure 5.72 The ladder is held in position by firefighters on the ropes while a timber brace is put in place.

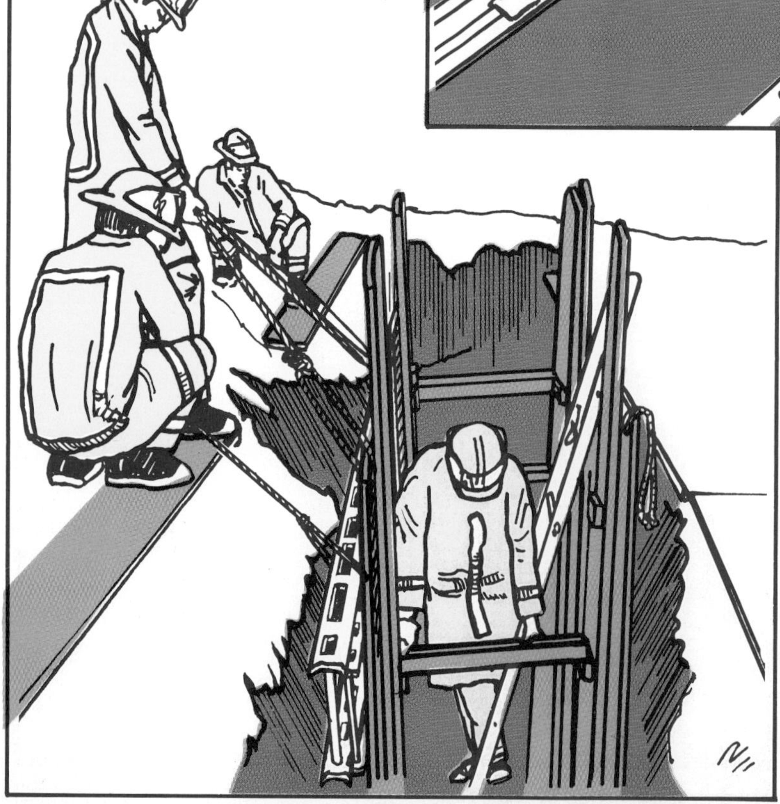

Figure 5.73 Shoring panels are put in place at either end as shown, then the brace and the plank are removed and another shoring panel put in place.

Both Sidewalls Collapsed

When both sidewalls of the ditch have collapsed, two 24 foot (7 m) extension ladders are utilized.

Step 1: Planks are secured to each ladder near both ends instead of in the middle (two required for each ladder) (Figure 5.74).

Step 2: Panels will be placed in the ditch on either side of the areas where the walls have collapsed. They are braced and then the ladders are lowered into place (Figure 5.75).

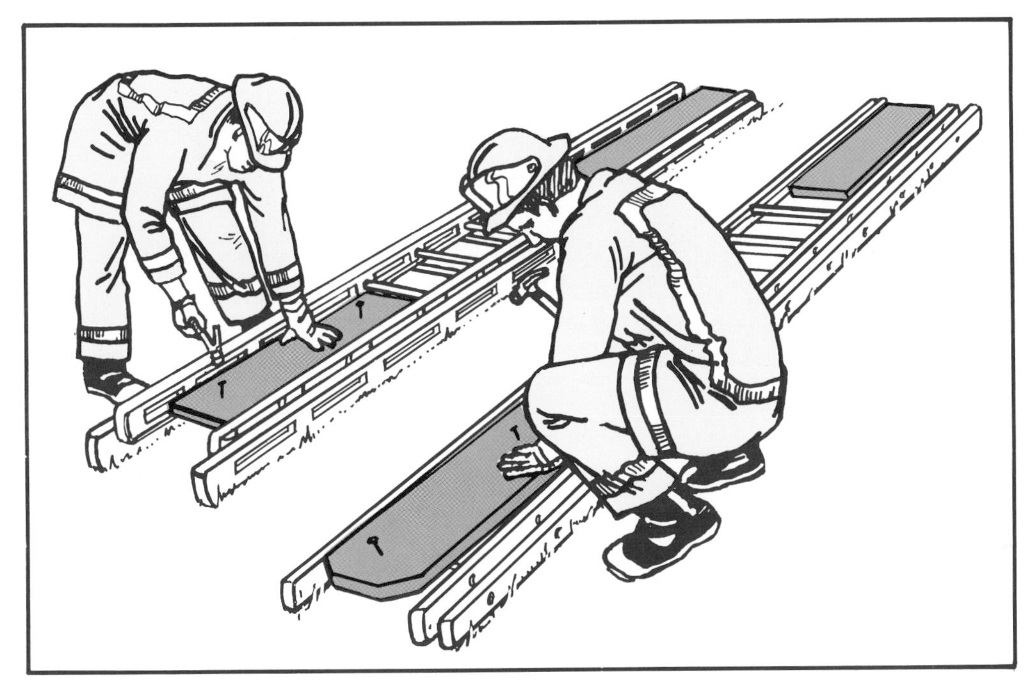

Figure 5.74 Planks are secured to each extension ladder near both ends.

Figure 5.75 The ladders are lowered into place after shoring panels have been placed on both sides of the caved-in area.

Step 3: Bracing will then be installed from one ladder to the other where planks are nailed; then panels are installed in the collapsed area (Figure 5.76).

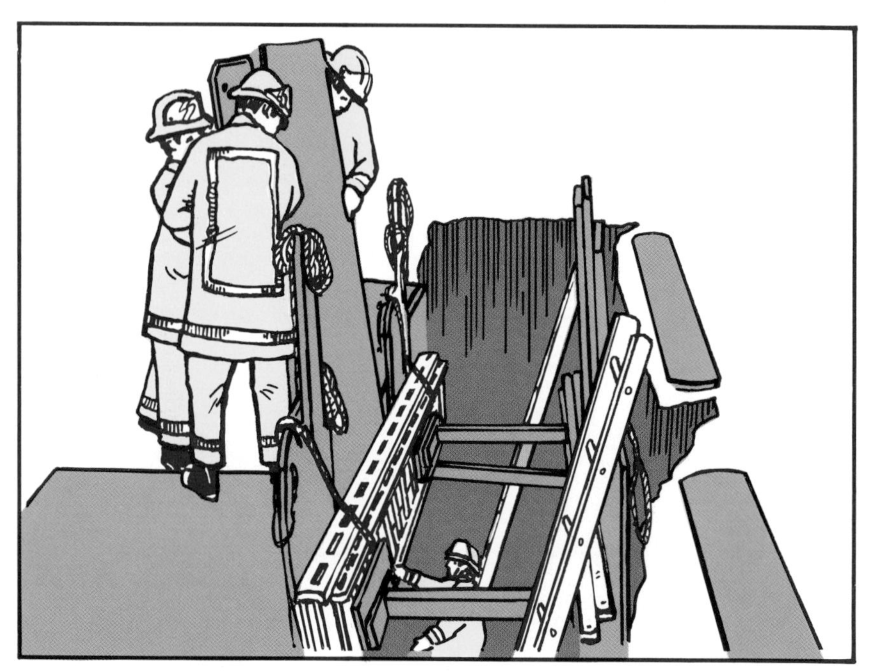

Figure 5.76 Bracing is then installed between planks. Then panels are installed in the collapsed area.

PROTECTING THE TRENCH DURING INCLEMENT WEATHER

It may be desirable to cover the collapsed area during inclement weather. A combination ladder used as an A-frame, two pike poles, a plank, and a tarp are employed for this purpose. The A-frame ladder is used to support the center of the tarp as in Figure 5.77.

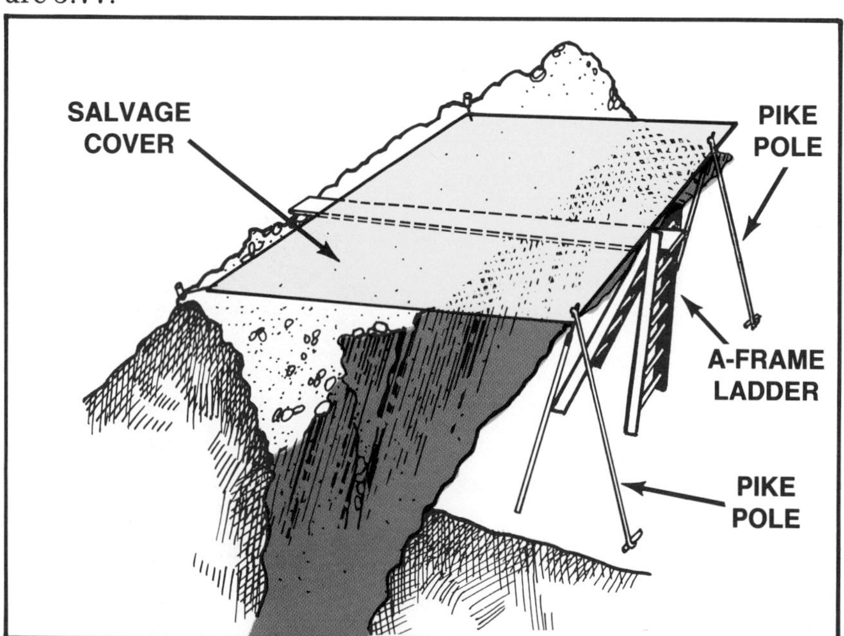

Figure 5.77 A combination ladder in the A- frame configuration serves as the basis for a shelter that is erected over a collapse area in bad weather.

SUPPORTING UNBROKEN UTILITY LINES

When utility lines cross a trench and the cave-in has not broken them, it is important to support them so that further operations do not cause them to break. A single ladder can be used for this purpose. It is laid across the trench over the utility lines and a rope is used to tie the pipes to the ladder so that the ladder is supporting them (Figure 5.78).

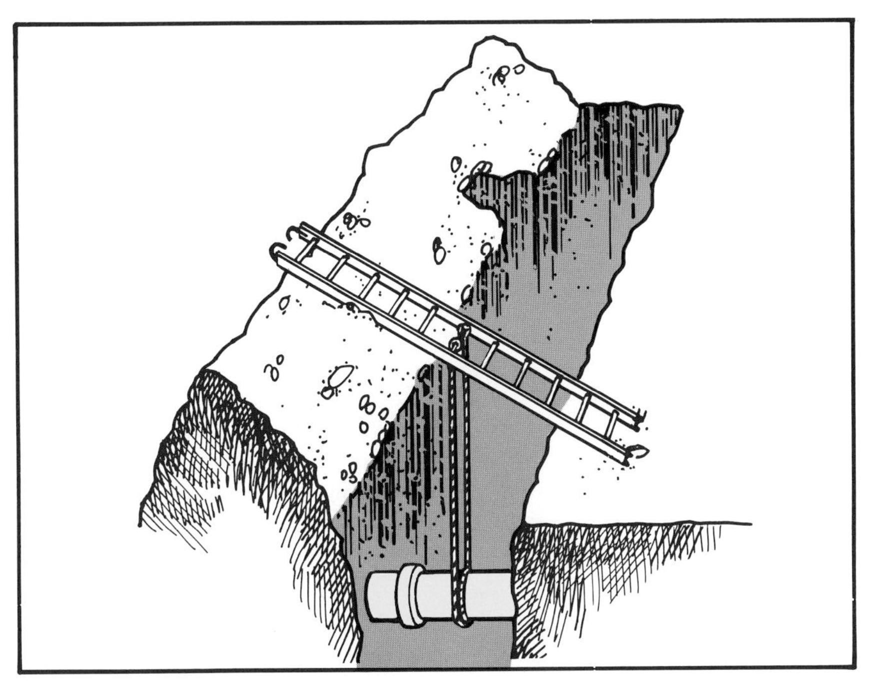

Figure 5.78 A ladder is laid across the trench right over the water, gas, or sewer line. A rope is tied at midpoint to support the utility line.

5.7 True or False.

	True	False
1. During trench rescue, when shoring up a one side-wall collapse, a plank is secured to both ends of an extension ladder.	☐	☐
2. To support utility lines that cross a trench near a collapse area, dirt should be backfilled at that point.	☐	☐

Making Devices for Hoisting

Because of obstacles or space constraints, it may be necessary to move an injured person or a load straight up. A block and tackle used in conjunction with riggings such as jib arms, gin poles, and A-frames can be used to do this. Fire department ground ladders are adaptable for use as "poles" for rigging these devices.

JIB ARMS

The block and tackle is rigged on one end of a single ladder. That end of the ladder is laid out of a window so that there is no more than 18 inches (457 mm) between the windowsill and the point where the block and tackle is secured. The other end of the ladder is secured by ripping up flooring and lashing it to floor joists (Figure 5.79).

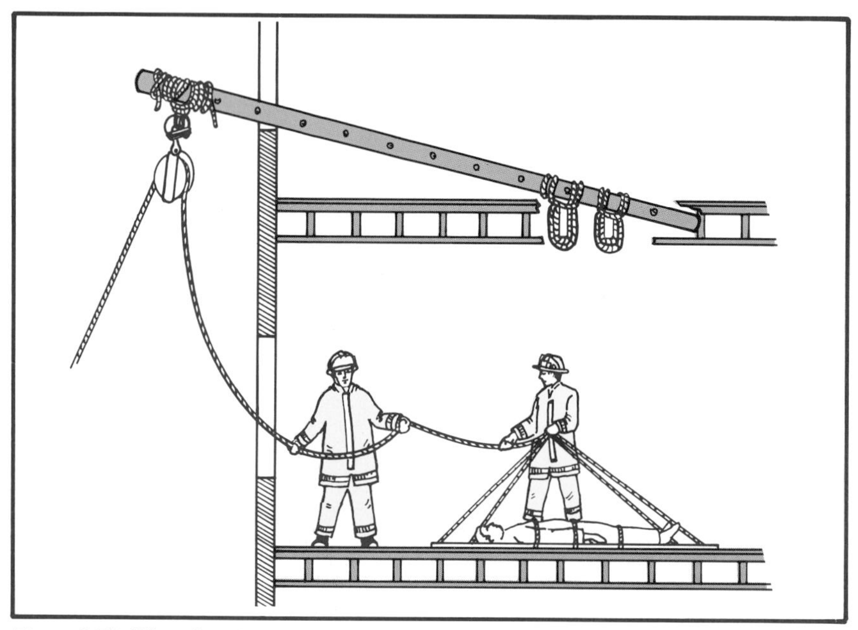

Figure 5.79 When there is a way to anchor one end, a ladder can be used as a jib arm.

GIN POLE

A gin pole is a single standing pole that is angled over the load and supported by rope guy lines. The hoisting rope is threaded down the pole to keep the pole in compression and to minimize shear tension (Figure 5.80).

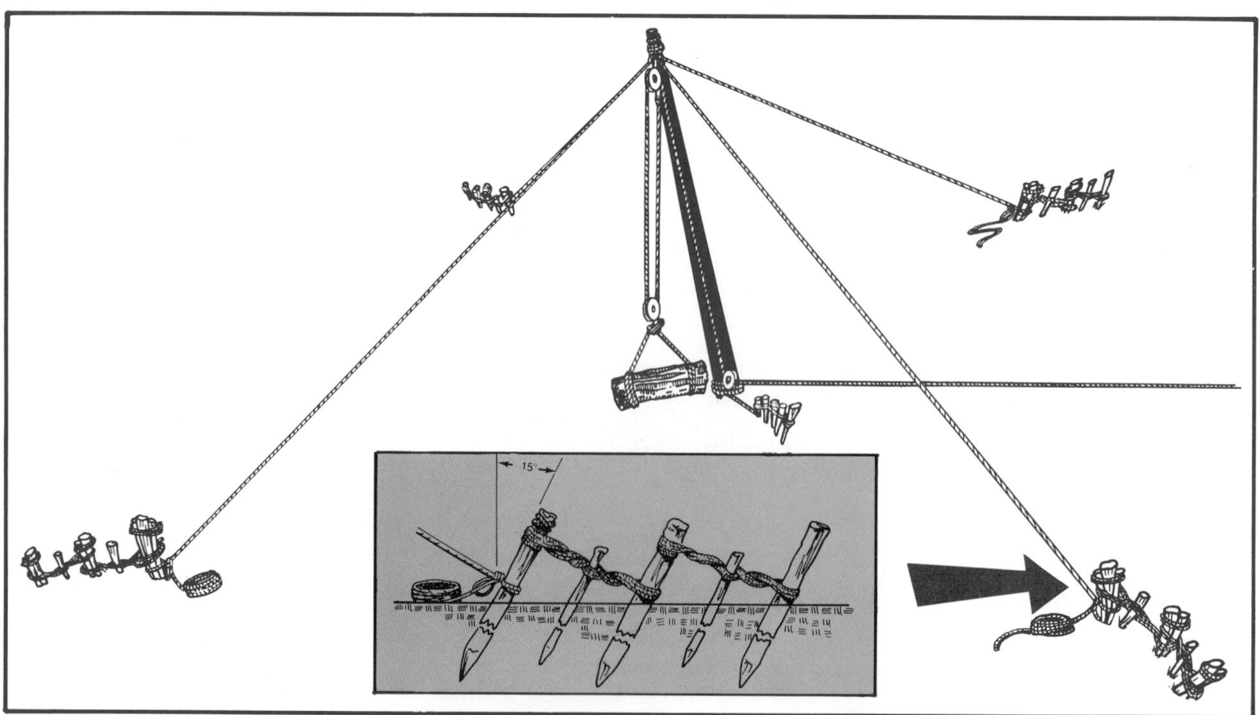

Figure 5.80 A standard gin pole.

A straight ladder makes an excellent gin pole. A rope, block and tackle, and four tie-down stakes are required along with the ladder. The ladder is laid at the selected site with the butt next to the hoisting area. The rope is played out and doubled to form two equal lengths of line with the bend at the tip of the ladder. Approximately three feet (1 m) from the bend, on each of the two lines, a standing bight is made that is large enough to slip over the tips of the beams. The loop between the bights is turned to form a figure eight and the slack pulled to the outside loop, which is dropped over the top of the ladder. The line on the opposite side of the ladder is grasped and pulled out, and the figure eight turned. After the procedure has been completed three times, all of the lines on the opposite side of the ladder are pulled forward and the clasp of the standing pulley is slipped over them. The hoisting end of the block and tackle rope is then run to the butt of the ladder and under the bottom rung. The guy line tie-down stakes are driven into the ground a ladder's length from the butt and half a ladder's length to the sides. The ladder is then raised into position and the guy ropes are secured to the stakes. Stakes are driven into the ground adjacent to the ladder's butt. The beams are lashed to the stakes. Figure 5.81 shows the completed gin pole assembly.

5.7 1. False
2. False

Figure 5.81 A roof ladder being used as a gin pole.

Alternate Method Using a Fire Department Pumper

An alternate rigging arrangement, using a fire department pumper to support the gin pole, has been found by many firefighters to be quicker and easier to rig. It also requires fewer special items of equipment and so may be more practical for the general fire service.

A pumper's tailboard is used as footing for the ladder butt, and the rear of the apparatus body, where it abuts the tailboard, is used as a brace to keep the ladder butt from slipping. The pump intake ports (large suction inlets) are used as anchors for tying off the guy lines. This eliminates the need for stakes, or binding lines, and saves time. It also simplifies the training process.

To utilize this method, a pumper is positioned so that the tailboard faces the site. A folded salvage cover or other padding is placed on the center of the tailboard with one edge turned up against the back of the apparatus body. The purpose of this padding is to prevent the ladder butt spurs from gouging the surfaces they are in contact with during this operation.

A single or roof ladder is rigged, as previously described, except that it is not lashed to any stakes. The ladder is placed on the padded area of the tailboard, then raised into position (Figure 5.82). The guy ropes are tied off to the large suction intake ports of the apparatus's fire pump (Figure 5.83). Care must be taken to balance the tension on the two ropes. The pumper is then maneuvered into position (Figure 5.84).

Figure 5.82 A roof ladder is placed on the tailboard where it abuts the rear of the apparatus body. The salvage cover keeps the butt spurs from scarring the rear of the apparatus.

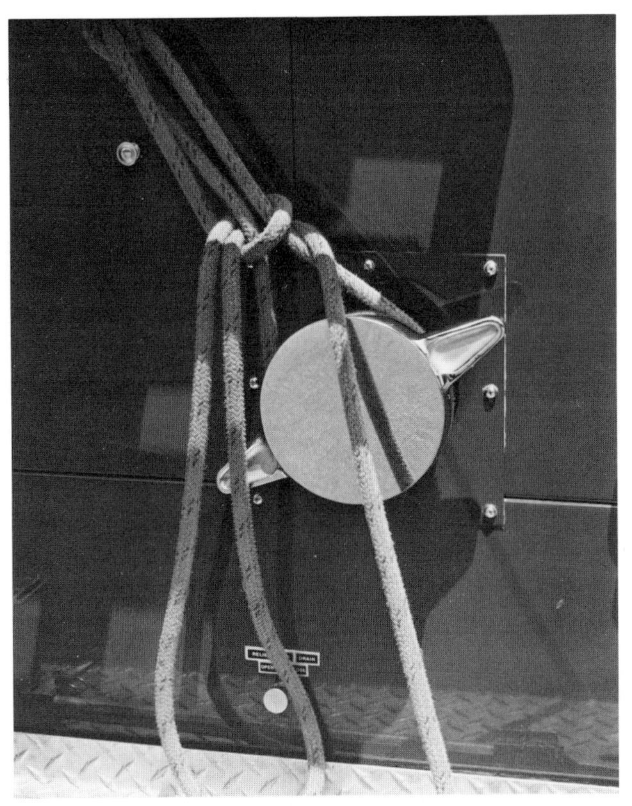

Figure 5.83 Guy lines are tied to the pumper intakes.

Figure 5.84 The pumper is backed to center the block and tackle over the person or object to be lifted.

Alternate Procedure when a Block and Tackle are not Available

Two ladder belts can be used with a rope when a block and tackle are not available. The hooks on the ladder belt are used instead of the pulleys of the block and tackle (Figures 5.85 and 5.86).

Figure 5.85 Safety belts being used in lieu of pulleys.

Figure 5.86 The safety belt is placed around the second rung and buckled. The hook is used as a sliding surface for the rope.

A-FRAME

The A-frame is particularly suited for raising loads out of trenches, wells, or other similar excavations. The A-frame is easiest to make when two single ladders of different widths are used. The steps for rigging the A-frame are as follows:

Step 1: Both ladders are placed on the ground on beams. The tips are brought together and the top rungs are lashed together to form the top of the A.

Step 2: A rope is used to form guy lines needed to minimize side-to-side movement. A bowline-on-a-bight is tied in the center of the rope. The loops should be long enough to pass over the ladder beams at the top of the A.

Step 3: The block and tackle is attached to the bight of the bowline and the A-frame is raised over the objective (Figure 5.87).

Figure 5.87 A bowline-on-a-bight is used to hold the two ladders together at the top and to support the block and tackle.

Step 4: A ladder is placed flat on the ground at right angles to the A-frame and the sway lines are secured to either end; or instead of a ladder, stakes are used to tie off sway lines.

Step 5: A rope hose tool or other piece of line is used to form the cross leg of the A or the ladder butts are staked and lashed so that the A-frame will not spread apart when the load is picked up (Figure 5.88).

Figure 5.88 The ladders must be lashed to stakes or otherwise secured so that they do not spread apart when the load is applied.

Step 6: A tag line is attached to the stretcher or victim to be lifted. When the victim clears the top of the excavation, slack can be given on the hauling line and tag line taken in, bringing the load to a safe location.

NOTE: Extension ladders should only be used in the retracted position, as loads would otherwise be limited to the strength of the ladder pawls.

5.8 Fill in the blanks.

1. Ground ladders can be used to make three types of devices to lift persons straight up, these are ________________, ________________, and ________________.

5.9 True or False.

	True	False
1. The alternate method of making a gin pole utilizing a pumper is more practical because it eliminates the need for stakes and lashing to them.	☐	☐

Extending Reach from Aerial Platforms

The following procedure can be used to extend the reach of aerial platforms that fall short of reaching persons who are in danger and cannot be reached otherwise. It should only be performed by well-trained crews. Single ladders with nonskid feet will help keep the butt from slipping. Extension ladders should only be used if no other option exists.

Step 1: Once the apparatus is positioned, a ground ladder is raised into the basket (Figures 5.89 and 5.90) and tied off. The tie is made with a rope hose tool or a bowline

Figure 5.89 A roof ladder is placed butt first into the aerial tower apparatus basket.

knot around a structural member of the basket or the nozzle piping (Figure 5.91). The butt is heeled by placing it against the back of the basket (Figure 5.92).

CAUTION: Some basket railings may not be strong enough to support a ladder tied to it as in the illustration. When this is the case, the ladder is tied to piping or other substantial structural members. Each apparatus should be evaluated ahead of time and a policy established that details how and to what the ground ladder is to be secured for this evolution.

Figure 5.90 The roof ladder is tilted up to approximately the proper angle of inclination.

5.8 jib arm, gin pole, A-frame

5.9 1. True

Figure 5.91 The ladder is lashed to the basket railing. Note that on some apparatus the basket railing is not substantial enough and an alternate lashing arrangement will be necessary.

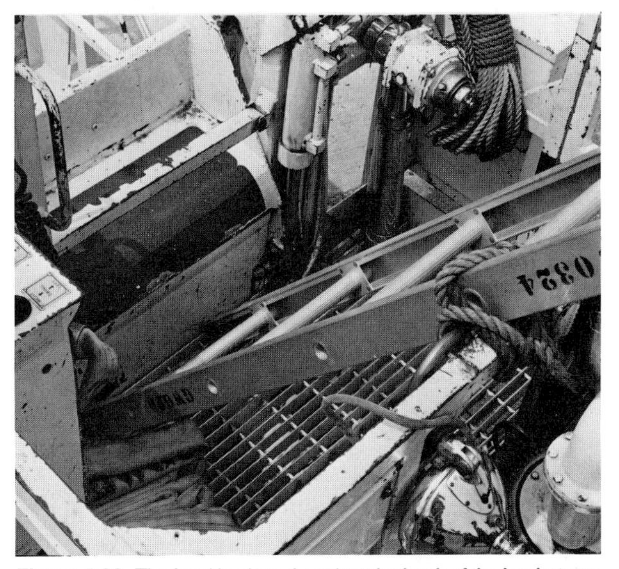

Figure 5.92 The heel is placed against the back of the basket or a similar heeling arrangement is made.

Step 2: One of the firefighters in the basket steadies the ladder while the platform is raised into the working position (Figure 5.93).

Step 3: The ladder is adjusted so that both beams rest evenly against the wall (Figure 5.94).

Step 4: One firefighter steadies the ladder while the other firefighter ascends it (Figure 5.95).

Figure 5.93 The aerial platform is raised into position. If necessary, the second firefighter steadies the roof ladder.

Figure 5.94 The aerial platform is adjusted so that both beams of the roof ladder rest evenly upon the windowsill or roof edge.

Figure 5.95 One firefighter climbs while the other steadies the roof ladder.

SAFETY CONSIDERATIONS

- Longer ladders present greater handling difficulties.
- The ground ladder must be supported against the building before climbing.
- The basket must not be moved after the ladder is in position.
- Lifelines or life belts must be used by firefighters.
- NEVER EXCEED maximum load ratings or operating capabilities (reach or angle) of elevating platforms or tower ladders.

5.10 True or False.

	True	False
1. The procedure for extending the reach of an aerial platform should only be performed by well trained crews.	☐	☐
2. Control of the aerial platform is routinely maintained within the platform during periods when a ladder used to extend the reach is being climbed.	☐	☐
3. When the tip of a ground ladder in the basket of an aerial platform is against the wall of a building, its weight is not considered in figuring the maximum loading of the aerial platform.	☐	☐

Using a Roof Ladder as a Pompier Ladder for Rescue

When a ground ladder comes one floor short in its reach, a roof ladder can be used as a pompier ladder to reach the next floor.

NOTE: This is not a recommended practice: it should only be used in extreme emergency for access to perform rescue.

Step 1: One firefighter ascends the first ladder and enters the highest window which it reaches. (Ascent may be via interior stairs if conditions permit). This firefighter will assist in placing the roof ladder and will secure it while it is being climbed.

Step 2: A second firefighter opens the hooks of the roof ladder and carries it up the first ladder with the hooks facing outward. The firefighter in the window helps steady the roof ladder while the firefighter on the ladder leg locks in or secures the life belt, strap, or rope hose tool (Figure 5.96a on next page).

Figure 5.96a The roof ladder is passed to the firefighter in the window.

Figure 5.96b Both firefighters pass the roof ladder upward to the window above.

Figure 5.96c The roof ladder hooks are placed over the windowsill.

Figure 5.96d The firefighter at the window holds the roof ladder securely in place while the firefighter on the extension ladder transfers to the roof ladder.

Figure 5.96e The firefighter climbs to the window above. Note that the key to the safety of this operation is the firefighter in the window opening holding the roof ladder securely against the side of the building.

5.10 1. True
2. False
3. False

Step 3: Both firefighters work together to pass the roof ladder upward to the window above. The roof ladder is then turned so that the hooks are inward (Figure 5.96b) and they are placed over the windowsill (Figure 5.96c).

CAUTION: It may be necessary to break out window glass to insert the roof ladder hooks, in which case extreme care must be taken to prevent injury from falling glass.

Step 4: The firefighter at the window grasps the roof ladder by the beams and uses body weight to hold it against the side of the building. Smaller statured persons should place both feet against the windowsill and brace themselves for added leverage.

Step 5: The firefighter on the other ladder transfers to the roof ladder (Figure 5.96d) and climbs to the window above (Figure 5.96e).

Using a Ladder for Emergency Ventilation

When it is necessary to ventilate quickly, such as clearing a building of smoke for search and rescue, ladders may be utilized to break out windows without a firefighter having to ascend the ladder.

The ladder is raised in line with the window and the tip is dropped against the window glass (Figure 5.97).

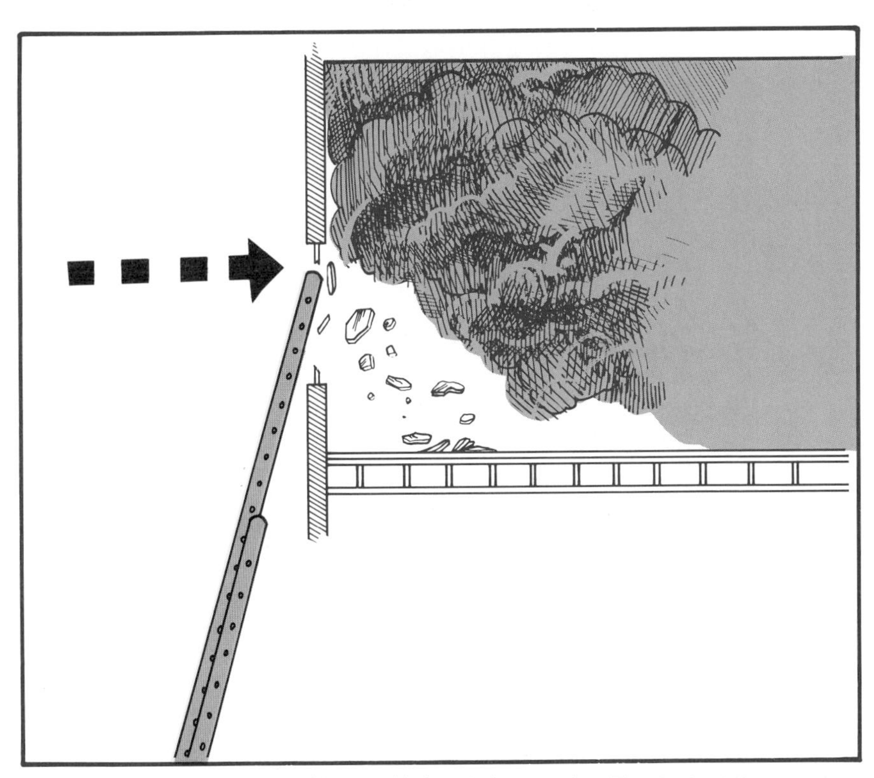

Figure 5.97 The ladder is raised in line with the window opening. The tip should be even with the top pane. It is dropped against the window to break out the glass.

CAUTION: Firefighters must be alert for the hazard of falling glass, particularly glass sliding down the beams.

The ladder is then carried vertically to the next window and the process is repeated. In this way, many windows can be ventilated in a short time.

NOTE: This procedure may damage the ladder beams near the top, particularly wood and fiber glass models, and repair will be necessary.

FIRE FIGHTING OPERATIONS AND USES

Fire fighting operations and uses for ground ladders are many and varied and are limited only by the imagination. The better known ones are detailed on the following pages.

Directing Fire Streams from Ground Ladders

When a fire stream is going to be directed onto a fire from a ladder, it is necessary to secure the hoseline to the ladder at the vantage point where the operation will take place.

Step 1: A dry hoseline with nozzle attached is advanced up a ladder to the desired vantage point. The nozzle and approximately two feet (.6 m) of hoseline are extended between two rungs so that the hoseline is draped over a rung, as shown in Figure 5.98.

Figure 5.98 The nozzle and approximately two feet (.6 m) of hoseline are draped over the ladder rung opposite the window opening.

CAUTION: The nozzle must be CLOSED during this part of the evolution.

Step 2: A rope hose tool or similar device is secured to the hoseline a rung or more below where the hoseline drapes over the rung (Figure 5.99).

Step 3: The hook end of the rope hose tool is wrapped around the rung below the one that the hoseline is draped over. Then it is passed over the hoseline. If there is still adequate slack, it is passed over the rung again on the other side of the hoseline. Then the hook is put over the rung that the hoseline is draped over. If there isn't enough slack, no second loop is made around the rung and the hook is placed over the rung that the hoseline is draped over (Figure 5.100).

Figure 5.99 A rope or chain hose tool is secured to the hoseline at the point shown.

Figure 5.100 The hose tool is used to secure the hoseline to the ladder rungs.

Step 4: The firefighter then descends several rungs and the hoseline is charged (Figure 5.101).

NOTE: The firefighter stays clear of the point where the hoseline bends over the rung because when the hoseline is charged there is a tendency for it to kick back and some adjustment of the lashing may be required. If the firefighter is standing behind it, there is a possibility of being knocked off the ladder.

Step 5: The firefighter then assumes a position on the ladder to operate the nozzle and leg locks in (Figure 5.102).

Figure 5.101 The firefighter stays clear of the area where the hose bends as the hoseline is charged.

Figure 5.102 The firefighter climbs into position to operate the nozzle, leg locks, and proceeds to direct the stream into the window opening.

5.11 True or False.

	True	False
1. After securing a hoseline to a ladder for directing a fire stream, the correct procedure is for the firefighter to leg lock in at nozzle level, grasp the nozzle, and call for water.	☐	☐

Hoisting Ladders

Ladders are sometimes needed on roofs or upper floors of buildings. Hoisting with a rope is probably the easiest and quickest way to accomplish this task. There are two methods and the one used is a matter of departmental policy.

The first method may be used on any ladder. Its primary advantage is in providing a fulcrum whereby the ladder can be swung into a window or over the cornice or parapet of a roof.

Step 1: Tie a large bowline knot and place the loop from the underneath side between the rungs of the ladder about one-third the length of the ladder from one end (Figure 5.103).

Step 2: Pull the loop and knot to the end of the ladder and place the loop over the beam ends (Figure 5.104).

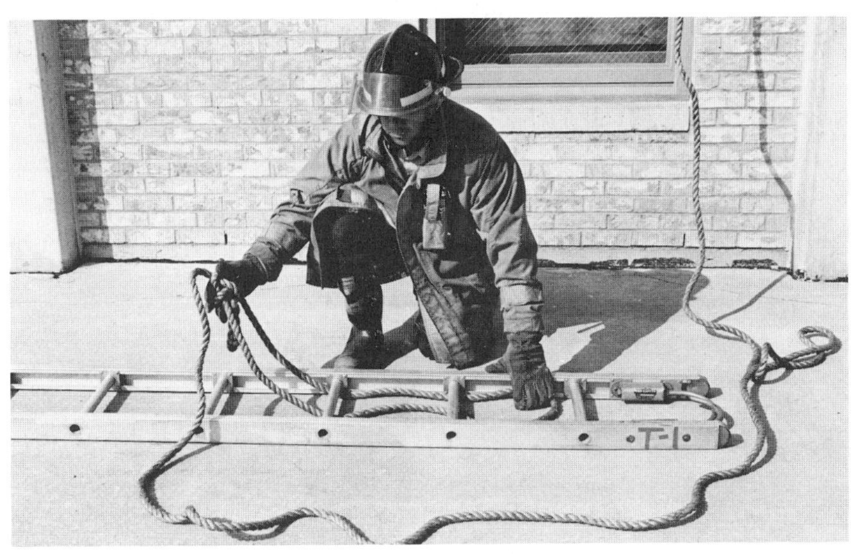

Figure 5.103 A bowline knot loop is stretched under several rungs in the manner shown.

Figure 5.104 The loop is pulled to the end of the ladder and it is placed over the beam ends.

Step 3: Complete the tie by pulling on the standing part of the rope (Figure 5.105).

Step 4: Firefighters in the building take up the slack line. The firefighters on the ground assist in lifting the ladder to vertical.

Step 5: The ladder is then hoisted up the outside of the building (Figure 5.106).

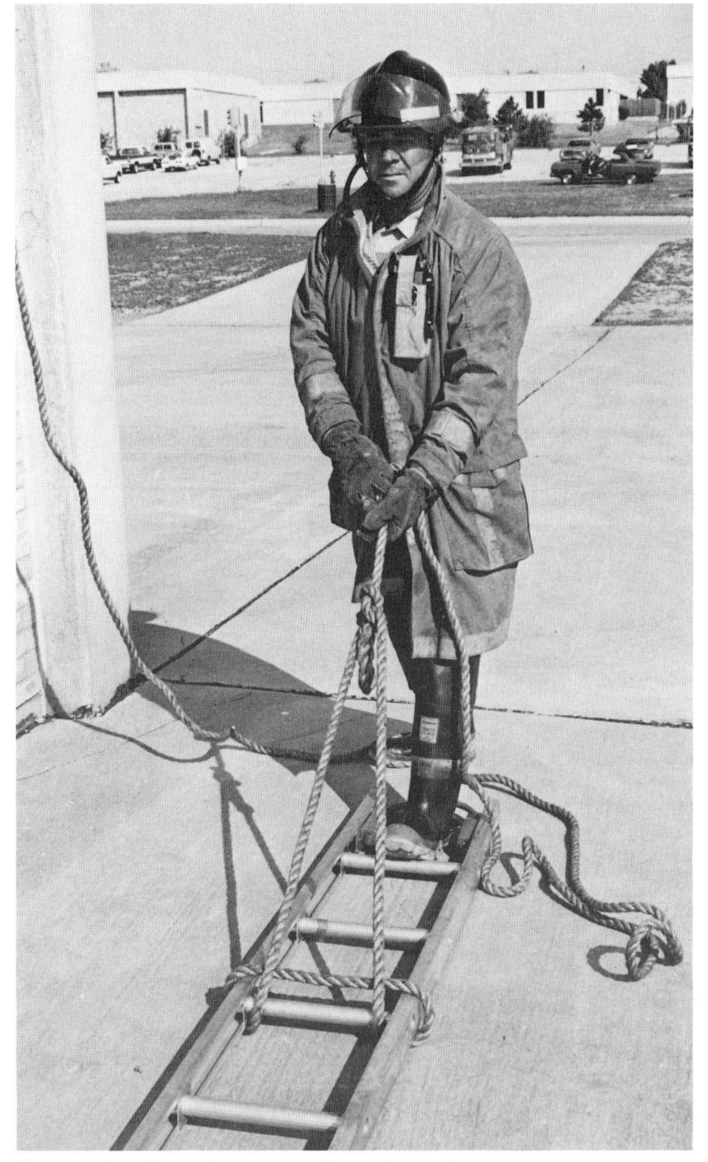

Figure 5.105 The tie is completed by pulling on the standing part of the rope.

Figure 5.106 The ladder is hoisted up the outside of the building. Note that the tie being behind the ladder keeps the tip tilted out so that it will not snag on protrusions.

NOTE: As a result of the rope loop being fed from underneath the ladder, the standing part of the rope and the bowline are between the ladder and the building as it is hoisted. This tilts the tip outward as the ladder is hoisted so that it will not snag on windowsills, etc. This arrangement, and tying it one-third of the way

5.11 1. False

down the ladder, projects the tip above the roof edge or windowsill when it reaches the point where the hoisting firefighters are located. This provides a short length of ladder that can be grasped for leverage when the ladder is pulled over the roof edge or windowsill.

The same tie can be used to lower a ladder except that the standing part of the line should be outside the ladder. The procedure swings the butt outward to keep it from catching on obstructions (Figure 5.107).

Figure 5.107 When a ladder is lowered, the tie is made on the other side so that the heel will kick outward.

The second method is as follows:

Step 1: The ladder is placed on the ground on one beam. The end of the rope is threaded between two rungs about one third the length of the ladder from one end (Figure 5.108).

Step 2: The end of the rope is then carried to the other end of the ladder where a clove hitch and a safety are tied around the upper beam (Figure 5.109).

Step 3: As the firefighters above pull on the standing part of the line the firefighters on the ground guide the ladder into a hoisting position (Figure 5.110).

Step 4: The ladder is then raised to where it is needed (Figure 5.111).

Figure 5.108 The ladder is resting on one beam. The end of the rope is threaded between two rungs about one-third the length of the ladder from one end.

Figure 5.109 The end of the rope is strung to the other end of the ladder where a clove hitch and safety are tied around the upper beam.

Figure 5.110 As the firefighters above hoist the ladder, those on the ground guide it.

Figure 5.111 The ladder is then hoisted up the outside of the building.

5.12 True or False.

	True	False
1. The method of hoisting where the bowline and the standing part of the rope are between the ladder and the building has the advantage that it tilts the tip outward when being hoisted so that it will not snag on protrusions.	☐	☐

Bridging for Fire Fighting

BRIDGING BETWEEN BUILDINGS, OVER TRENCHES, DITCHES, ETC.

The same procedures for bridging are used as when rescue is the objective (See page 305). It is still important to remember that this is not a recommended ladder use and should be undertaken only if necessary. The weight of a hoseline on the ladder must be counted in considering weight loading of ladders used in this manner.

BRIDGING OVER HOLES IN FLOORS

Holes sometimes burn through floors at locations needed for access, such as just inside the front door. A single or roof ladder can be laid horizontal through the doorway and across the hole so that firefighters can crawl in (Figure 5.112). This arrangement can be improved if there are any boards available that can be laid flat on the ladder, making a continuous walkway (Figure 5.113).

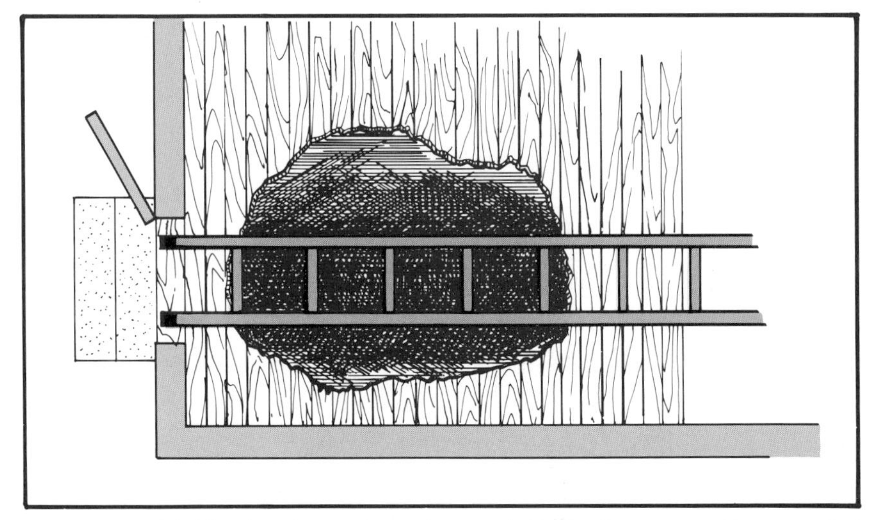

Figure 5.112 A ladder placed over a burned out portion of the floor can provide access that would otherwise not be possible. The same arrangement can be used over floor pits such as in older garages.

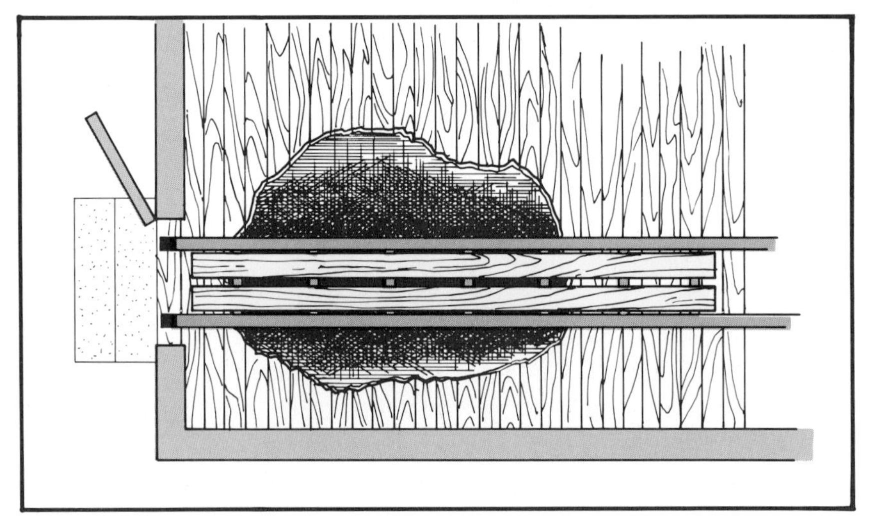

Figure 5.113 A plank laid on the ladder will make accessibility easier.

BRIDGING TO REACH BUILDINGS UNDER CONSTRUCTION

The same process can be used to gain access to buildings under construction that have not been backfilled around basement walls (Figure 5.114).

5.12 1. True

Figure 5.114 Bridging across excavations around buildings under construction may be the only means of access.

BRIDGING OVER FENCES AND WALLS

This procedure requires two ladders or an extension ladder that has been disassembled to make two single ladders. A standard pumper complement of a 24-foot (7 m) two section-extension ladder and a 14-foot (4 m) roof ladder is frequently used.

When the specific point of bridging is not critical, a point adjacent to a fence post is recommended. The procedure is as follows:

Step 1: The larger or heavier ladder, if there is one, is raised into place against the fence.

Step 2: One firefighter climbs this ladder and leg locks in; the second firefighter passes the other ladder, butt first, to the firefighter on the first ladder.

NOTE: If only one firefighter is available, the second ladder is raised butt upward beside the first ladder before the firefighter ascends it.

Step 3: The firefighter on the ladder grasps the second ladder and slides or passes it over the top of the fence or wall.

Step 4: The second ladder butt is then lowered to the ground. The ladder is shifted so that one of its beams abuts a beam of the first ladder; then the two are lashed together (Figure 5.115).

When low fences or barriers are encountered, the combination ladder can be used. It is opened into the A-frame configuration and then picked up and placed over the fence or wall.

The telescoping beam combination ladder is unique in the way it is handled when forming an A-frame for bridging a fence.

Figure 5.115 Two short ladders or a disassembled extension ladder can be lashed together to bridge a fence or wall. A rope hose tool is a convenient way that the sections can be secured together.

Step 1: Both halves are extended (telescoped) to form a single ladder (Figure 5.116).

Step 2: The single ladder is raised and placed against the fence (Figure 5.117).

Step 3: A firefighter climbs to the hinge point and grasps a rung of the upper half with one hand. The other hand is used to release the hinge lock (Figure 5.118).

Step 4: The upper half of the ladder is allowed to swing downward (Figure 5.119).

Step 5: When the upper half beams reach the ground on the other side of the fence, the hinge lock is set and the A-frame ladder which has been formed is ready for use (Figure 5.120).

Figure 5.116 Both halves are telescoped to form a long "single" ladder.

Figure 5.117 The ladder is placed against the fence or wall. If it is used to bridge a fence, a position near a post is preferred.

Figure 5.118 One firefighter heels it while the other climbs to the hinge point, grasps a rung of the upper half, and releases the hinge latch.

Figure 5.119 The upper half is lowered to the ground on the other side of the fence.

Figure 5.120 When the upper half is in place, the hinge lock is operated to hold it in place.

5.13 True or False.

	True	False
1. Bridging between buildings or over ditches is a routine fire fighting procedure.	☐	☐
2. Bridging a low fence or barrier can be accomplished by using a combination ladder as an A-frame.	☐	☐

5.14 Fill in the blanks.

1. When bridging a fence, the ladders are normally placed adjacent to a ______________.

2. A ladder placed horizontally over a hole in the floor will be easier to traverse if ______________.

BRIDGING TO KEEP HOSELINES CLEAR OF ROADWAY

Live fire training or situations such as dump or debris fires will occasionally occur where the only water source is on the other side of a road and it is desirable to keep the road open. Since there is time, ladders can be lashed together to form a bridge over the highway that will permit all but high-bodied vehicles to pass without driving over the hoseline. A 24-foot (7 m) extension ladder and a roof ladder are usually used, but three single ladders can be used. If the 24-foot (7 m) extension ladder is used, the two sections are separated so that there are two equal length ladders for the sides of the bridge structure. The other ladder is used for the span section.

The ladders are lashed together while lying on their sides on the ground (Figure 5.121). Dry hoselines are attached with rope hose tools or hose straps and then the assemblage is raised over the roadway (Figure 5.122). Firefighters will be necessary to stop vehicles with high bodies and assist motorists, as one-way traffic may be necessary.

Figure 5.121 The three ladder sections are lashed together while lying on one beam on the ground. Guy ropes are attached.

Figure 5.122 A dry hoseline is attached with hose straps and then the assembly is moved into place and tilted up over the roadway. The guy ropes are tied off to fixed objects.

Improvising a Long Roof Ladder

When a roof ladder is not long enough to reach from a roof peak to the eaves, such as would be the case with the church shown in Figure 5.123, a second roof ladder or a single ladder can be lashed to the first ladder to obtain the needed extra length. A solid beam roof ladder and a wider, truss beam second ladder are

Figure 5.123 The distance between the roof peak and the eaves of this church is so great that one roof ladder would not reach.

used (Figure 5.124). When a single ladder is available that will reach by itself but cannot be secured because it does not have hooks, the roof ladder can be lashed to the single ladder (Figure 5.125).

Figure 5.124 The narrower solid beam roof ladder is lashed to the wider truss beam ladder. Note that the assembly is placed on the roof in such a manner that the narrow ladder is underneath.

Figure 5.125 A long truss beam single ladder is tied to a short roof ladder. The roof ladder must be narrower so that it will nest inside the truss beam ladder allowing the assembly to lie flat on the roof.

CAUTION: The rungs of both ladders must be securely tied together with rope, rope hose tools, or straps. When rope is used it should be a lifeline. Twist ties (using a Spanish windlass) are used as little as possible.

5.13 1. False
2. True

5.14 1. post
2. a plank is laid on the ladder

Using Ladders to Support Smoke Ejectors

Smoke ejectors can be hung from or supported on ladders in a variety of ways, as outlined below.

- In front of window openings (Figure 5.126)
- In door or archway openings (Figures 5.127 and 5.128)
- In stairwells (Figure 5.129)

Figure 5.126 A ladder is raised with the tip placed over the window. Then the smoke ejector is suspended from a rung.

Figure 5.127 A ladder is placed in front of a doorway. A smoke ejector is suspended from a rung similar to the way it is placed in front of a window.

Figure 5.128 An A-frame ladder is set up in a doorway. Smoke ejectors are suspended from it.

Figure 5.129 A smoke ejector suspended from a ladder in a stairwell.

- Over floor openings (Figure 5.130)
- Over scuttle or flat roof openings (Figure 5.131)
- Over openings in sloped roofs (Figure 5.132)
- Over window wells (Figure 5.133)

Ladders used for this purpose will require considerable scrubbing to remove spot deposits; tar remover or a similar safety solvent will be required.

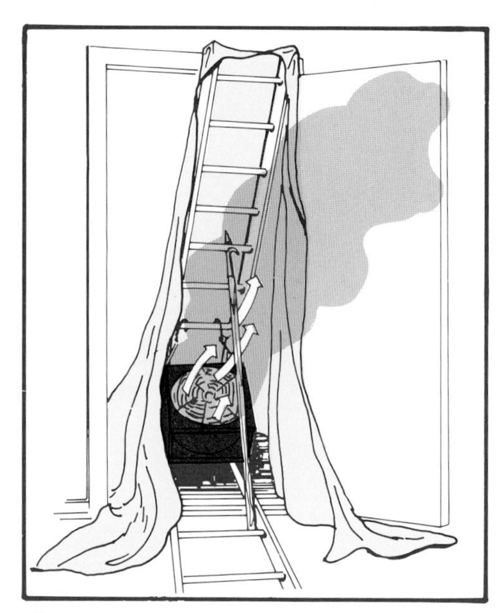

Figure 5.130 A ladder is laid flat over the hole in the floor. Another ladder is positioned from the edge of the hole to the doorway. The smoke ejector is placed at an angle where the two ladders meet, its top hung from the second ladder, its bottom resting on the first ladder.

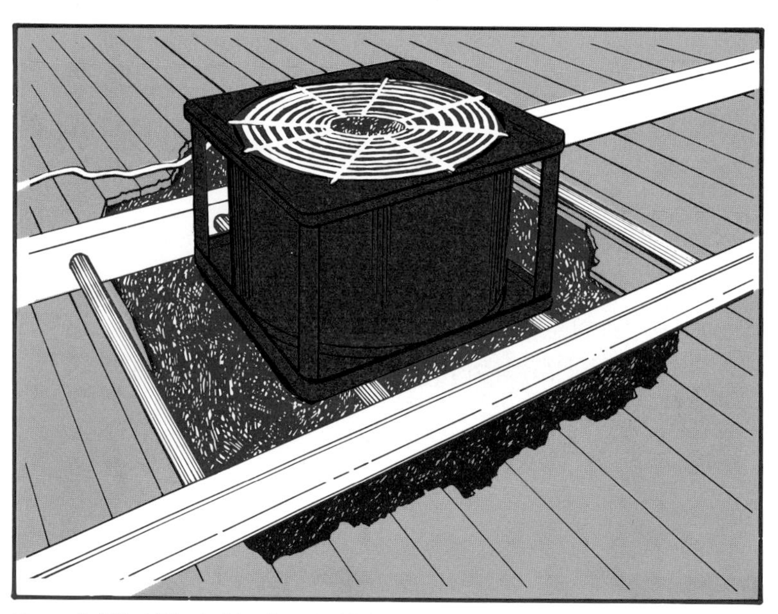

Figure 5.131 All the ladder does in this instance is provide a means of locating the smoke ejector over the opening.

Figure 5.132 The roof ladder provides support for the smoke ejector as well as a way of keeping it on a sloped roof.

Figure 5.133 Ladders are also used to support smoke ejectors over window well openings.

Keeping a Suction Strainer off the Bottom by Using a Ladder

When drafting water, it is important to maintain 18 inches (450 mm) between the openings of the strainer and the bottom. A single or roof ladder can be used for this purpose. The suction hose with strainer attached is brought through between the two bottom rungs, or the second and third rungs from the bottom, depending on the steepness of the bank. This tilts the strainer toward horizontal and keeps it off the bottom (Figure 5.134).

Figure 5.134 The end of the hard suction is threaded through a ladder to keep the suction strainer off of the bottom.

5.15 True or False.

	True	False
1. When a ladder will not reach a roof peak, a second roof or single ladder can be lashed to the first ladder to provide the extra length needed.	☐	☐
2. When a ladder is used to support smoke ejectors, the resulting soot that builds up on the ladder should be removed by wiping the ladder with gasoline, kerosene, etc.	☐	☐
3. When a ladder is used with the end of the suction hose inserted through two lower rungs during drafting, the purpose of the ladder is to give a firefighter footing so that the individual may climb into the water to reach the strainer and keep it clear of debris.	☐	☐

Making a Water Chute with a Tarp and a Ladder

In salvage operations, ladders can be used to support water chutes for removing water from a building. To do this, a salvage cover is partially unfolded in order to roll the edges, which are placed inside the ladder beams. Note that if the rolled edges are placed underneath the cover, the weight of the water in the trough tends to tighten the rolls. On ladders with high beams, the rolls do not have to be inverted since the beams will hold them (Figures 5.135 thru 5.138).

Figure 5.135a Salvage cover edges are rolled to form a chute. *Courtesy of the Chicago Fire Department.*

Figure 5.135b The rolled salvage cover is placed on a ladder which provides the support to carry its and the water's weight. *Courtesy of the Chicago Fire Department.*

Figure 5.136 Ladder supporting a water chute. *Courtesy of Chicago Fire Department.*

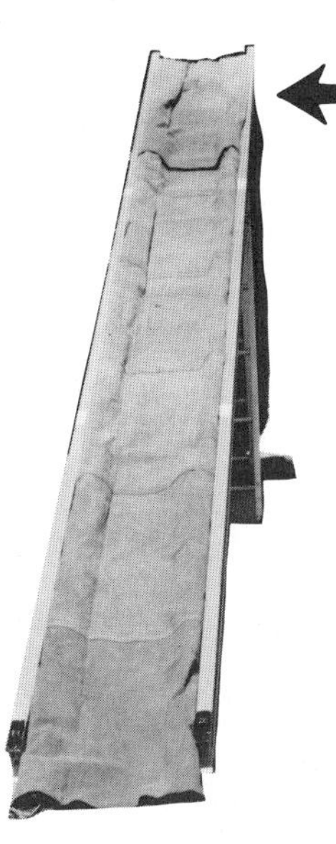

Figure 5.137 Two ladders are used in this instance, one to elevate one end of the ladder which is supporting the water chute.

Figure 5.138 An A-frame ladder supports a pike pole which in turn supports another ladder upon which a water chute has been placed.

Using a Ladder as a Battering Ram

Due to stresses developed when ladders are used as battering rams, they should only be employed in this manner as a last resort. Any ladder used as a battering ram must be service tested before being returned to normal service. Extension ladders should only be used when single ladders are not available because the fly section may move on impact, presenting a clear danger to firefighters' hands and fingers. Breaching objectives should always be struck with the ladder butt.

Step 1: The ladder is held flat at arm's length. The firefighters position themselves standing opposite each other, grasping the rungs palms down, two rungs between hands. The hands should be placed as in Figure 5.139.

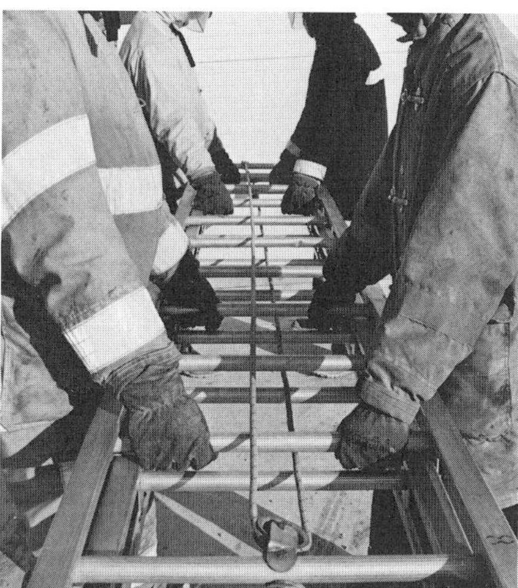

Figure 5.139 Gripping the ladder for use as a battering ram.

Step 2: At the "ready" command, the ladder is drawn back until the forward arm is across the firefighter's body.

Step 3: At the "strike" command, the firefighters thrust forward together, striking the objective solidly (Figure 5.140).

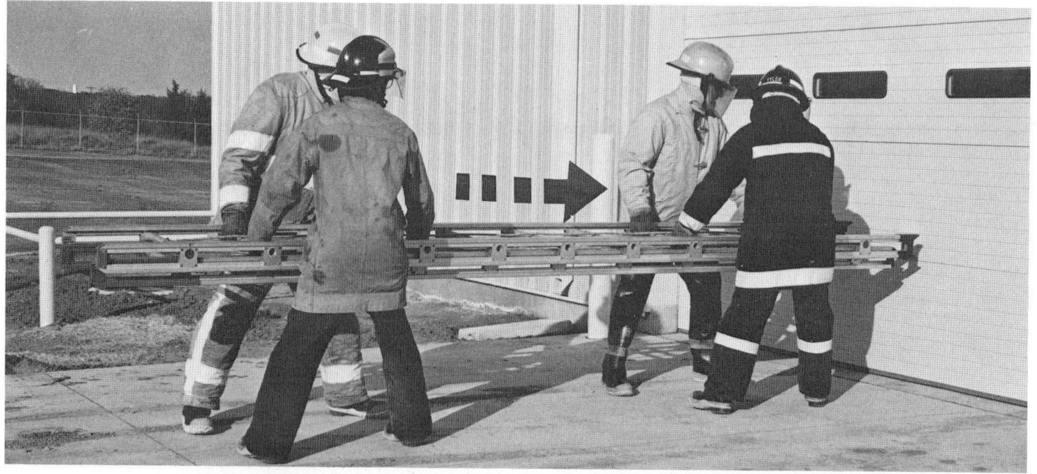

Figure 5.140 All bring the ladder forward in unison to strike the door.

NOTE: Care should be taken to avoid striking persons standing near the ladder tip during the back stroke.

Where the objective is narrow the ladder may need to be used beam over beam (Figure 5.141).

5.15 1. True
2. False
3. False

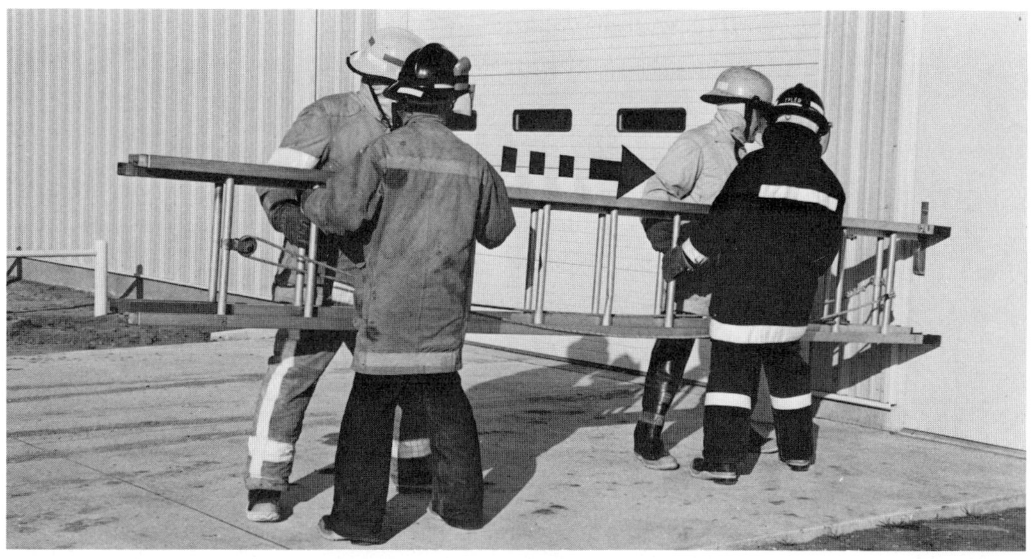

Figure 5.141 In narrower spaces the ladder is turned on edge.

Using a Roof Ladder and Rope for Remote Control and Advancement of a Nozzle (Also Called an Improvised Cellar Pipe)

It may be desirable to extend a nozzle into a basement through a hole cut in the floor above, or into an area through a doorway, without firefighters having to enter. A ladder and a utility rope can be used to do this and provide some control of the stream direction.

Step 1: A dry hoseline with nozzle attached is laid out on the ground. A roof ladder is laid atop the hoseline, butt end toward the nozzle. The nozzle is extended up through the ladder between the second and third rungs and the hoseline is lashed to the third rung.

Step 2: The midpoint of the rope is tied to the nozzle near the tip. If a shutoff nozzle is used, the shutoff is also tied in the open position (Figures 5.142 and 5.143).

Figure 5.143 The hoseline has been tied to the third rung. The nozzle rests on the second rung. A utility rope has been secured to the nozzle in such a way that it will hold the shutoff in the open position. This rope will be used like reins to control some of the nozzle movement.

Figure 5.142 A ladder and hoseline rigged as an improvised cellar pipe.

Step 3: The assembly is then extended through the opening, keeping the hoseline straight in line with the ladder. The hoseline is charged simultaneously with its advancement. The rope reins are used to elevate the stream (Figure 5.144).

Step 4: The ladder is tilted or rocked to change the direction of the stream (Figure 5.145).

Figure 5.144 Pulling back on the reins lifts the nozzle. Slacking off allows the nozzle's weight to lower it.

Figure 5.145 The ladder is tilted for sideways movement of the stream.

ALTERNATE METHOD

Another method of rigging ladders and a hoseline allows deeper penetration through a door opening or from a more remote location. It will also allow a hoseline to be advanced into a building from across a trench, ditch, or space between buildings. Two roof or single ladders of different widths are rigged together. A two and one-half inch (64 mm) or three inch (76 mm) hose, with a 300 gpm (1325 Lpm) to 500 gpm (1900 Lpm) nozzle attached may be used.

Step 1: Place the tip end of a roof ladder between the top two rungs of a wider roof or single ladder, hooks open and turned upward. Align the rungs. Place a straight bar between the underside of the beams of the first roof ladder and the upper surface of the second ladder's beams near its tip. A second straight bar or similar tool is placed between the underside of the beams of the second ladder and the upper surface of the beams of the first roof ladder near the hooks. This allows the stresses between the two ladders to be transferred from beam to beam and avoids stressing the rungs. A short length of rope is used to secure the two bars to each other. The ladders are lashed together with a rope hose tool (Figures 5.146 and 5.147).

Figure 5.146 The tip of a roof ladder with the hooks open is spliced to a wider ladder. Straight bars or similar tools are used to tilt up the roof ladder.

Figure 5.147 Rope hose tools can be used in lieu of a short length of utility rope.

Step 2: Stretch a two and one-half inch (64 mm) or three-inch (76 mm) hoseline, with nozzle attached, along both ladders to the butt of the ladder which is angled upward. Use another straight bar or similar tool to angle the nozzle upward. The straight bar is laid across the beams, under the nozzle tip, to do this. The hoseline is secured to the upper rungs of the ladder with rope hose tools or a short length of rope (Figure 5.148 on next page).

The completed assembly is shown in Figure 5.149.

Figure 5.148 A hoseline is strung along both ladders. The nozzle is lashed to the butt of the narrower ladder. A straight bar or similar tool is used to tilt the nozzle upward.

Figure 5.149 Completed assembly for the alternate method of rigging a hoseline and ladders for remote advancement.

Making a Ladder Pipe Applicator

This procedure may be used on fires above the reach of conventional elevated streams. However, since firefighters must work directly under the fire floor, it should only be used in buildings with protected construction (i.e., protected steel framework with poured floors). A short extension ladder, a ladder pipe, ladder pipe control handle, two ladder pipe remote control ropes, two short ropes, and two short pike poles are needed. An in-line gate valve is needed for the hoseline supplying the ladder pipe. It is located where the last section of hose couples up. The steps needed to rig and use this assembly are

Step 1: Place the short extension ladder on the floor, pull the fly section out a distance of four rungs, and set the pawls. Place the ladder tip on the windowsill. Clamp the ladder pipe, complete with control handle and attached hoseline, to the center of the two top rungs. Attach ladder pipe remote control ropes; one to the eye on the ladder pipe control handle, the other to the eye on the stream shaper ring. Secure the hoseline to the ladder with a hose strap or similar device (Figure 5.150).

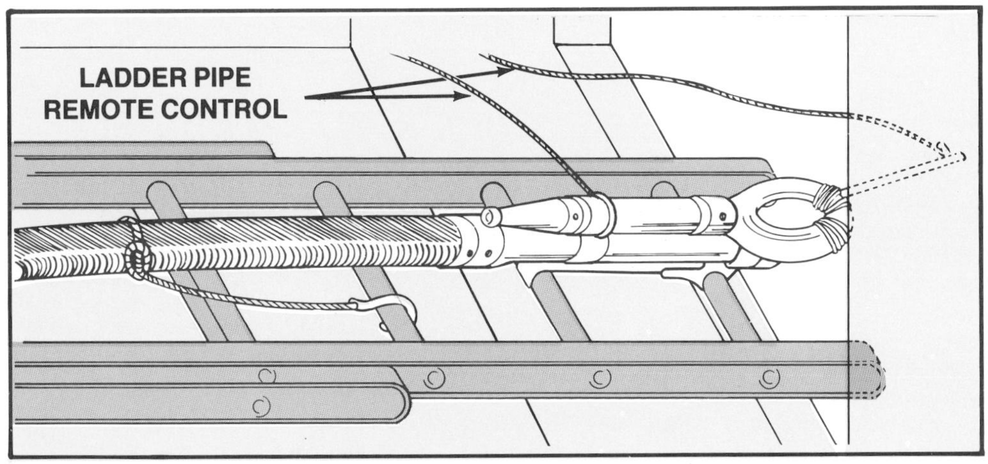

Figure 5.150 The fly of a short extension ladder is extended at least four rungs. A ladder pipe with remote control ropes is attached to the top two rungs. A rope hose tool or hose strap is used to secure the hoseline to a ladder rung.

Step 2: Using an appropriate bridging maneuver, position the assembly to extend out of the window. Adjust the fly extension if necessary. Be sure that the pawls are latched (Figure 5.151).

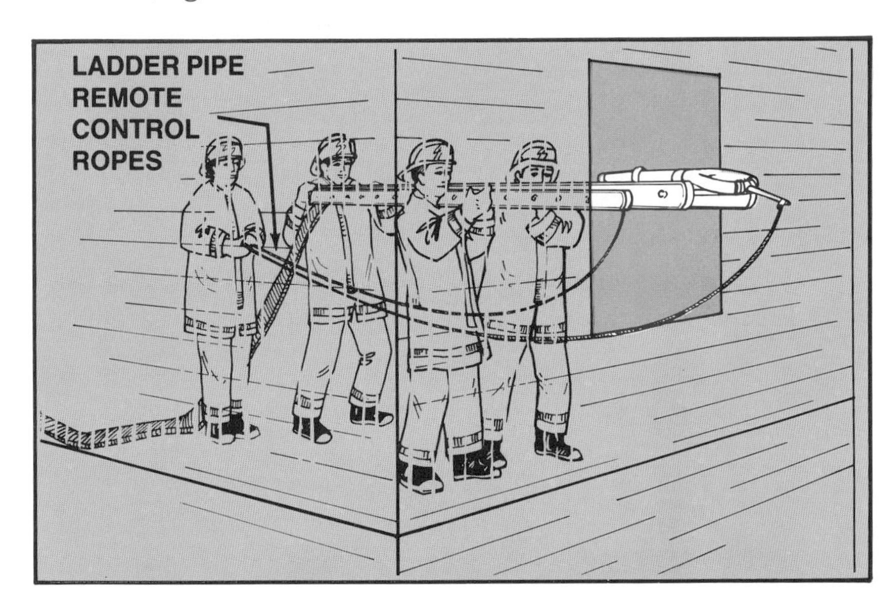

Figure 5.151 The same carry as for bridging or raising under obstructions is used to pick up the assembly and extend it out of the window.

Step 3: The assembly is held in place while a short pike pole is placed across the inside of the window opening near the top. The pike pole and the ladder are lashed together using a short section of rope. A half hitch is tied around the rung and two clove hitches with binders are used around the pike pole, one at either side of the window opening. A second pike pole is placed across the outside of the window opening near the sill. It is secured by the same method (Figure 5.152).

Step 4: Charge the hoseline slowly until stability of the apparatus is established. Manipulate the stream with the control ropes (Figures 5.153 and 5.154).

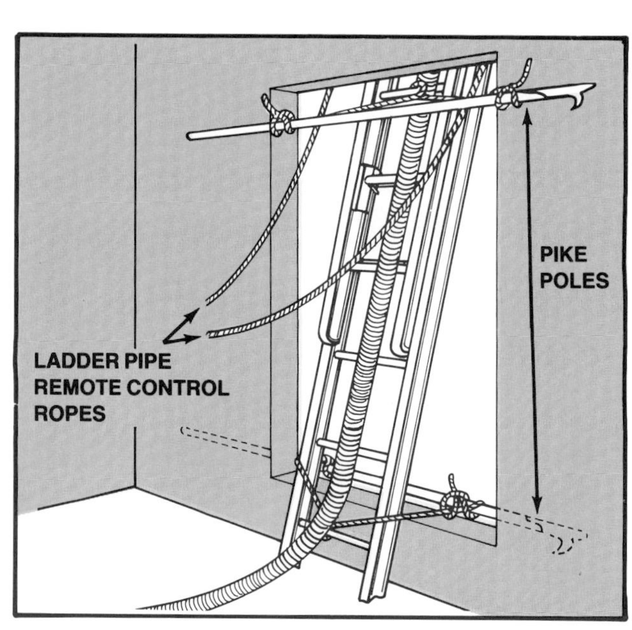

Figure 5.152 The ladder is secured in place using two short pike poles and two short lengths of rope.

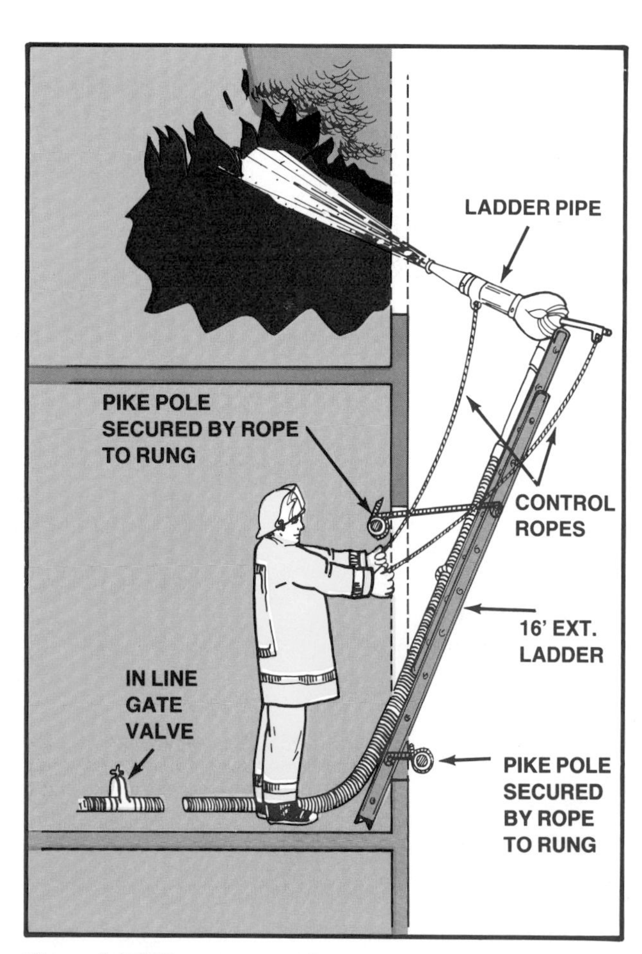

Figure 5.153 Remote control lines are used to move the ladder pipe up and down into the window opening of the floor above.

Figure 5.154 Remote control ladder pipe assembly in operation.

Making an Under-a-Pier Applicator

Fires under piers are difficult to control because of the problem of getting a stream onto the fire. One method of directing a large caliber stream under a burning pier uses two single or roof ladders of different widths, a ladder pipe, hose straps or rope hose tools, short sections of rope, and two half-inch (15 mm) reinforcing rods, as follows:

Step 1: A roof or single ladder is laid on the pier. The ladder pipe with a hoseline coupled to it is attached to the two top rungs. (If a roof ladder is being used, the two bottom rungs). The remote control rope is attached to the stream shaper ring.

Step 2: The ladder is turned over and the ladder pipe control handle is attached. The other remote control rope is attached to the eye on the handle.

Step 3: One firefighter holds the two remote control ropes while others pick up the ladder and extend it downward over the side of the pier or through a sizeable hole cut in the pier deck. The hoseline and ladder pipe must be on the side toward which the stream will be directed.

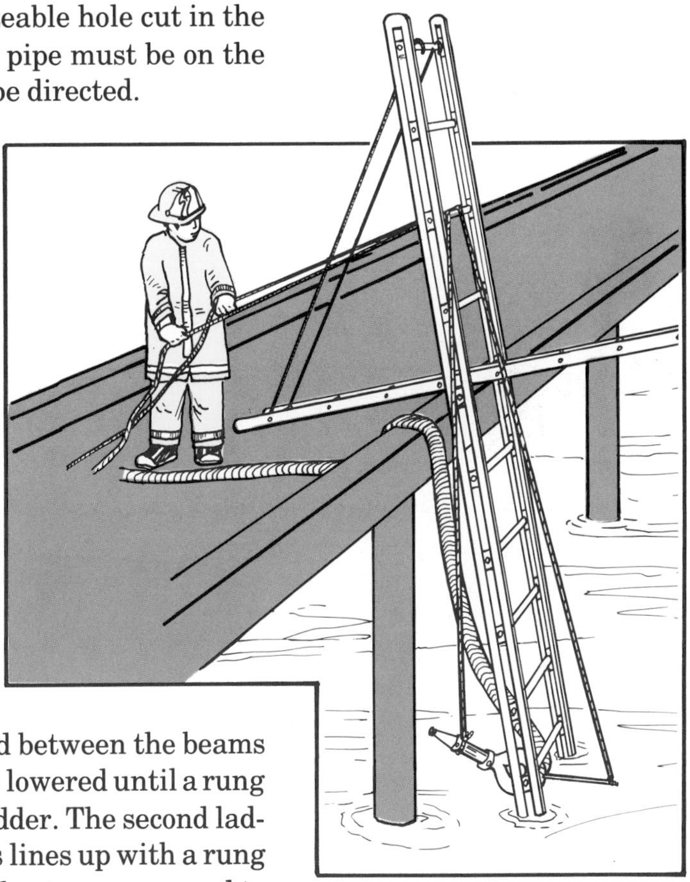

Figure 5.155 An under a pier applicator rigged with two ladders, rope, a ladder pipe, and two metal rods.

Step 4: A short, narrower ladder is inserted between the beams of the first ladder. The first ladder is lowered until a rung contacts the beams of the second ladder. The second ladder is adjusted until one of its rungs lines up with a rung of the first ladder. A hose strap or short rope are used to join the rungs of the two ladders (clove hitch and binder).

Step 5: The vertical ladder's angle is secured with the reinforcing rods, whose ends are bent into rungs and held by tying with a short piece of rope.

The completed assembly is illustrated in Figure 5.155.

5.16 True or False.

	True	False
1. When a ladder is used as a battering ram, the butt of the ladder should be used to strike the objective.	☐	☐
2. Buildings with protected steel framework and poured floors prohibit the use of the ladder pipe applicator.	☐	☐

Using Ladders to Construct a Catch Basin

A triangular shaped catch basin, made by lashing two ladders and a pike pole together to form a framework, then lined with a salvage cover so that it will hold water (Figure 5.156), has several uses.

- It provides a way for a pumper to get water from a hydrant having different threads than those of the hose fittings carried by the pumper.
- It permits use of hydrant when damaged hydrant outlets prevent the pumper from hooking up.
- It serves instead of a folding portable tank when rural tanker shuttle operations are necessary.

Figure 5.156 Two ladders and a pike pole are lashed together in a triangular framework and a salvage cover is placed between the sides to form a catch basin.

When a catch basin is used at a hydrant, it is placed in front of one of the hydrant outlets. The side formed by the pike pole should be turned so that it is not in line with the hydrant outlet. The hydrant outlet should be gated so that the flow can be controlled. This is done to prevent overflow and to keep the velocity of the stream of water from causing the salvage cover to come loose from the pike pole as illustrated in Figure 5.157.

Figure 5.157 If filled from a hydrant, care must be taken so that the velocity of the stream doesn't hit the pike pole side of the catch basin because this may pull the salvage cover loose.

When a catch basin is used during a tanker shuttle operation, care must be taken so that water being dumped does not flow against the pike pole side with any significant force. In all cases, the pumper obtains water by drafting from the catch basin. The pumper's hard suction hose should not be laid across the pike pole side of the catch basin because the weight may bend or break the pike pole.

Making a Dam Across a Stream Using a Ladder and a Salvage Cover

Sometimes streams are fast running but too shallow for drafting. When a floating dock strainer is available, a ladder and salvage cover can be used to dam the stream and raise the water level to permit drafting.

Step 1: A salvage cover is spread on the ground. The ladder is placed at one of the long sides (Figure 5.158).

Figure 5.158 A salvage cover is spread flat on the ground beside the stream. A ladder is placed atop it along one edge. Note how shallow the stream is and that the stream lacks sharply defined banks.

Step 2: The ladder is rolled up in the cover, leaving about four feet (1 m) of cover free to form a flap (Figure 5.159).

Step 3: The ladder and cover assembly is placed across the stream, preferably at a point where the stream bottom is level (Figure 5.160).

Step 4: The flap is stretched upstream and anchored with rocks or by straight bars stuck through the grommets (Figure 5.161).

Step 5: It is sometimes necessary to do some chinking at the ends to prevent serious leakage.

Step 6: It may be necessary to support the ladder near its midpoint to prevent the weight of the water that builds up from bowing the ladder (Figure 5.162).

Figure 5.159 The ladder is rolled up in the salvage cover. The process is halted at least four feet (1 m) from the far edge so that a flap is left.

Figure 5.160 The assembly is picked up and placed across the stream with the flap extending out toward the upstream side. Some tucking is done at the ends.

Figure 5.161 The flap is anchored in place.

5.16 1. True
2. False

Figure 5.162 Pike poles are used to help support the ladder. After a short period, the water should rise sufficiently to take draft.

Review

Answers on page 388

True or False.

	True	False
1. It is better to evacuate persons through stairways, connecting buildings, elevating platforms, and aerial ladders than down ground ladders.	☐	☐
2. Persons being evacuated down a ground ladder are generally sufficiently scared of the fire that they will not need assistance climbing down the ladder.	☐	☐
3. When a firefighter assists a person down a ladder, the firefighter's knee is placed between the victim's legs to provide support in case the person slips or becomes unconscious.	☐	☐
4. When bringing an unconscious victim down a ladder, the victim's feet are placed outside the beams to prevent entanglement.	☐	☐
5. Bridging should be done with extension ladders because of the added security offered by the ladder pawls.	☐	☐

Fill in the blanks.

6. The four methods used to place a single ladder across an open space (bridging) are:
   (1) ____________________
   (2) ____________________
   (3) ____________________
   (4) ____________________

Check the correct response.

7. Using a ladder as a fulcrum to lower an injured person requires a minimum of ____________.
   - ☐ A. Three firefighters
   - ☐ B. Four firefighters
   - ☐ C. Five firefighters

8. In the ladder fulcrum method of lowering an injured person, the person is maintained in a ____________ position.
   - ☐ A. Horizontal
   - ☐ B. Sitting Up
   - ☐ C. Vertical

Short Essay.

9. Describe how a firefighter can make a stretcher using a single or roof ladder. ____________________
   ____
   ____
   ____

10. After prying an overturned vehicle with a ladder, what action should be taken as far as further use of the ladder? ____________

____

____.

Check the correct response.

11. The ____________ ladder is used as part of the rigging to cover a collapsed area of a trench during rescue operations in inclement weather.

- ☐ A. Single
- ☐ B. Roof
- ☐ C. Pompier
- ☐ D. A-frame
- ☐ E. Extension

Short Essay.

12. List five safety considerations when using a single or roof ladder to extend the reach of an aerial platform.

(1) ____

(2) ____

(3) ____

(4) ____

(5) ____

Fill in the blanks.

13. A ________________ ladder may be used as a pompier ladder for rescue in extreme emergencies.

14. Firefighters breaking windows with ladders should be alert for ____.

Check the correct response.

15. Two ladders, a pike pole, and a salvage cover are used to construct a catch basin when ____________.

- ☐ A. Hydrant threads are different than hose and appliance carried on apparatus.
- ☐ B. Hydrant threads are damaged.
- ☐ C. A folding tank is not available for rural tanker shuttle operations.
- ☐ D. All of the above.

True or False.

	True	False
16. Dams constructed of a ladder and a salvage cover are used to slow down the flow of streams so that a pumper can draft from them.	☐	☐

Appendix A

________________ Fire Department

GROUND LADDER TESTING AND REPAIR RECORD

ifsta

1. MFGR:	2. MFGR'S Model or Code#:	3. MFGR'S Serial#:	4. FD ID#:

5. Date Purchased	6. Date Placed in Service	7. Unit or Location to Which Assigned:

8. Type: ☐ Single ☐ Roof ☐ Extension ☐ Pole ☐ Folding ☐ Combination ☐ Pompier

9. Length:	10. Construction Materials: ☐ Wood ☐ Metal ☐ Fiberglass	11. Beam Type: ☐ Solid ☐ Truss

12. Certified as Meeting NFPA Standard 1931: ☐ Yes ☐ No Edition Year:

13. Reason for Test: ☐ Annual Service Test ☐ Suspected Damage, Overload, Unusual Use ☐ Exposed to Heat ☐ Retest After Repair

14. Test Date:	15. Person(s) Performing:

16. Heat Sensor Label Check: ☐ Label Unchanged ☐ Label Changed / Heat Exposure Indicated ☐ No Label Present

17. ☐ Horizontal Bending Test Performed Weight Used: Amount of Deformation: ☐ Passed ☐ Failed

18. ☐ Hardware Test Performed Weight Used: ☐ Passed ☐ Failed → Location and Part Failing:

19. ☐ Roof Hook Test Performed Weight Used: ☐ Passed ☐ Failed → Location and Part Failing:

20. ☐ Pompier Ladder Test Performed Weight Used: ☐ Passed ☐ Failed → Location and Part Failing:

21. ☐ Hardness Test Performed ☐ Instrument Calibrated Before Test ☐ Instrument Calibration Verified Immediately After Test

INSTRUMENT USED: ________________ Min. Acceptable Reading for this Instrument: ________________

☐ Passed ☐ Failed- Location of Failure: ________________ Failure Reading: ________________

22. ☐ Eddy Current Test Performed Performed By: ________________ Firm Name: ________________

☐ Passed ☐ Failed- Location of Failure: ________________

23. ☐ Liquid Penetrant Test Performed Performed By: ________________ Firm Name: ________________

☐ Passed ☐ Failed- Location of Failure: ________________

24. Status of Ladder as Result of Test: ☐ In Service ☐ Out of Service for Further Testing ☐ Out of Service for Repair ☐ Destroyed ☐ Other

25. Repair Notes: (Date and Initial Entries)

____

____

Remarks: (Use Section Number)

____

____

____

____

____

Signature of Person Responsible For Test

Index

A

A-frame for hoisting, 338, 339
Aerial platforms, extending reach from, 340-342
Angle of inclination
  checking for proper, 259-261
  definition, 14
Auditorium raise, 239-241

B

Battering ram, 366-367
Beam
  definition, 14
  construction, 40
Beam and hardware test, 76
Beam Cantilever bending tests, 75
Bed section, 15
Bedded position, 15
Bridging
  for fire fighting, 353
  for rescue 305-314
Butt, 15
Butt spur, 68
Butt spur slip test, 75

C

C-channel beams, 40
Carrying ladders, 114-171
  one-firefighter, 114-130
  two-firefighter, 130-144
  three-firefighter, 145-157
  four-firefighter, 157-163
  five-firefighter, 164-167
  six-firefighter, 167-168
  carrying other ladders, 169-170
  special carry for narrow passageways, 171
Catch basin, 374
Cellar pipe, improvised, 367-369
Certification, 30, 72
Cleaning, 82
Climbing, 259-266
  checking for proper angle of inclination, 259-261
  pompier ladders, 280-282
  techniques, 262-266
  to place a roof ladder, 270-280
Combination ladders
  carry, 190
  construction, 56
  definition, 12, 13
  raises, 231-237
Components, 40-59
Composite ladders, 38, 52
Compression test, 76
Conscious persons, assisting down ladders, 292, 293
Construction
  certification, 30
  features of, 38-73
  materials of, 31-37
Cyclic rung-pawl test, 76

D

Damming a stream, 375-377
Deflection test, 74
Design and construction
  design verification testing, 74-77
  features of construction, 38-73
  materials used, 31-38
  workmanship, 28-30
Design verification testing, 74-77
  beam and hardware test, 76
  beam cantilever bending test, 75
  butt spur slip test, 75
  compression test, 76
  cyclic rung-pawl test, 76
  deflection test, 74
  horizontal bending test, 74
  ladder section twist test, 75
  multisection extending force test, 76
  pawl tip load test, 76
  pompier ladders, 77
  roof hook strength test, 76
  rung bending strength test, 74
  rung-to-beam shear strength test, 74, 75
  rung torque test, 75
  side sway test, 75
  single pawl load test, 76
Designated length, 16
Distance from building, 176, 177
Dome raise, 239-241

E
Eddy current test, 93
Electrical hazards, 184
Extending reach of aerial platforms, 340-342
Extension ladders
  carries, 114-163
  construction, 40-52
  definition, 12
  inspection, 78-81
  raises, 196-209

F
Factory raise, 242-244
Features of construction, 38-73
  hardware and accessories, 59-73
  loading, 39
  major components, 40-59
  width, 38, 39
Fences, bridging over, 355-357
Fiber glass ladders
  construction, 52
  definition, 37
Five- and six-firefighter pole ladder raise, 219-229
Five-firefighter carries, 164-167
  flat arm's length method, 166-167
  flat-shoulder method, 164-166
Fly position, 184
Fly section, 16
Folding ladders
  carry, 169
  construction, 53-54
  definition, 11
  raise, 230, 231
Foot pads, 68
Four-firefighter carries, 157-163
  arm's length on-edge method, 163
  flat arm's length method, 162
  flat-shoulder method, 157-162
  low-shoulder method, 163
Four-firefighter pole ladder raises, 212-219
Four-firefighter raises, 207-209
Fulcrum, 315-319

G
Gin poles, 334-337

H
Halyard
  construction, 65
  definition, 16
  tying, 255-259
Halyard anchor, 65
Hardness service testing, 89
Hardness tester, 90
Hardware and accessories, 59-73
  butt spurs, 68
  foot pads, 68
  halyard/halyard anchor/pulley, 65, 66
  labels, 71-73
  levelers, 70, 71
  mud guard, 70
  pawls, 59-64
  protection plates, 70
  roof ladder hooks, 65
  staypoles, staypole spurs, and toggles, 69
  stops, 67
  tie rods, 68, 69
  toe rods, 69
Heat sensor label, 73-78
Heeling, 252, 253
History, 2-4
Hoisting devices, 333-339
Hoisting ladders, 349-352
Holes, bridging over, 354, 355
Horizontal bending test, 74
Hotel raise, 242-244

I
Ice rescue, 326, 327
Identification number, 17
Inside width, 18
Inspection of ladders, 78, 79
Introduction, 1-5

J
Jib arms, 333, 334

L
Labels, 71
Ladder drawbridge, 311-314
Ladder float drag, 328
Ladder pipe applicator, 370-372
Ladder reach, 108
Ladder requirements, 5
Ladder section twist test, 75
Ladder terms, 14-21
  angle of inclination, 14
  beam, 14
  bed section, 15

bedded position, 15
butt, 15
designated length, 16
dogs, 16
fly section, 16
halyard, 16
heel, 16
identification number, 17
inside width, 18
maximum extended length, 18
nesting, 18
outside width, 18
pawls, 19
rail, 19
retracted, 19
rungs, 20
side rail, 20
stripping ladder, 20, 21
tip, 21
Ladder types
combination, 12, 13
extension, 12
folding, 11
pole, 12
pompier, 14
roof, 10
single, 9, 10
Ladders carried on aerial apparatus, 102
Ladders carried on pumpers, 101
Ladders carried on other apparatus, 105
Ladders, 31-38
composite, 38
fiber glass, 37
metal, 31
wood, 34
Leg locking, 267-270
Levelers, 70
Lifting and lowering, 113, 114
Liquid penetrant testing, 93
Loading, 39
Location on apparatus, 99-107
aerial apparatus, 102-105
pumpers, 100-102
other apparatus, 105-107
Long roof ladder, improvising, 359-361
Lowering extended ladders to below grade, 301-305
Lowering persons, support for, 321, 322

M

Maintenance, 79-81
Manufacturers' identification label, 71
Materials used, 31-38
composite, 38
fiber glass, 37
metal, 31-34
wood, 34-37
Maximum extended length, 18
Metal ladders, 31, 40
Methods of mounting, 99-107
aerial apparatus, 102-105
pumpers, 100-102
other apparatus, 105-107
Mud guard, 70
Multiple ladder carry, two-firefighter, 143, 144
Multisection rung-pawl test, 76

Narrow passageway carry, 171
NFPA Standard 1901: *Standard for Automotive Fire Apparatus,* 100, 102, 105
NFPA Standard 1931: *Standard on Design and Design Verification Tests for Fire Department Ground Ladders,* 5, 16, 17, 30, 38, 45, 50, 53, 55, 69, 71, 72, 74-77, 108
NFPA Standard 1932: *Standard on the Use, Maintenance, and Service Testing of Fire Department Ground Ladders,* 5, 47, 77-81, 84-93
Nesting, 18

O

Obstructions, raising ladders under, 245-252
two-firefighter single or roof ladder, 245-248
three-firefighter single or roof ladder, 248
three-firefighter alternate, 249-251
four-firefighter extension ladder, 252
One-firefighter carries, 114-131
arm's length method, 127-131
high-shoulder method, 121-126
low-shoulder method, 114-120
One-firefighter raises, 194-200
short single and roof ladders, 194
all others, 195-200
Outside width, 18

P

Parallel raise, 192, 193
Pawl tip load test, 76
Pawls, 19, 59-64
Pivoting, 185-187
one-firefighter, 185, 186

two-firefighter, 186, 187
pole-ladders, 255
Placement, 172-177
Pole ladders
carries, 157-169
definition, 12
raises, 209-229
Pompier ladders
carry, 70
climbing, 280-282
construction, 55
definition, 14
raise, 237, 238
Positioning, 172-177
locations to avoid, 175, 176
objectives, 172
proper distance from building, 177
responsibility for, 172
use affecting placement, 172-175
other factors, 175
Protection plates, 70
Prying, 328
Pulley, 65
Purpose, 4

R

Raising ladders, 183-239
one-firefighter raises, 194-200
two-firefighter raises, 201-203
three-firefighter raises, 204-206
four-firefighter raises, 207-209
four-firefighter pole ladder raise, 212-219
five-and six-firefighter pole ladder raises, 219-229
combination ladder raises, 231-237
folding ladder raises, 230, 231
pompier ladder raises, 237, 238
special raises, 239
Reach of ladders, 108
Removing ladders from apparatus, 110-112
Repairing, 82-84
Requirements, 5
Rescue operations
assisting persons down ladders, 292, 293
bridging, 305-314
bringing unconscious persons down, 293-297
emergency ventilation to assist rescue, 345, 346
extending reach of aerial platforms, 340, 342
float drag, 328
ice rescue, 326, 327
ladder fulcrum, 315-319
ladder slide, 319-321
ladder sling, 321, 322
ladder stretcher, 323-325
ladders for shoring, 328-332
lowering extension ladders to below-grade locations, 301-305
making devices for hoisting, 333-339
removing unconscious leg-locked firefighters, 298-301
supporting broken utility lines during rescue, 333
using a roof ladder as a pompier ladder, 343-345
Requirements, ladders, 5
Review questions
chapter one, 23, 24
chapter two, 94-96
chapter three, 178-180
chapter four, 285-287
chapter five, 378, 379
Review questions, answers, 386-388
Right angle raise, 192, 193
Roadway, bridging over, 358, 359
Rolling, 191
Roof hook strength test, 76
Roof ladder hooks, 65
Roof ladders
carrying, 120, 135, 136
climbing to place, 270-280
definition, 10
improvising long, 359-361
Rung bending strength test, 74
Rungs
construction 45, 50
definition, 20
rung to beam shear strength test, 74
torque test, 75

S

Safety, 283, 284, 343
Safety belts, 266, 267
Scope, of manual, 4
Securing ladders, 252-259
heeling, 252-254
tying in, 254-255
tying the halyard, 255-259
Selecting ladders, 107-110
Serial number, 72

Service testing, 84-93
methods, 85-93
what constitutes failure, 85
when to test, 85
Shifting in a vertical position, 187-190
one-firefighter, 188
two-firefighter, 188, 189
shifting pole ladders, 189, 190
Shoring trench side with ladders, 328-332
Side sway test, 75
Single ladders
carries, 114-143
construction, 39-52
definition, 9, 10
raises, 194, 201-204
Single pawl load test, 76
Six-firefighter carries, 167, 168
flat arm's length method, 168
flat-shoulder method, 167, 168
Smoke ejectors, supporting on ladders, 361-363
Solid beam construction, 40, 46
Staypole spurs, 69
Staypoles, 69
passing unattached, 210, 211
Stops, 67
Streams, directing from ladders, 346-348
Strength service testing, 86
Stretchers made with ladders, 323-325
Stripping ladder, 20, 21
Suction strainer, supporting with ladder, 363

T

T-channel beams, 44
Test Records, 93
Testing
design verification, 74
service, 84-93
Three-firefighter carries, 145-157
arm's length on-edge method, 157
flat arm's length method, 153-156
flat-shoulder method, 145-153
low-shoulder method, 156
Three-firefighter raises, 204-206
Tie rod, 68
Tip, 21
Toe Rod, 69
Toggles, 69
Two-firefighter carries, 131-143
arm's length on-edge method, 140
hip or underarm method, 136
low-shoulder method, 131
multiple-ladder carry, 143
Two-firefighter raises, 201-203
beam method, 202, 203
flat method, 201, 202
Truss beams, 43, 47
Tubular beams, 42
Tying the halyard, 255-259
Tying the ladder in, 254, 255
Types of ladders, 9-14
combination, 12, 13
extension, 12
folding, 11
pole, 12
pompier, 14
roof, 10
single, 9, 10

U

Unconscious firefighter, removing from leg lock, 298-301
Unconscious persons, bringing down ladders, 293-297
Under-a-pier applicator, 373, 374

V

Ventilation with a ladder, 345, 346

W

Walls, bridging over, 355-357
Water chutes, 364, 365
Weight comparison of ladders, 32
Width, 38
Wood ladders, 34, 46
Working from ladders, 266-270
leg lock method, 267-270
safety belt method, 266, 267
Workmanship, 28-30

Review Answers

Chapter 1 Review Answers

1. C
2. A
3. B
4. C
5. B
6. True
7. False
8. False
9. True
10. True
11. Angle of Inclination
12. False
13. True

Chapter 2 Review Answers

1. label, attached to ladder
2. will not
3. hardness test, strength service test
4. discoloration of the fiber glass
5. 300 pounds (136 kg)
   750 pounds (340 kg)
6. hook, finger, torsion spring, five
7. True
8. False
9. False
10. False
11. False
12. Car wax
13. To maintain reliability
14. False
15. False
16. True
17. ⅜ inch (10 mm), 825 pounds (374 km)
18. Dry, spray
19. safety solvent
20. True
21. True
22. True
23. True
24. False
25. overloading, impact, unusual, heat

Chapter 3 Review Answers

1. repetitive, practical training
2. flat or vertical racking, rear or side removal
3. True
4. False
5. 35 foot (12 m)
6. leg muscles
7. low-shoulder, high-shoulder, arm's length
8. False
9. True
10. C
11. 
12. 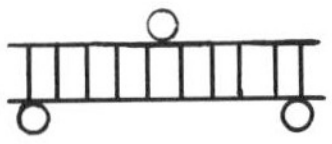
13. 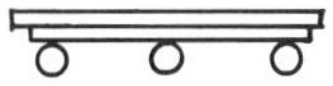
14. 
15. 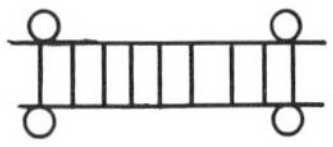
16. 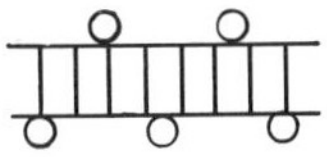
17. 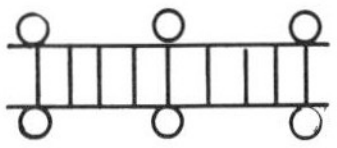
18. False
19. True
20. butt end
21. C

Chapter 4 Review Answers

1. E
2. A
3. C
4. A
5. B
6. B
7. use body weight to shove the ladder into the building
8. False
9. True
10. True
11. C
12. C
13. A
14. bounce, sway
15. use of leg muscles
16. use of leg lock or safety belt
17. danger of eye injury
18. C
19. B
20. False
21. True
22. False
23. False
24. True

Chapter 5 Review Answers

1. True
2. False
3. True
4. True
5. False
6. three-firefighter shoulder, three-firefighter arm's length, hoisting, drawbridge
7. B
8. A
9. coat or salvage cover placed on the ladder
10. service tested
11. D
12. (1) ground ladder must be supported against building before climbing
    (2) longer ladders present greater handling difficulties
    (3) the basket must not be moved after the ladder is in position
    (4) lifelines or life belts must be worn by firefighters
    (5) never exceed maximum load ratings or operating capabilities (reach or angle) of apparatus
13. roof
14. falling glass
15. D
16. False

IFSTA MATERIALS

FIRE SERVICE ORIENTATION & INDOCTRINATION

History, traditions, and organization of the fire service; operation of the fire department and responsibilities and duties of firefighters; fire department companies and their functions; glossary of fire service terms.

FIRE SERVICE FIRST AID PRACTICES

Brief explanations of the nervous, skeletal, muscular, abdominal, digestive, and genitourinary systems; injuries and treatment relating to each system; bleeding control and bandaging; artificial respiration, cardiopulmonary resuscitation (CPR), shock, poisoning, and emergencies caused by heat and cold; fractures, sprains, and dislocations; emergency childbirth; short-distance transfer of patients; ambulances; conducting a primary and secondary survey.

ESSENTIALS OF FIRE FIGHTING

This manual was prepared to meet the objectives set forth in levels I and II of NFPA, *Fire Fighter Professional Qualifications, 1981.* Included in the manual are the basics of fire behavior, extinguishers, ropes and knots, self-contained breathing apparatus, ladders, forcible entry, rescue, water supply, fire streams, hose, ventilation, salvage and overhaul, fire cause determination, fire suppression techniques, communications, sprinkler systems, and fire inspection.

IFSTA'S 500 COMPETENCIES FOR FIREFIGHTER CERTIFICATION

This manual identifies the competencies that must be achieved for certification as a firefighter for levels I and II. The text also identifies what the instructor needs to give the student, NFPA standards, and has space to record the student's score, local standards, and the instructor's initials.

FIRE SERVICE GROUND LADDER PRACTICES

Various terms applied to ladders; types, construction, maintenance, and testing of fire service ground ladders; detailed information on handling ground ladders and special tasks related to them.

FIRE HOSE PRACTICES

Construction, care, and testing of hose and various fire hose accessories; preparation and manipulation of hose for rolls, folds, connections, carries, drags, and special operations; loads and layouts for fire hose.

SALVAGE AND OVERHAUL PRACTICES

Planning and preparing for salvage operations, care and preparation of equipment, methods of spreading and folding salvage covers, most effective way to handle water runoff, value of proper overhaul and equipment needed, and recognizing and preserving arson evidence.

FORCIBLE ENTRY, ROPE AND PORTABLE EXTINGUISHER PRACTICES

Types of forcible entry tools and general building construction; use of tools in opening doors, windows, roofs, floors, walls, partitions and ceilings; types, uses, and care of ropes, knots, and portable fire extinguishers.

SELF-CONTAINED BREATHING APPARATUS

This manual is the most comprehensive text available on self-contained breathing apparatus. Beginning with the history of breathing apparatus and the reasons they are needed, to how to use them, including maintenance and care, the firefighter is taken step by step with the aid of programmed-learning questions and answers throughout to complete knowledge of the subject. The donning, operation, and care of all types of breathing apparatus are covered in depth, as are training in SCBA use, breathing-air purification, and recharging cylinders. There are also special chapters on emergency escape procedures and interior search and rescue.

FIRE VENTILATION PRACTICES

Objectives and advantages of ventilation; requirements for burning, flammable liquid characteristics and products of combustion; phases of burning, backdrafts, and the transmission of heat; construction features to be considered; the ventilation process including evaluating and size up is discussed at length.

FIRE SERVICE RESCUE PRACTICES

IFSTA's *Rescue* has been enlarged and brought up-to- date. Sections include water and ice rescue, trenching, cave rescue, rigging, search-and-rescue techniques for inside structures and outside, and taking command at an incident. Also included are vehicle extrication and a complete section on rescue tools. The book covers all the information called for by the rescue sections of NFPA 1001 for Fire Fighter I, II, and III, and is profusely illustrated.

THE FIRE DEPARTMENT COMPANY OFFICER

This manual focuses on the basic principles of fire department organization, working relationships, and personnel management. For the firefighter aspiring to become a company officer and the company officer who wishes to improve management skills this manual will be invaluable. This manual will help individuals develop and improve the necessary traits to effectively manage the fire company.

FIRE CAUSE DETERMINATION

Covers need for determination, finding origin and cause, documenting evidence, interviewing witnesses, courtroom demeanor, and more. Ideal text for company officers, firefighters, inspectors, investigators, insurance and industrial personnel.

PRIVATE FIRE PROTECTION & DETECTION

Automatic sprinkler systems, special extinguishing systems, standpipes, detection and alarm systems. Includes how to test sprinkler systems for the firefighter to meet NFPA 1001.

INDUSTRIAL FIRE PROTECTION

Devastating fires in industrial plants do occur at a rate of 145 fires every day. *Industrial Fire Protection* is the single source document designed for training and managing industrial fire brigades.

This text is a must for all industrial sites, large and small, to meet the requirements of the Occupational Safety and Health Administration's (OSHA) regulation 29 CFR part 1910, Subpart L, concerning incipient industrial fire fighting.

HAZ MAT RESPONSE TEAM LEAK & SPILL GUIDE

A brief, practical treatise that reviews operations at spills and leaks. Sample S.O.P. and command recommendations along with a decontamination guide.

FIRE SERVICE INSTRUCTOR

Characteristics of good instructor; determining training requirements and what to teach; types, principles, and procedures for teaching and learning; training aids and devices; conference leadership.

PUBLIC FIRE EDUCATION

A valuable contribution to your community's fire safety. Includes public fire education planning, target audiences, seasonal fire problems, smoke detectors, working with the media, burn injuries, and resource exchange.

FIRE PREVENTION AND INSPECTION PRACTICES

Fire prevention bureau and inspecting agencies; fire hazards and causes; prevention and inspection techniques; building construction, occupancy, and fire load; special-purpose inspections; inspection forms and checklists, along with reference sources; maps and symbols; records and reports.

WATER SUPPLIES FOR FIRE PROTECTION

Importance, basic components, adequacy, reliability, and carrying capacity of water systems; specifications, installation, maintenance, and distribution of fire hydrants; flow test and control valves; sprinkler and standpipe systems.

FIRE APPARATUS PRACTICES

Various types of fire apparatus classified by functions; driving and operating apparatus including pumpers, aerial ladders, and elevating platforms; maintenance and testing of apparatus.

FIRE STREAM PRACTICES

Characteristics, requirements, and principles of fire streams; developing, computing, and applying various types of streams to operational situations; formulas for application of hydraulics; actions and reactions created by applying streams under different circumstances.

FIRE PROTECTION ADMINISTRATION

A reprint of the Illinois Department of Commerce and Community Affairs publication. A manual for trustees, municipal officials, and fire chiefs of fire districts and small communities. Subjects covered include officials' duties and responsibilities, organization and management, personnel management and training, budgeting and finance, annexation and disconnection.

FIREFIGHTER SAFETY

Basic concepts and philosophy of accident prevention; essentials of a safety program and training for safety; station house facility safety; hazards en route and at the emergency scene; personal protective equipment; special hazards, including chemicals, electricity, and radioactive materials; inspection safety; health considerations.

FIRE PROBLEMS IN HIGH-RISE BUILDINGS
Locating, confining, and extinguishing fires; heat, smoke, fire gases, and life hazards; exposures, water supplies and communications; pre-fire planning, ventilation, salvage and overhaul; smokeproof stairways and problems of building design and maintenance; tactical checklist.

AIRCRAFT FIRE PROTECTION AND RESCUE PROCEDURES
Aircraft types, engines, and systems, conventional and specialized fire fighting apparatus, tools, clothing, extinguishing agents, dangerous materials, communications, pre-fire planning, and airfield operations.

GROUND COVER FIRE FIGHTING PRACTICES
Ground cover fire apparatus, equipment, extinguishing agents, and fireground safety; organization and planning for ground cover fire; authority, jurisdiction, and mutual aid, techniques and procedures used for combating ground cover fire.

FIRE SERVICE PRACTICES FOR VOLUNTEER AND SMALL COMMUNITY FIRE DEPARTMENTS
A general overview of material covered in detail in *Forcible Entry, Ladders, Hose, Salvage and Overhaul, Fire Streams, Apparatus, Ventilation, Rescue, Inspection,* and *Self-Contained Breathing Apparatus, and Public Fire Education.*

INSTRUCTOR GUIDE SETS
Available for *Forcible Entry, Hose, Salvage and Overhaul, Fire Streams, Apparatus, Ventilation, Rescue, First Aid, Inspection, Aircraft, and for the slide program Fire Department Support of Automatic Sprinkler Systems.* Basic lesson plan, tips for instructor, references.

TRANSPARENCIES
Multicolored overhead transparencies to augment *Essentials of Fire Fighting* are now available. Since costs and availability vary with different chapters, contact IFSTA Headquarters for details. Units available:

Fire Behavior; Portable Extinguishers; Ropes and Knots; Hose Tools and Appliances; Handling Hose; Handling Ground Ladders; Ventilation; Fire Streams; Ladder Carries and Raises; Forcible Entry; Salvage and Overhaul; Prevention and Identification; Ground Cover Fires; Communications; Water Supply; Automatic Sprinkler Systems; Rescue; Protective Breathing Apparatus.

FIREFIGHTER VIDEOTAPE SERIES
Video programs for reinforcement of basic skills and knowledge on a variety of fire fighting topics. Excellent for use with *Essentials* or *Volunteer* to review and emphasize subjects. Titles available: The Anatomy and Behavior of Fire, Fire Safety, Protective Breathing Apparatus, Fire Hose and Nozzles — Part 1, Fire Hose and Nozzles — Part 2, Ventilation, Sprinklers, Ladders, Forcible Entry, Rescue, Ropes and Knots, Salvage, Fire Alarm and Communications, General Qualifications, First Aid, Inspection — Part 1, and Inspection — Part 2.

SLIDES

2-inch by 2-inch slides that can be used in any 35 mm slide projector; supplements to respective manuals and sprinkler guide sets.

Sprinklers
Module 1: Introduction to Automatic Sprinkler Protection
Module 2: Types of Sprinkler Systems
Module 3: Maintenance and Inspection of Sprinkler Systems
Module 4: Components of Water Supply Systems
Module 5: Testing and Analysis of Water Supply Systems
Module 6: Factors Affecting the Adequacy of Sprinkler and Water Systems

Smoke Detectors Can Save Your Life
Matches Aren't For Children
Public Relations for the Fire Service
Public Fire Education Specialist (Slide/Tape)
Salvage*

*The complete package consists of the slides, instructor's manual, and instructor's guide sets.

MANUAL HOLDER
The fast, efficient way to organize your IFSTA manuals. Those attractive heavy-duty vinyl holders have specially designed side panels that allow easy access to all of your IFSTA manuals. Manual holders stand unsupported and will hold up to eight manuals.

IFSTA BINDERS
Heavy-duty three ring binders that will allow you to organize and protect your IFSTA manuals. Available in two sizes, 1½ inch and 3 inch.

GUIDE SHEET BINDERS
Free with purchase of complete guide set. Binders also available separately.

WATER FLOW TEST SUMMARY SHEETS
50 summary sheets and instructions on how to use; logarithmic scale to simplify the process of determining the available water in an area.

PERSONNEL RECORD FOLDERS
Personnel record folders should be used by the training officer for each member of the department. Such data as IFSTA training, technical training (seminars), and college work can be recorded in the file, along with other valuable information. Letter size or legal size.

Ship to: Date ____________

Name ____________ Customer Number ____________

Organization ____________ Phone ____________

Address ____________

City ____________ State ____________ Zip ____________

ifsta

Send to
Fire Protection Publications
Oklahoma State University
Stillwater, Oklahoma 74078
(405) 624-5723
Or Contact Your Local Distributor

ORDER FORM

IFSTA MANUALS

WRITE THE NUMBER OF COPIES OF EACH MANUAL NEXT TO ITS TITLE

Title	No. Of Each
Indoctrination	____
First Aid	____
Essentials	____
500 Competencies for Essentials	____
Ladders	____
Hose	____
Salvage and Overhaul	____
Forcible Entry	____
Self-Contained Breathing Apparatus	____
Ventilation	____
Rescue	____
Company Officer	____

Title	No. of Each
Fire Cause Determination	____
Private Fire Protection	____
Industrial Fire Protection	____
Haz Mat Leak and Spill Guide	____
Instructor	____
Public Fire Education	____
Fire Prevention/ Inspection	____
Water Supplies	____
Fire Streams	____
Apparatus Practices	____
Fire Protection Administration	____

Title	No. of Each
Safety	____
Aircraft	____
High-Rise	____
Volunteer	____
Ground Cover	____
Manual Binder	____

SLIDES

Title	No. of Each
Salvage	____
Public Fire Education Specialists	____
Smoke Detectors Can Save Your Life	____

Title	No. of Each
Matches Aren't For Children	____
Public Relations for the Fire Service	____

VISUAL AIDS

Multicolored overhead transparencies to augment each chapter of *Essentials of Fire Fighting* are now available. Also available are slide programs for each of the major fire pump manufacturers and slides for sprinkler systems. A new addition, The firefighter videotape series is available for both Beta and VHS formats. Since costs and availability vary with different sets, contact Fire Protection Publications for details.

OTHER MANUALS AND MATERIALS MAY BE ORDERED BELOW:

QUANTITY	TITLE	LIST PRICE	TOTAL

All Foreign Orders must be prepaid in U.S. currency and include 20% shipping and handling charges.

Obtain postage and prices from current IFSTA Catalog or they will be inserted by Customer Services.

Note: Payment with your order saves you postage and handling charges when ordering from Fire Protection Publications.

Payment Enclosed ☐ Bill Me Later ☐

Allow 4 to 6 weeks for delivery.

SUBTOTAL $ ____________

Discount, if applicable $ ____________

Postage and Handling, if applicable $ ____________

TOTAL $ ____________

FOR ORDERS
TOLL FREE NUMBER — 800-654-4055

Oklahoma, Hawaii, and Alaska call collect.

COMMENT SHEET

DATE ________________ NAME ____

ADDRESS ____

ORGANIZATION REPRESENTED ____

CHAPTER TITLE ____ NUMBER ________

SECTION/PARAGRAPH/FIGURE ____________________________________ PAGE ________

1. Proposal (include proposed wording, or identification of wording to be deleted), OR PROPOSED FIGURE:

2. Statement of Problem and Substantiation for Proposal:

RETURN TO: IFSTA Editor
Fire Protection Publications
Oklahoma State University
Stillwater, OK 74078

SIGNATURE ______________________

Use this sheet to make any suggestions, recommendations, or comments. We need your input to make the manuals the most up to date as possible. Your help is appreciated. Use additional pages if necessary.